Handbook of Mobile Communication Studies

Handbook of Mobile Communication Studies

edited by James E. Katz

The MIT Press
Cambridge, Massachusetts
London, England

For information about special quantity discounts, please e-mail special_sales@mitpress.mit.edu.

This book was set in Stone Serif and Stone Sans on 3B2 by Asco Typesetters, Hong Kong.
Printed and bound in the United States of America.

Library of Congress Cataloging-in-Publication Data

Handbook of mobile communication studies / edited by James E. Katz.
    p.   cm.
Includes bibliographical references and index.
ISBN 978-0-262-11312-0 (hardcover : alk. paper)
1. Cellular telephones—Social aspects. 2. Wireless communication systems—Social aspects.
3. Interpersonal communication—Technological innovations—Social aspects. 4. Communication
and culture. I. Katz, James Everett.
HE9713.H36   2008
303.48′33—dc22   2007001992

10   9   8   7   6   5   4   3   2   1

Rick Rescorla

Morristown, New Jersey
Born 5.27.39
Died 9.11.01

Contents

Acknowledgments

Many of the chapters in this volume were drawn from a conference held on May 21, 2005, at Rutgers University in New Brunswick. The twin pillars of this conference ("Mobile Communication and the Network Society") were Manuel Castells, who framed issues regarding mobile communication, and Kenneth Gergen, who analyzed mobile communication's evolving social consequences. Their participation, along with many others, is a deeply significant tribute to the international attention that scholars are increasingly devoting to mobile communication.

An edited book is the work of many hearts and minds, and I am highly indebted to the considerable number of researchers who have directly and indirectly contributed to the volume. I can thank only a small proportion of them here. They include my graduate students who so generously shared their skills and time: Kalpana David, Miles Cho, Chih-Hui Lai, Katie Lever, Emily Liang, Dan Su, and Shenwei Zhao. Ronald E. Rice and Satomi Sugiyama provided good advice at many junctures. Yi-Fan Chen was a vital participant in every stage of the project, and I owe her a special debt of thanks.

My gratitude goes to William Mitchell of MIT for his encouragement and to Manuel Castells of USC for his collegiality and congeniality. Manuel not only generously contributed to the book's development but also provided its afterword. Mary Curtis and Irving Louis Horowitz tendered valuable advice at every stage of the book project.

Thanks also go to my colleagues Leopoldina Fortunati, Richard S. Ling, Scott Campbell, Jonathan Donner, and Howard Rheingold. Jan Ellis, Akiba Cohen, Gerard Goggin, Pui-Lam "Patrick" Law, Chantal de Gournay, Hui-Min Kuo, Steve Love, Boxu Yang, Mike Noll, and Kristóf Nyíri also have been important sources of inspiration. Dean Gustav Friedrich kindly cosponsored our organizing conference. To all these fine scholars, I tender my heartfelt thanks.

1 | Introduction

James E. Katz

Hello, Neil and Buzz, I am talking to you by telephone from the Oval Room at the White House, and this certainly has to be the most historic telephone call ever made.

—U.S. President Richard M. Nixon to Apollo 11 astronauts Neil Armstrong and Edwin "Buzz" Aldrin, July 20, 1969 (White House History Association 1997)

Whether or not this ringing up of the astronauts was the most historic of telephone calls, it did set a long-distance record, and a mobile one at that. President Nixon's 1969 call to the moon capped an era of increasingly distant and mobile interpersonal communication technology and services that began in 1876 when the first voice message to be electrically transmitted occurred, which was of course Alexander Graham Bell's cry, "Watson, come here; I need you."

The calls of Nixon and Bell contrast to one made on April 3, 1973, which was the first public mobile telephone call ever made. On that spring day, an inebriated Martin Cooper of Motorola placed a call from a Manhattan street to Joel Engel, his relentless New Jersey–based rival at Bell Labs. Thus was opened the era of the portable cellular telephone. As Cooper recalls the story, he and Engel had been competing abrasively for years. Cooper says he would often call Engel to annoy him or just to brag. When Engel answered Cooper's call at his Bell Labs office, he had no idea that Cooper was using the occasion to show off the newly developed Motorola technology. Cooper began the conversation by crowing, "Guess who this is, you sorry son-of-a-bitch." Engel—having been pestered by Cooper for many years, and not appreciating the circumstances of Cooper's call—apparently did not respond directly to Cooper. Rather Engel said (perhaps to someone else in his office), "It's him again" (Orenstam 2005). The call apparently was to Engel an irritating interruption.

Nixon's call was ponderous but important as a crafted effect and symbol. Bell's call was a spontaneous expression of need. Cooper's call, though, was to break some important news and, perhaps more importantly, to brag. Nixon's and Bell's calls were different from Cooper's in another way: unlike the recipients of their calls, the recipient of Cooper's call was initially unaware of the weightiness of the historic moment.

Additionally, the Cooper-Engel exchange adumbrates the countless billions of mobile phone calls to follow: misinterpretation and social competition. Moreover, it seems that during that mobile-originated call Engel was having to juggle his interaction with the caller and a physically co-present colleague.

All three archetypical calls—which among them have served to convey important information, act as public symbols, seek assistance, and engage in piquant competition—are themes commonly found in the chapters presented in this volume. More specifically, the book's aim is to present a collection of analyses of the mosaic of mobile personal communication, and most especially the mobile phone. It provides telescopic and microscopic views of phenomena that are transforming the structure of global organization as surely as the intimacy of ordinary life. Authors seek to understand how mobile communication reconstructs the organization and content of daily life in a variety of contexts. In doing so, the contributors reveal a range from critical to quotidian, and from humdrum to surprising instances of communication.

Many of their efforts yield intriguing points of comparison with the three kinds of telephone calls sketched at the chapter's outset. As was the case with President Nixon's call to fully mobilized astronauts, some chapters highlight astonishing and amazing qualities of new mobile communication. Other chapters are aligned more closely with the unanticipated interaction and social dependency emblematic of Mr. Bell's urgent cry for help. His was a spontaneous cry seeking intervention by a distant party (in this case only slightly distant, but sufficiently so as to make the cry historic). Yet other chapters hone in on the type of call that was prototyped by the call of Cooper's mobile phone call to Bell Labs's Engel. Mr. Cooper's call was one of social competition; it carried news phrased in banal terms but was in fact of transcendent importance. Billions of calls daily, historic or not, are now integral parts of the never-ending struggle through which people communicate their thoughts and interact with other people, and even with machines.

This handbook aims to convey in manageable form recent thinking and research on the social aspects of mobile communication. No handbook—especially one that addresses a subject as vast as the way three billion or so people use an endlessly flexible technology—can include all related topics. Neither can those that are included be necessarily dealt with to the depth that one might wish. These are the cruel realities that face any editor, and must be resolved in a way that cannot always be to everyone's satisfaction.

My choice of topics has been guided by the overarching idea that mobile communication has become mainstream even while it remains a subject of fascination in usage configurations and social consequences. As such, the handbook aims at examining the way mobile communication is fitting into and altering social processes in many places around the globe and at many levels within society. In essence, then, it presents a series of analyses of how the reality of being mobile and in communication with distant information and personal resources affects daily life. Of course, with more than a third of all humans in the world operating under such conditions, it is hard to make precise claims that are at once manifestly universal and useful. Yet, as the chapters in this volume demonstrate, there are some remarkably consistent changes in personal routines and social organization as a result of literally putting mobile communication resources into the hands of people.

The contributors show how mobile communication profoundly affects the tempo, structure, and process of daily life. Topics discussed include

- who is integrated into mobile communication networks and why.
- how social networks are created and sustained by mobile communication.
- how mobile communication fits into an array of communication strategies including the Internet and face-to-face.
- the way traditional forms of social organization are circumvented or reinvented to suit the needs of the increasingly mobile user.
- how quickly miraculous technologies become ordinary and even necessary.
- how ordinary technology becomes mysterious, extraordinary, and even miraculous.
- the symbolic uses of mobile communication beyond mere content.
- the uses of mobile communication in political organizing and social protest, and in marshaling resources.

The chapters in this volume cut across vast social issues and geographic domains. They highlight both elite and mass users, utilitarian and expressive uses, and political and operational consequences. The chapters also have foci that range from individual to collective issues, and from industrialized to rapidly (or slowly) developing societies. The themes also cut across psychological, sociological, and cultural levels of analysis. At their heart, though, is an enduring theme of how mobile communication has affected the quality of life in both exotic and ordinary settings. Mobile communication is now a mainstream activity in all human activities, and is increasingly sharing if not (yet) predominating life's center stage in both intrusive and subtle ways.

The volume has four main themes, with chapters drawing out each of them. These themes are digital divides and social mobility, sociality and co-presence, politics and social change, and culture and imagination. The book concludes with a few comments by the editor and an afterword by Manuel Castells. In terms of specific coverage, it is useful to sketch here the volume's chapter contents.

1. It begins with the present chapter wherein major issues are briefly sketched, followed by substantive chapters.
2. Lara Srivastava shows on a global and comparative scale the enthusiastic embrace of the cell phone by people from all regions and strata. In terms of speed and breadth of adoption, the mobile phone is without historical parallel. Recounting the enormous worldwide success enjoyed by mobile communication, she indicates that mobile

communication has done more to overcome the digital divide than the PC, the device that had been the favored hope of international aid agencies.

3. Jonathan Donner investigates Manuel Castells's assertion that the Informational Society has brought about a "Fourth World" of marginalized peoples who are bypassed by information technologies and excluded from significant participation in the economic and social organization of the new millennium. Dr. Donner finds that mobile communication will narrow the gap between the Fourth World and more developed countries, but cannot in themselves close it entirely. He also considers if a new smaller digital divide will open within Fourth World villages between the "less poor" and "poorest poor," an important issue that requires monitoring in the years ahead.

4. Ragnhild Overå describes the context of Ghana and shows how improved access to telephones, and especially mobile phones, affect daily life and economic opportunities for Ghana's traders in agricultural produce. She focuses on the way in which traders change social, economic, and spatial practices when they acquire mobile phones, and shows how some resources are increased but in a far-from-equitable way. Ms. Overå also traces gender implications of the distribution of new communication technology.

5. Pui-lam "Patrick" Law and Yinni Peng discuss mobile phone use among the migrant workers in southern China and how it has aided the formation of free-floating networks. The rapid and extensive penetration of mobile phones among the migrant workers in Guangdong since early 2000 has expanded contacts with their fellow villagers who have been widely scattered. Mobile phones can also for the first time extend and continue networks developed originally in the factories. With these expanded and flexible networks, migrant workers are more resourceful in getting job market information. It also helps them demand greater rights and lead more autonomous lives.

6. Judith Mariscal and Carla Marisa Bonina trace the dramatic rise of mobile phones in Mexico, partly attributable to prepaid subscription availability. While there are gender and economic status–based digital divides among subscribers in Mexico, it remains noteworthy that more than a quarter of Mexico's lowest economic segment has mobile phones. In a survey conducted for this book, the authors discovered that young women have a perceived heavy dependence on the mobile, and that young teens say they find little difference between face-to-face and mobile communication, suggesting that the traditional distinction older people have between real and virtual is being erased. The mobile is becoming a central feature for the people of Mexico.

7. Jan Chipchase points out that relative to developed markets, emerging markets have vast numbers of textually nonliterate people. Effective use of growing numbers of mobile phone features in these regions requires an understanding of textual features such as prompts, contact management, and synchronous communication. This presents the textually nonliterate with a severe challenge. Mr. Chipchase argues that resolving this challenge would benefit the nonliterate poor in developing countries, and give them unprecedented opportunities in a host of other and even novel service settings.

8. Patricia Mechael shows how despite the rapidly growing importance of mobile communication in healthcare, there are many difficult and often nuanced problems in its application. Although these obstacles are not going to stunt use, particularly in the developing world, her ethnographic report provides an important complement to the system-centered approaches that often characterize donor approaches to creating health care infrastructures.

9. Lourdes M. Portus explores how Filipino urban poor residents in a squatter community give meanings to the mobile phone and how they negotiate its acquisition and use. She finds that despite grinding poverty, the mobile phone is increasingly critical to social role fulfillment and one's self-perception; it can also reduce or exacerbate interpersonal tensions. Dr. Portus uses her focus group interviews to show that ordinary "talk" about the mobile phone reflects the values and socioeconomic outlook of its users and would-be users.

10. Sherry Turkle explores communication mobility and its implications for those who live in the upper strata of the advanced industrial societies. The professionals who "phone it in" are scrutinized from a sociological view. So too are today's teens who apparently experience Paris as an opportunity for texting rather than a romantic or political engagement. Mobile connectivity while away from home leaves little time to inquire, as perhaps both Adolph Hitler and Jacques Chirac did in their day, "Is Paris Burning?" She also examines the digital dustbin awaiting those who are no longer wanted in the brave new technological world. In the process of her examination, she raises important questions about the use of human beings as the opportunities grow to leave emotional work to the infinite patience of robotic pets.

11. Christian Licoppe examines ringtones as a form of social exchange. He also highlights how the choices of musical introductions are done with careful forethought of the likely effect they will have on both the individual recipient and the anticipated and potentially unwilling ambient audience. Professor Licoppe underscores the "connected presence" dimension of ringtones, and that although ringtones are seemingly personal statements, they are also a means of maintaining bridges with others.

12. Scott Campbell sees that the recent explosion of mobile phones and other wearable personal communication technologies (PCTs) calls for a reconceptualization of the relationship between technology and body. PCTs are now worn on the body and are clearly often regarded as part of one's personal statement or fashion. He applies a theoretical framing that Mark Aakhus and I developed, called *Apparatgeist* (Katz and Aakhus 2002), to predict accurately what relational uses of the mobile phone are linked to perceptions of PCTs as fashion, while logistical and safety/security uses are not. This suggests that theories focusing on the unique aspects of mobile communication may lead to valuable new insights.

13. Richard S. Ling examines societal cohesion and social ritual in light of increasingly mobile communication. He extends to mobile telephony the analyses originally

developed by Durkheim, Goffman, and Collins. He asks if there is something special about face-to-face interaction, for which nothing can substitute, no matter how "rich" the medium. Ling suggests that interaction can become just as fulfilling distantly as when it occurs face-to-face, and thus there are no enduring barriers to developing a seamless continuum of social interaction across media.

14. Naomi S. Baron finds that a variety of technologies—landline telephones, e-mail, instant messaging (IM), and mobile phones—are increasingly enabling users to regulate interpersonal access in terms of speech or textual acts. Using the metaphor of differentially "adjusting the volume" of individual linguistic interactions, Baron explores the social consequences of technologically empowering individuals to manage the terms of linguistic engagement.

15. Peter B. White and Naomi Rosh White look at tourists' experiences of communicating using mobile and fixed telephones in New Zealand. Drs. White and White find that tourists use the voice and texting functions of telephones to create a sense of co-presence with people from whom they are temporarily separated by long distances. Such people were eager to be seen by group members as being continuously engaged in their social groups and relationships even while acknowledging the distance separating them from those back home. There were many strategies addressing mobile communication to control the frequency and nature of contact. The findings of Drs. White and White have intriguing implications for the psychology of travel, tourism, and leisure.

16. The chapter by Kakuko Miyata, Jeffrey Boase, and Barry Wellman presents one of the first studies to collect information about social networks and e-mail use over time, in this case in Japan. Japan is particularly interesting because both mobile and PC-based telecommunication are popular there. As such, it offers a unique opportunity to map out the adoption of *keitai* (Internet-enabled mobile phones) to communicate with close friends and family. This chapter probes how it is used currently and how it appears to affect relationships, especially in terms of change over time. It examines the prospect that as the first generation of *keitai*-savvy young adults matures, they may continue to make *keitai* their primary means of electronic communication. The answer to this question raises the prospect that there will be a gradual demise of PC e-mail in Japan.

17. Howard Rheingold explores how mobile phones and SMS (short message service) technology in conjunction with Internet tools have become powerful influences on public demonstrations and elections as well as in spreading information that has been officially suppressed. Ghana and Korea are among the many countries where these mobile technologies have affected voter turnout as well as the way the elections themselves turned out. Also, Mr. Rheingold notes, these tools can be appropriated for violent protest and terrorism.

18. Ilpo Koskinen focuses on multimedia services, or MMS. In terms of this technology, he surveys their social uses including moblogs (Web sites accessible with mobile

phones), citizen journalism, and mobile mass media. Professor Koskinen argues that mobile multimedia primarily add social, sensual, and emotional elements to mobile telephony. He says that although moblogs may break the boundary of private and public content and enable citizen journalism, they will more likely just become another means of sharing private photo albums over the Internet. He believes, as does Kenneth Gergen, that mobile multimedia will lead to society being increasingly grouped into monadic clusters that turn their attention away from civil concerns.

19. Mohammad Ibahrine examines mobile communication in the Arab world, especially the way it disrupts traditional structures and methods of regulating interpersonal communication. Dr. Ibahrine discusses the production and distribution of mobile media content by the Arab masses as well as the changing patterns of social and cultural understanding and practices within a large communicative context between individuals and communities. Dr. Ibahrine also examines the sociopolitical mobilization of some segments of the "Arab street" via mobile hyper-coordination.

20. On-Kwok Lai discusses the ways in which location remains important in Japan despite much heralded expectations that with new telecommunication services such considerations would become minimized. Dr. Lai details how cultural, demographic, policy, and technological factors comingle to yield services of particular utility to the young and the aged in Japan.

21. Shahiraa Sahul Hameed examines mobile communication in three spheres of Singaporean life—education and youth, antisocial behaviors and security, and religion. Incidents involving the use of mobiles and text messaging have gained public attention; the widespread debate they provoke says much about the interaction between culture and technology. The incidents also make problematic certain taken-for-granted norms. Ms. Hameed finds that the mobile is an important agent of social change but also presents an opportunity to reinforce standards of conduct.

22. Kenneth J. Gergen reviews structural changes of the past half century in the character of democratic process, with special attention to the increasingly important function of mobile communication. Dr. Gergen holds that mobile communication has become increasingly important because it helps give rise to democratic participation lodged between the overarching structure of government and the local community, an area he terms *Mittelbau*. On the local level, mobile communication alters civil society by giving rise to monadic clusters, which are small groups linked in close and continuous communication. Such clusters lead on the one hand to political disengagement, and on the other to political polarization. Professor Gergen also sees that mobile communication is contributing to an erasure of the individual as the locus of political decision making ("the heart of democracy"), and opening the door to a view of democratic process as relational flow.

23. Gustavo Mesch and Ilan Talmud discuss how culture modulates the use of information and communication technologies (ICTs) among Israeli teens, contrasting those from Jewish and Arab ethnic groups. The Jewish adolescents used communication

technologies in ways that do not depart greatly from patterns in other Western societies. Their choices among cell phone, instant messaging, and e-mail depends largely on cost considerations and is made to expand and maintain social ties. Network expansion was achieved through meeting new buddies online, but with the goal of moving quickly to cell phone calls and face-to-face meetings. Contrarily, Arab adolescents used cell phones to maintain local ties when face-to-face communication is an unrealistic alternative. As well, many ties with the opposite gender were created by seeming "mistakes"; these are kept secret and tend to remain virtual to avoid violating norms.

24. Jonathan Donner, Nimmi Rangaswamy, Molly Wright Steenson, and Carolyn Wei analyze the way mobile phone technology acts as both a change agent and a site where existing tensions in Indian middle-class families are played. It also offers a snapshot of the aspirational consumption that is characteristic of the new middle class in India. Three case studies relate mobiles to family financial decisions, romantic relationships, and domestic space. The studies show that whereas elements of autonomy and individuation do arise from mobile phone use, the adoption of mobiles as a family process more accurately describes its diffusion in middle-class India.

25. Thomas Molony discusses the mainstreamed situation in the East African country of Tanzania. He notes that the recent impression of ICTs in Africa is of countless motivated individuals using mobile phones and the Internet to pull themselves out of poverty. This view should be countered by an understanding of the ordinary adoption of mobile phones. Using data from a series of semistructured interviews conducted in Tanzania in 2005 and 2006, he concludes that there is an important informal economy of the acquisition and sale of mobiles. While potential economic benefits of ownership are valued, Dr. Molony says, to users the social networking aspects are probably of paramount importance.

26. Gerard Goggin presents an overview of cultural studies of mobile communication. He sketches some of the main research findings in this area. As part of his analysis, Goggin shows that while the early absence of cultural studies of mobile technology is slowly being rectified, there are some important topics left unexplored. He uses the balance of his chapter to identify pressing needs for research on cultural aspects of the mobile.

27. James E. Katz, Katie M. Lever, and Yi-Fan Chen discuss the mobile music phenomenon. They show how the technology is designed and used to provide a soundscape (or audio ecology) for one's life; it is important to users not only for its entertainment capabilities but also as a form of environmental control. It allows users to control interaction with others, especially those who might wish to gain "face time" with the technology's possessor. Style and self-expression dimensions are important in guiding acquisition behaviors and display behaviors. But perhaps the most surprising finding is that for a substantial minority of users, the technology seems to be employed to

build bridges and make social connections for potential new friends and to solidify bonds among current members of one's social circle. As sharing capabilities of the devices grow, so will users avail themselves of its bridging aspects. Formerly a technology of personal isolation, like so many others before it, the mobile as well seems destined to be expanded into a technology of conviviality.

28. Bart Barendregt and Raul Pertierra analyze the religious and supernatural aspects of mobile communication technology, primarily in Indonesia and the Philippines. Belief in the ability of mobile phones to communicate with the supernatural is widespread, and local media feature the uncanny experiences users have had with their phones. Asian phone ghost stories could be considered a metacomment on modernity itself and the mobile phone becomes one of modernity's main icons. In keeping with cultural views of technology, Pertierra and Barendregt assert that popular beliefs about the supernatural meld with people's orientations toward technology and its uses.

29. Madanmohan Rao and Mira Desai survey the mobile environment in India, one of the world's fastest-growing mobile markets. They address three sets of impacts: civic, media, and social. E-government services for citizens, as well as political mobilization and communication by citizens, is increasing, due largely to SMS functionality. Media impacts include use of mobile services such as SMS polling for TV programming and gaming, and surveys of questions on issues such as appropriate social behavior. Finally, they examine social impacts of mobiles when extended to family relations, dating, and space-time negotiation based on an original survey conducted in Mumbai. They conclude that mobiles will play a critical role in India's modernization. The discussion covers broader implications of the technology in day-to-day life.

30. Sophia Krzys Acord and James E. Katz interrogate the world of mobile-mediated gaming, comparing it with traditional games and nongaming mobility. While a large and lively area, in contrast to many other mobile applications, gaming has not grown at anywhere near the expected rate. They identify three modalities of gaming, each with varying normative consequences for social relationships, individual psychology, and public space.

31. Youn-ah Kang looks at users of an online world to compare those who have both mobile and PC-based access versus PC-based alone. Ms. Kang finds significant differences between the two groups, with the mobile users much more active and involved. Using both qualitative and quantitative methods, she believes the uses of mobile technology are leading to greater convergence of online and offline worlds. It seems that such convergence is likely to lead to greater social integration, just the opposite result of what pessimists often fear about results of the growth of communication technology.

32. In the concluding chapter, rather than providing a summary of summaries, I identify issues that I see emerging from the studies presented in this handbook, and which I believe are likely to not only remain consequential but also grow in importance.

33. The afterword by Manuel Castells presents his summative view as to the long-term and profound impact of mobile communication. His four points serve to not only highlight and recapitulate central points that may be discerned among the chapters here, but cast in clarity a long-term research agenda that may be pursued to better understand human potential as mediated through personal, powerful communication technology. Readers would be well served by attending closely to his statement.

Before turning to the specific chapters, it is useful to highlight some points that underlie analytical and case-study chapters. They are presented here to obviate repeating them in specific chapter contexts.

First, the mobile device is seldom used as an isolated technology. In the developed world, it exists in relation to technologies such as the landline phone, physical cable infrastructure, and the Internet. Even in the developing world, there are many modalities for communication, and sometimes they also interact across platforms. As shown by Dr. Ibahrine, for instance, this is the case in Lebanon (and formerly in Saudi Arabia) where one could use one's mobile to vote favorably for talent performance being shown on television. Overå shows how public payphones are used to supplement mobile phones, and Mesch and Talmud analyze the interplay and substitutability of the Internet and mobile phone.

Second, the emphasis here is on using mobiles for conventional interpersonal communication in various settings. This is but one of many possible emphases. Everywhere, the mobile is a multipurpose device, the uses of which extend far beyond voice or text messaging. It has become a music platform, calendar, watch, alarm clock, calculator, and game player. It is also a PC terminal for interacting with the Web and Internet. TV viewer, social date-finding service, game center, and medical data repository are among its popular uses. It is a portal to advertising, health monitoring, and, seemingly, the supernatural. Mobiles are environmentally interactional and useful for a host of geolocational services. There are some exciting innovative uses such as "smart mobbing," distributed mobile games, "art happenings," and "symphonies" (generally performed by the young).

This book's central purpose is to look at the ordinary life of the mass of users. It is less engaged with analyzing emerging experiments of those walking (and talking) on the wild side. So while the innovative and unusual topics mentioned above are probed to some extent, the emphasis is on everyday uses by everyday people, and what these uses mean for social organization and interpersonal communication.

Third, there are a good many people who are not mobile phone users, though each day there are many fewer of them. Some of these people desperately want mobile devices but cannot afford or otherwise obtain or use them. There are a variety of reasons for this, including accidents of geography and location as well as one's physical or fiscal condition. And it is worth pointing out, as seen for example in chapter 6 on Mexico, where at any given time there may be a good number of former users or "mobile phone dropouts." Most of these dropouts will be future users and they usually become dropouts because of temporary financial setbacks (Rice and Katz 2003). Yet that being recognized, it is important to point out that others directly reject the technology. This may be because they do not want to be beholden to others ringing them up at times unknown, or because of cultural or religious reasons.

Fourth, the analyses in the chapters show that sometimes there are surprising and novel uses for mobile devices, and that these uses are not readily foreseeable even though they may have dramatic consequences. Still more noteworthy, these uses are often undertaken by those who ordinarily are not associated in the public mind with innovative technology. These include religious groups and diviners, matchmakers and con artists. The perceived value and symbolism of having a mobile also extends far beyond the mobile's functional use. These perceptions may vary widely by particular subculture, too, as seen in chapter 9 by Portus on the Philippines, for instance.

Finally, this volume is not aimed at dissecting mobile communication uses in business, commercial, or professional settings. Nor does it aspire to analyze regulatory or macroeconomic policies. Finally, we do not aim to report on avant-garde experiments and imaginative one-off innovators. All of these are fascinating areas, but fall outside the scope of this book. Rather, its interest is in the social side of human life, what is known as "studies." Contributors seek to understand the cultural, familial, and interpersonal consequences of mobile communication in a global context. They are exploring contextualized issues against cultural and national backdrops. They in essence pinpoint how mobile communication has become part of mainstream human existence, how major cultural and social interaction patterns are being readjusted, and what newly created structures and processes result therefrom. That is ample enough for even the most ambitious single-volume study.

References

Orenstam, N. 2005. Doctor Cellphone. http://www.valleyofthegeeks.com/Features/Cooper.html.

Rice, R. E., and Katz, J. E. 2003. Comparing Internet and mobile phone usage: Digital divides of usage, adoption and dropouts. *Telecommunications Policy* 27(8–9): 597–623.

White House History Association. 1997. *Jump-starting the space race*. http://www.whitehousehistory.org/04/subs/04_a02_e05.html.

Digital Divides and Social Mobility

2 The Mobile Makes Its Mark

Lara Srivastava

The startling rise in mobile phone usage over the past decade has a number of important consequences and implications for both economic and social life. This chapter explores current mobile growth, implications for developing countries, and future scenarios.

One of the most striking characteristics of the world we live in today is the increasing use of technology for accessing information and mediating communication. The phenomenal spread of mobile and Internet technologies and applications are unprecedented in any other domain of human activity. Information and communication technology (ICT) has been touted as a key element of economic growth over the past fifteen years, and is maintaining its lead as the fastest-growing service sector, outstripping growth in basic services such as health, housing, and food (see figure 2.1).

Even the rapid expansion of the Internet was exceeded by the lightning speed of the mobile phone's dramatic uptake. One might find a mobile phone in remote villages of the developing world—the same is not true of the Internet. During the 1990s, both technologies grew at similar rates, albeit with a two-year lag, that is, the Global System for Mobile Communications (GSM) having been launched in 1991 and the first generation of Internet browsers (e.g., Mosaic) introduced in 1993. At the turn of the century, the growth of the Internet slowed down while mobile growth surged ahead. In particular, between 2000 and 2003, around twice as many new mobile cellular subscribers as new Internet users have been added worldwide (figure 2.2). It was at the end of 2002 that mobile technology truly entered the mainstream, with the number of mobile lines overtaking the number of fixed lines on a global scale. This occurred not only in industrialized nations but also in the developing world, where the lack of fixed-line infrastructure, the relatively low cost of deployment, and the advent of low-cost prepaid services stimulated the rapid adoption of mobile services. Mobile communication has overtaken fixed across geographic parameters such as countries, across sociodemographic ones such as gender, income, or age, as well as across economic parameters such as service cost or national GDP.

Changes in the proportion of households' expenditure by category

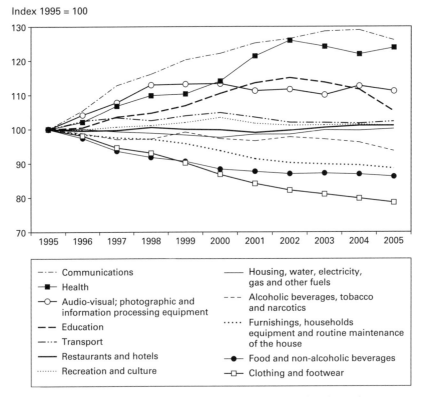

Notes: "Communications" includes telecommunication equipment and services and postal services. New Zealand and Turkey not included in calculations.

Source: OECD, SNA database.

Figure 2.1
Household expenditures by category, 1995–2005.

In January 2006, there was already more than one mobile phone for every three inhabitants on the planet, with the total number reaching 2.17 billion. In January 2007, the number of mobiles in the world doubled the number of fixed lines, at 2.6 billion. Despite the downturn of the economy over the past few years and the burst of the dot-com bubble, mobile communications are continuing to grow rapidly, with new services such as mobile gaming and television being adopted around the world.

It is not surprising, therefore, that the current age has oft been described as a mobile and wireless one. Access to information and communication is no longer limited to fixed locations. There is an increasing dependence on mobile networks—with the loss

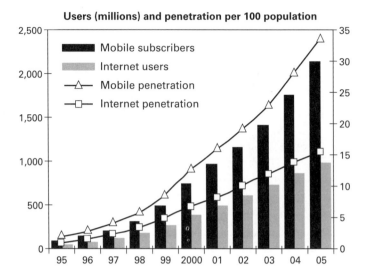

Figure 2.2
Mobile and internet users worldwide, 1995–2005.

of personal mobile phones causing panic and disruption in daily lives. Moreover, many people use their mobiles even when a fixed line is available (and even when it is cheaper). There is also an increasing migration to data services, with the phenomenal growth of messaging services (such as SMS) being an important and telling example. People often prefer to send a message before engaging in a call, or as a faster more intimate alternative to e-mail. The reason for the great and unexpected popularity of SMS seems to be its frequently lower cost relative to a call, and the convenience of communicating with a busy or otherwise occupied party. Individual households in a number of economies are using wireless LAN routers to enable further mobility for their networked devices. Always-on access to information and communication and anytime-anywhere communications are the hallmarks of the present age.

Second-generation (2G) mobile network technologies still dominate, with the vast majority of the world using GSM. In the 2G world, a number of other mobile standards existed, from time division multiple access (TDMA) in the United States to personal digital cellular (PDC) in Japan, making international roaming outside the GSM regions difficult, if not impossible. Despite the success of mobile telephony, data transmission speeds for 2G networks were too slow in most cases to allow efficient Internet access over the portable devices (with the notable exception of Japan's i-mode). In an effort to address the fragmentation of the 2G market and to enable the next generation of multimedia mobile connectivity, the International Telecommunication Union (ITU)

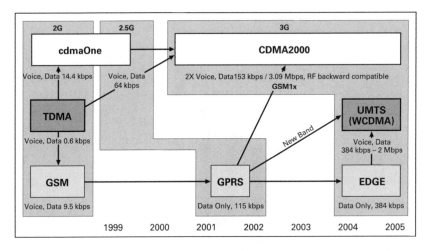

Figure 2.3
From 2G to 3G.

began work in the late 1980s on a new, global standard for mobile communications. This work culminated in the development of the IMT-2000 (International Mobile Tele-communications-2000) standard.

The goal of the IMT-2000 project was to harmonize third-generation mobile network radio interfaces into a single global standard. The IMT-2000 family encompasses three different access technologies (code division multiple access, CDMA; TDMA; and frequency division multiple access, FDMA) and five different radio interfaces with full interoperability. However, most deployments have centered around two main interfaces, CDMA2000 and W-CDMA (Wideband Code-Division Multiple Access; also known in Europe as UMTS, the Universal Mobile Telecommunications System). China is now finalizing its homegrown standard, TD-SCDMA (Time Division-Synchronous Code Division Multiple Access), though licenses have yet to be allocated (figure 2.3).

The first 3G networks were launched in Korea and Japan in 2000 and 2001. Japan's NTT DoCoMo deployed a W-CDMA network and Korea's SK Telecom a CDMA20001x network. W-CDMA requires a complex network upgrade from 2G mobile networks such as GSM. On the other hand, the deployment of CDMA20001x is a much simpler and less costly jump from 2G networks (e.g., cdmaOne or TDMA). Data rates, however, are lower than W-CDMA (the enhanced data network CDMA20001x EV-DO offers higher data rates). This is the main reason for which CDMA2000 operators are ahead of their W-CDMA counterparts in rolling out networks and services.

Thus, after an initial inertia roughly equivalent to the GSM experience and that of many other innovations, faster 3G networks have begun to spread rapidly. There were more than sixty million 3G mobile broadband users at the beginning of 2006, in around sixty different economies, and representing just under 3 percent of total mobile users. By the end of 2006, this number jumped to 420 million. These numbers do not include the slower CDMA20001x deployments, which offer speeds of less than 256 kbit/s. The leading countries in terms of total subscribers as of January 2006 were Japan, Korea, Italy, United Kingdom, and the United States.

Not surprisingly given their head start, Korea and Japan are the biggest markets for 3G mobile broadband. In Europe, Italy has made great strides: At the beginning of 2006, it ranked as the third biggest market in terms of subscriber numbers, and second in terms of mobile broadband penetration (figure 2.4a–b).

In Europe, 3G W-CDMA services (known in the region as UMTS) were launched in 2003, a few years later than in Asia. The new entrant H3G took the market by storm in Italy, Great Britain, Austria, Denmark, and Sweden. Incumbent operators with 3G licenses, however, waited until 2004 (and even 2005) to begin offering 3G services. In January 2006, about 3.4 percent of all mobile subscribers were using a mobile broadband service. It is noteworthy that those European countries in which a new entrant has been allowed to establish itself were faster in launching 3G services, whereas incumbent operators have followed the rate of rollout stipulated in their licenses. As the only new entrant on the European market, H3G has been relatively more swift in providing services to consumers, including attractive flat-rate data packages.

Asia Ahead

Asia's position as the epicenter of mobile communications is undeniable. Over the past decade, it has maintained its lead in mobile ownership—in January 2006, the region accounted for 41 percent of the world's mobile subscribers, compared to 31 percent in Europe and 21 percent in the Americas. Despite a low overall penetration per capita, Asia is home to a number of success stories in mobile communications such as mobile Internet, 3G, and mobile TV. It was able to forge ahead with mobile data while Europe struggled with WAP, low data revenues, and handsets with limited functionality. Four of the five leading 3G economies (in terms of market size) are in Asia: Japan, Korea, India, and China.

In early 2006, Asia was home to over 52 percent of the world's mobile broadband subscribers and was the first to commercially launch high-speed services. NTT DoCoMo introduced its W-CDMA 3G services as early as October 2001, under the brand name FOMA (Freedom of Mobile Multimedia Access), of which there were twenty-six million subscribers in 2006, representing just over 50 percent of the subscriber base. Internet access over mobile phones, however, was not new to Japan, due to the overwhelming success of mobile browsing services first introduced as far back as 1999, and in December 2006, more than 85 percent of mobile users in the country (85.6 million) subscribed to a browser phone service. Video messaging was also

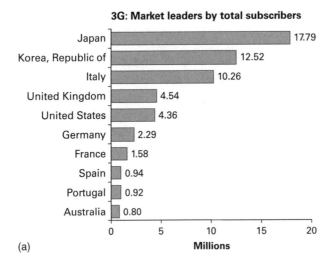

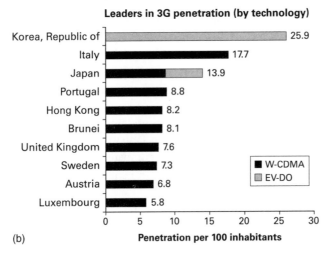

Figure 2.4a–b
Top 3G mobile markets.

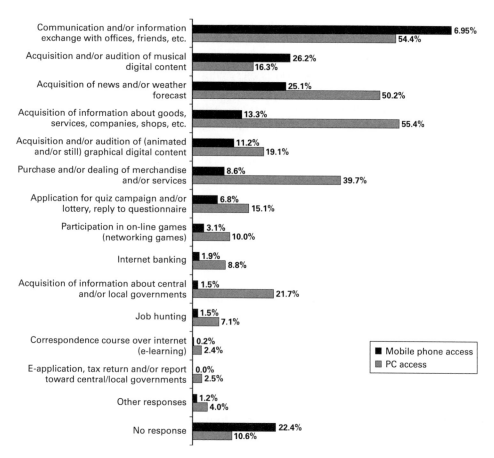

Figure 2.5

Activities using PC vs. mobile phones in Japan, 2005.

successful—and easy to use: with a built-in mobile video camera, users can take short videos (containing audio), attach it an e-mail called a "super mail," and send it to another handset. Recipients simply open the e-mail to view the video.

South Korea's SK Telecom was the first to launch CDMA20001x in October 2000. The country has the world's highest proportion of high-speed mobile handsets per capita—more than 80 percent. Service providers focus on residential rather than business users, and a third of the revenue generated by these stems from data transmission. In Korea and Japan, many of the applications, such as music and games, are designed to appeal to the youth market. Mobile phones are more popular than PCs for communication, and digital music downloaded over mobiles has now overtaken digital music downloaded by PCs (figure 2.5).

However, the unfolding of the mobile market in Japan and Korea is in contrast to most other Asian countries, many of which are developing or emerging economies. In China, where overall mobile penetration is still low (at just under 30 percent in January 2006), the main data application is messaging. The overwhelming popularity of SMS is due to its relatively low cost compared to a voice call. During 2005 alone, Chinese users sent about 300 billion SMS messages. Trials of the homegrown 3G standard, TD-SCDMA, are also underway. The conditions for rolling out the new 3G network are good, given the population density, a growing infrastructure base, and the popularity of SMS. However, the low penetration rate, the time needed to perfect the network standard, combined with misgivings about foreign investment have been delaying its progress.

Though the other giant in the region, India, though it figures in the top five mobile economies, it has yet to deploy mobile broadband. Over CDMA20001x networks, services on offer are for the most part limited mobility services. And the country's overall mobile penetration rate, at 8.15 percent in January 2006, is still under a third of China's rate. Still, and perhaps as a result, it remains the second-fastest-growing mobile market in the world. Many regulatory barriers remain, however, such as high license and spectrum fees, lack of opportunities for infrastructure sharing, and a persistently fragmented market based on "circles."

More Mobiles in the Developing World

Mobile communications have been a boon to the developing world. Before their deployment in rural areas, some citizens had to walk a fair distance before finding a telephone to contact family members or to conduct business. Mobile phones have served to extend access to information and communications, and have in this respect narrowed the digital divide, though much work still remains to be done. Mobile use in the developing world is more widespread than any other ICT, that is, personal computers or fixed-line telephones.

Even though industrialized countries account for the largest proportion of the global telecommunication sector (by value and by number of subscribers), much of the new growth is occurring in the developing world and that is where most of the potential for growth exists. In 1993, there were 3 million mobile phones and 141 million fixed lines in the developing world—ten years later, mobile phones had jumped to 608 million, while fixed lines trailed at 546 million. And in the period between 2000 and 2003, it was the developing world that accounted for almost 60 percent of the new growth in the market (new mobile users added).

The digital divide seems to be shrinking in some technologies, such as fixed lines and mobile phones, but expanding in other newer technologies including broadband

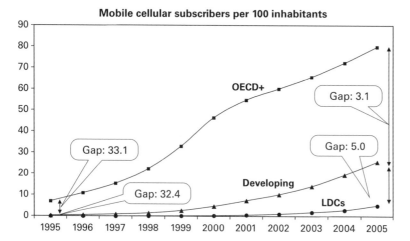

Figure 2.6
Mobile phone international digital divide, 1995–2005.

(figure 2.6). Mobile communications has been a particular success story, with the number of mobile cellular subscribers in developing countries rising from just 12 million in 1995 to more than 1.15 billion in 2005. This represents a compound annual growth rate of 58 percent per year. For mobile at least, the ratio between OECD+ countries and developing economies has decreased considerably, falling from 33.1 to 3.1. Among the least developed countries (LDCs), mobile subscribers outnumber fixed lines by seven to one.

There are a number of reasons for the rapid growth of mobile communications in developing countries. Most importantly, and as already mentioned, the systemic lack of fixed-line infrastructure means there is a significant latent demand for communication services. The low price of rolling out mobile networks allowed for deployment in rural and low-income areas. Unlike fixed networks, mobile networks are also more suitable to regions that are difficult to reach or have rugged and accidental terrain. Moreover, compared with personal computers or the Internet, mobile phones require a lower skills base and can be used by the financially disadvantaged and the illiterate. The advent of the prepaid business models reduced the cost of ownership, and in particular "first ownership," encouraging new users to adopt services on a trial basis (and until they could afford it). Mobile handsets also can be shared much more easily than Internet connections or personal computers. In some cases, mobile sharing has even become a business opportunity, creating jobs and extending access. Differences in the relative success of mobile communications in developing countries, for example, those oft remarked between India and China, depend on factors such as demographics (population density, income distribution by region), social norms and culture (e.g., urbanization), regulatory structure (e.g., competition), and economic factors (e.g., average household income).

The advent of mobile phones in underserved areas has facilitated social networks and has had a positive impact on social capital. Through mobile phones, citizens can communicate with a wider section of society, from family members to teachers and business associates. Many farmers in developing countries (e.g., Kenya) use mobile phones to check market prices for commodities and thus avoid exploitation by middlemen. Mobile phones are an important point of contact allowing people to participate in an economic system: for instance, a tailor in a small village of India can use a mobile phone to receive calls from potential clients even when on the road. The shop is always open. Mobile phones can also reduce the need for unnecessary travel to speak with business associates or purchase items that may be out of stock. Many women have gained economic freedom through mobile phones, setting up small businesses and even reselling mobile talk time.

The growth of the mobile industry in many developing countries has been limited not only by call charges but also by the high cost of handset ownership. A survey conducted by the GSM Association found that in sixteen of the poorest fifty countries, taxes represent 20 percent of the cost of owning a mobile phone. This constitutes a significant deterrent to the widespread adoption of the mobile. In an effort to address the problem, the GSM Association is promoting a low-cost (USD 30) handset to stimulate adoption in developing countries. Countries like India, where mobile overtook fixed in October 2004, have slashed their handset taxes, but must now turn their attention to persistently high communication tariffs. Though mobile phones have brought a number of benefits to the developing world, much work remains to be done on issues such as affordability and network coverage.

Mobile Magic to Come

Today's mobile phone is certainly for talking. But more and more, it is being used for data applications, multimedia, and entertainment. But what might the mobile phone of the future be? There are a number of new areas that mobile manufacturers are exploring, but two stand out: sensors and radio-frequency identification (RFID).

Reading with Mobiles

Simply put, radio-frequency identification enables the real-time identification of everyday objects and the collection and management of information regarding their status. Information embedded in tags the size of a grain of rice can include location, status, date of manufacture, date of expiration, owner, and so on. The tiny tags (stores of information) can be affixed to items such as medication, tires, and clothing, as well as fast-moving consumer goods such as toothbrushes and cosmetics. Information contained on the tags can be read by RFID readers in selected locations. Early applications of RFID made their appearance in transport and logistics, for example, highway toll collection and access control for secured buildings. But this decade has seen a significant rise in interest in the technology for a wide variety of applications such as retail and lifestyle. The potential uses of RFID in the retail industry are manifold. In the area of supply-chain management, stores can locate and replace inventory more efficiently and accurately using the technology. For shoppers, items tagged with RFID chips no longer need to be scanned individually at the checkout, making for a faster and more convenient shopping experience. The inclusion of RFID readers in mobile phones is the next natural step. Nokia introduced two GSM handsets with RFID-reading capabilities for businesses during 2004. Within a couple of years, the company intends to give consumers the ability to use RFID-enabled mobile phones to read information about consumer products sold in retail stores and is developing a prototype jointly with Verisign. Analysts predict that by 2015, six hundred million mobile phones will be RFID-enabled. In the area of standards-setting, the Near Field Communications Forum is looking at ways to use RFID to bridge the connectivity gap between all kinds of devices (such as the mobile phone) and electronic information transfer. Mobile phones can become an important platform for users to communicate with networked or smart objects and open a host of possibilities for location-based services. In the future, by pointing to objects tagged with RFID, users will be able to scan or download information about price, ingredients, and any promotions.

One of the first trials of mobile RFID shopping was run in Tokyo in 2003 using tags, rather than readers, affixed to mobile phones. The tags were used to locate customers as they passed readers installed in different areas of the mall. Through messaging and Internet services (i-mode), shoppers could receive information on their mobile phones about targeted promotions and entertainment options.

Since the loss (even temporary) of a mobile phone often causes panic or disruption in a user's daily life, scientists are beginning to find solutions to address our growing dependence on technology. The MIT Media Lab has recently designed a build-your-own bag for those who tend to leave behind mobile phones or keys, in particular when leaving home or the office. The bag is made of computerized fabric patches with radio receivers and antenna, which communicate with RFID tags affixed to a mobile phone or key ring. Furnished with a list of prespecified tagged objects, the bag's intelligent reader is automatically activated the moment it is picked up, initiating a verification of the contents of the bag against the preset list and alerting the user if something is missing. Mobile phones with the same reading capabilities as the MIT bag are also likely to be developed.

Sensing with Mobiles

The use of information stored on RFID tags in combination with sensor capabilities will no doubt extend communication and convenience even further. The main function of sensors is to measure specific phenomena through the collection of physical values such as pressure, shape, humidity, temperature, noise, and velocity. Making sensors mobile will give them more autonomy and collaborative potential with other objects, and make the mobile an even more personalized and responsive device than it is today. Like RFID, sensors will provide yet another mechanism for acquiring data and information about a user's environment and the objects in it, including the mobile phone itself. Both LG Electronics (Korea) and NTT DoCoMo (Japan) have released handsets equipped with fingerprint sensors to determine the identity of the device's owner. Fingerprint sensors can also be used to verify the age of users, thus enabling the limitation of access to adult content, gaming applications, and chat rooms. Though Asian manufacturers already have begun deploying such sensor phones, their release in Europe is expected to proceed more tentatively.

In addition to biometrics and security functions, the inclusion of sensors in mobile phones can make possible a host of other applications. The introduction of the camera phone has given mobiles the sense of sight, so to speak. And mobiles may soon develop complementary olfactory capabilities. Siemens is one company that has created a mobile phone with tiny sensors that alert users to bad breath and similar gases. Nanotechnology inside the mobile phone may enable it to detect chemical changes in the environment. Such phones could help joggers with determining the level of ozone, or act as breathalyzers for designated drivers. Thought is being given to the inclusion of sensors such as blood glucose meters for diabetics and peak flow meters for asthmatics. Early such prototypes are already on show. In this way, mobiles can not only monitor the status of a user's health and environment, but also act as command and control centers for an intelligent home or a smart vehicle.

Conclusion

Mobile communication is decidedly an integral part of personal, national, and economic life, facilitating business and increasing the conveniences of daily existence. Though it has not yet achieved universal service, its reach now extends to men, women, and children all over the world—in both developed and developing countries. We have seen systems of transportation evolve from the horse carriage to the rocket. Similarly, communication systems have grown from Morse's telegraph and Marconi's radio to the mobile of today, and the end to this development is not yet in sight. The difference is that few of us have ever been in a rocket but a large many have used a mobile phone. The mass adoption of mobile phones is unprecedented in the world of technology.

The mobile handset looks roughly the same as when it first appeared in our world. But that is where the comparison stops. The resemblance is deceptive, for not only is it vastly more advanced today but it promises to mutate into something quite different and for which no adequate nomenclature has yet been developed. From a mere telephone, it is becoming a powerful computing tool of which the telephony part is but a small subset. Much can be learned from tech-savvy economies such as Japan and Korea that are at the cutting edge of new services and applications, having turned a once simple device into a universal and multipurpose portal. But the impact of mobiles in the developing world is no less than spectacular: it has extended connectivity, stimulated businesses, and created new jobs.

It does not strain credulity to envisage a rapidly approaching time when the mobile will act as a command and control center, anticipating our daily needs in the home and on the road. Since humans began using tools, nothing resembling the mobile phone has been in their grasp. And this tool of tools continues to grow in power, scope, and utility. Perhaps it will never quite match the finger of Steven Spielberg's E.T., but it seems to approach gradually its mystical power.

3 | Shrinking Fourth World? Mobiles, Development, and Inclusion

Jonathan Donner

In his far-reaching trilogy, *The Information Age: Economy, Society, and Culture*, Manuel Castells (1996; 1997; 1998) describes how advances in information technology have combined with the strengthening of global capitalism to usher in a new paradigm of human organization, called the Informational Society. One of the trilogy's central assertions is that the transition to the Informational Society has brought about a "Fourth World" of marginalized peoples and regions bypassed by information technologies and excluded from participation in the networks of production, exchange, and consumption that are the primary mode of organization in the new millennium.

As the other chapters in this volume make clear, the postmillennial world continues to experience rapid technological change, this time brought about by a sudden proliferation of wireless, mobile, and personal communication devices. The highest rates of growth in mobile penetration are currently in the developing world, as millions of people purchase their first handsets.

The rapid uptake of mobiles in the developing world has not escaped the attention of the mass media, nor of policymakers. Recent research suggests that mobiles, like other ICTs, contribute to growth (Waverman, Meschi, and Fuss 2005) and foreign direct investment (Williams 2005), and make markets more efficient (Jensen 2007). Yet the case for the mobile's role in broad-based economic development is far from settled. In particular, questions remain at the micro-level about the actual household-, firm-, and village-level processes by which mobile telecommunications support growth and poverty alleviation. Similarly, the ongoing policy dialogue about the digital divide (Katz and Aspden 1997) has expanded to include mobile communication devices. Some argue that the devices' rapid uptake is evidence for a closing divide (*Economist* 2005); others are concerned that the gaps between the mobile haves and have-nots represent a new, untethered manifestation of the divide.

So for both intellectual and policy reasons, it is important to consider how the spread of mobile telephony affects economic development in the poorest communities on the planet. In considering the question, I use Castells's framework of the Fourth World as an organizing principle to re-examine two questions: What are the most important impacts of mobile technologies? And, how will they diffuse? I say *re-examine* since neither question is particularly new. However, the answers to both questions are different in the Fourth World, due to extreme economic scarcity and the lack of land-lines. Indeed, addressing these differences is a key to understanding how mobile telephony is likely to diminish, though not eliminate, the inequities and isolation that characterize the Fourth World.

Background

In the trilogy, Castells argues that in an informational age, synchronized and integrated networks of information, production, and exchange are the new and prominent feature of social organization. People, firms, and regions organize around these networks, which take on structures and goals of their own. These networks challenge (but do not replace) the nation-state, transform (but do not replace) stand-alone firms, and transform (but do not eliminate) human experience of space and time.

The reorganization of human activity around networks has brought about a bifurcation of the world into informational producers and vulnerable, replaceable labor (Castells 1998, p. 364). The logical extreme of this bifurcation is the "Fourth World," a term Castells uses not to identify a group of impoverished sovereign states (e.g., the "Third World" (Horowitz 1972)) but rather to identify a form of marginalized existence that exists across the globe in both rich and poor nations (Manuel and Posluns 1974). Put simply, the inhabitants of the Fourth World are the "structurally irrelevant" (Castells 1996, p. 135) in the current structure of the global economy; they are neither producers nor significant consumers within it. Their lives are difficult; disease, illiteracy, crime, and vulnerability to environmental harm compound to create what Jeffrey Sachs (2005) calls "Poverty Traps." Residents of the rural Fourth World are isolated by geography and economic activity. Those of the urban Fourth World—whether in the slums of the developing world or the decaying former industrial neighborhoods of North American and European cities—live immediately adjacent to the worlds of information production (Santos 1979) but remain excluded and do not participate in the informational exchanges that sustain and define their cities.

Thus, a community's use of ICTs is one of the central determinants of its participation in the informational system, or its relegation to the Fourth World. Addressing Africa's "technological apartheid," Castells argues, "Information technology, and the ability to use it and adapt it, is the critical factor in generating and accessing wealth, power, and privilege in our time" (Castells 1998, p. 92).

When originally completed in 1998, the first editions of *The Information Age* did not cover mobile telephony in any detail. However, the recent review book *Mobile Communication and Society* (Castells et al. 2007) addresses this gap by reviewing many studies to date of mobile communication. The review is broad and, although it includes a rich discussion of mobiles in the developing world, it does not specifically revisit the concept of the Fourth World.

Thus, this brief chapter must tread lightly, linking two concepts—the existence of a Fourth World within an Informational Society, and more recently, mobile communication—both of which have been covered in great detail by Castells and his colleagues, and that have been and will continue to be debated extensively around the world. Nevertheless, I think the Fourth World is an important concept to be brought directly to bear on our understanding of the significance of mobiles within human society. By framing isolation and exclusion in informational terms, rather than strictly regional, national, or cultural ones, it allows for a sharper identification of the relationships among connectivity, ICTs, networks, and prosperity. This chapter focuses on implications of these relationships for the emerging discussion around mobile theory (represented by this volume), and does not comment directly on the larger debates on the nature and structure of the Fourth World nor the Informational Society as a whole.

This analysis focuses on the rural and urban poor to the exclusion of the more prosperous users living in the developing world. There are important unanswered questions about how the usage patterns of emerging urban middle classes (and elites) in the developing world mirror or differ from their counterparts in high-income nations. However, the Fourth World is at its core defined by socioeconomic marginalization, and studies of mobile use among communities facing severe economic constraint (Donner 2004; Gamos 2003; Goodman 2005; Molony 2006; Samuel, Shah, and Hadingham 2005; Souter et al. 2005) remain rare.

An absence of such studies is understandable—mobiles were developed for and first deployed in more prosperous parts of the world (Dholakia and Zwick 2004). Indeed, the scholarship about the social significance of mobile technologies is itself primarily a product of the knowledge centers of the Informational Society. Yet to the common archetypes of the perpetually connected teenager and the data-devouring mobile information worker, researchers need to add the users who live on the margins of the Informational Society. After all, many have elected to spend the equivalent of USD 40—a substantial amount of their total annual income—for a handset and prepaid card. My purpose then is to revisit two of the broad concerns from the mobile society literature, exploring the impacts of mobile technologies and the patterns of their spread through society as they relate to inhabitants of the Fourth World. The discussion will highlight two additional concerns left unresolved by the evidence to date.

What Are the Most Important Impacts of Mobile Telephony?

As mentioned, the recent review by Castells et al. (2007) addresses the various impacts of widespread mobile use: on youth culture, on language, on politics, and on human experiences of space and time. Another approach, offered by Katz and Aakhus (2002), suggests an overarching theme of *Apparatgeist*—a universal spirit embodied in mobile technologies that, by reducing the costs of communication and by increasing individual control over the time, location, and content of communication, tends to encourage individualism and self-expression. Insofar as there is a bias in this general form of inquiry, the bias is toward finding ways in which mobile technologies enable social change (Murdock 2004). Like this chapter, many of these works focus on the significance of a single technology (mobile telephones) for society. In doing so, changes associated with the new technology are often framed, implicitly or explicitly, as challenging, improving, or replacing other communication technologies, particularly the landline telephone and the Internet. These comparisons break down, however, in the context of the Fourth World, where the other reference technologies are simply absent.

Some researchers have begun to tackle how these changes might manifest in the developing world. They focus on how users can draw on mobile technology to redistribute political power, giving the previously disenfranchised a voice in the dialogue (Obadare 2004; Pertierra et al. 2002; Rafael 2003). Mobiles also carry significant symbolic power, representing modernity, prosperity, and individuality (Donner 2004; Varbanov 2002). Migrants in China are purchasing mobiles priced far beyond their means (Castells et al. 2006). In Africa, as elsewhere, some people without the means to purchase a mobile may make a false one out of wood or carry a broken one just for display. Indeed, for those in the Fourth World living on a few dollars a day, exposure to extensive advertising may cause more unfulfilled want than satisfied need (Alhassan 2004). Yet while these questions of individuality, identity, and political motivation are important to understanding the use of mobiles in the developing world, they do not directly engage the economic-equity nexus central to the dialogue on ICTs and economic development.

To focus on the qualities that are unique to the mobile phone is to risk obscuring their most important function for many users in the developing world: *affordable, basic, person-to-person connectivity*. Though mobiles offer SMS, roaming, ringtones, and other new services, they are often functionally equivalent to landlines in terms of their economic and social utility (Hamilton 2003; Hodge 2005). As the surprisingly deep demand for mobile services in sub-Saharan Africa has made clear, the desire of people to use telecommunications services for communication is strong indeed (Gamos 2003).

There are three primary components to this argument for substitutability between landlines and mobiles. First, basic connectivity for businesses, education, and social and governmental institutions is important. For example, my work in Rwanda suggests that small-business owners use mobiles to acquire new customers and new suppliers, to learn about job opportunities, and to obtain critical information about market prices (Donner 2006). More broadly, callers can use mobiles or landlines to break through isolation to call an ambulance (Souter et al. 2005), consult an agricultural expert, coordinate remittances (Tall 2004), or even check on a loved one living far away.

Second, many users do not require the kind of broad, intercity roaming that GSM and CDMA systems deliver. Qiu's (2005) research on the little smart system in China makes this clear—little smart provides low-cost, limited mobility handsets to more than eighty-five million users in smaller cities and rural areas throughout China. Similar systems have deployed in India (O'Neill 2003). In these cases, the "mobile" and roaming elements of the system are not as important as the fact that their wirelessness reduces the cost of basic connectivity to end users.

Finally, shared-access approaches, such as the Grameen phone program (Bayes 2001) and the phone shop franchises in South Africa (Reck and Wood 2003), challenge our focus on mobility and perpetual contact. Shared mobiles are neither "anytime" nor "everywhere" for their users. Yet they are very popular and critical to providing access to rural areas and poor communities (Donner 2005).

If we pursue this line of functional equivalence, then Saunders, Warford, and Wellenius's (1994) review of the importance of telecommunications in economic development remains an excellent guide to our understanding of the impact of mobiles in the Fourth World. These impacts include better market information; improved transport efficiency and more distributed economic development; reduction of isolation and increase in security for villages, organizations, and people; and increased connectivity to (and coordination with) international economic activity. Beyond economic impacts, research on the social impacts of the landline phone (de Sola Pool 1977) on urbanization and geographic specialization are similarly helpful.

What Pattern Will Mobile Diffusion Take?

It is a standard refrain that mobile CDMA and GSM networks are now spreading much faster than the landline copper networks, providing coverage to rural areas where landline coverage is prohibitively expensive. The world will have three billion mobile users long before it has three billion landlines, perhaps by 2010 (*Wireless Intelligence* 2005). To examine how this particular diffusion pattern will impact the Fourth World, let us first address the twin challenges of access and affordability (relative to other options), which will remain the primary barriers to mobile adoption among the poor citizens of the Fourth World.

Access to mobile and GSM signals continue to improve. More than 80 percent of the world's population live under or nearby a mobile signal (World Bank Global ICT Department 2005). The majority of that last 20 percent without access to a signal lives in the most isolated, least cost-effective, most sparsely populated, or most politically restricted places to serve. If a place lacks access to a mobile signal, it is much more likely that its inhabitants live in the Fourth World.

Affordability remains a distinct challenge for those in the Fourth World, even for those who live under a signal. The price of the handset continues to drop, with Motorola announcing in 2005 its intention to develop a USD 30 handset (Baines 2005). Even less expensive handsets are not out of the question. Nevertheless, more than two billion people live on a dollar or two a day, and cannot easily afford a mobile handset, nor the airtime to use it most fully. As discussed in the previous section, shared access models will continue to be quite popular among this large group.

What of those in the Fourth World who do elect to purchase a mobile, or visit a GSM shared phone kiosk? To some optimists, the mere potential to connect to anyone anytime, enabled by the mobile, might represent an escape from membership in the Fourth World. Yet it seems more prudent to reiterate that like other ICTs (Grace, Kenny, and Qiang 2001), mobiles can enable economic activity that leads to growth, but do not directly create that growth. In other words, one should view mobiles (or telephones in general) as a necessary but not sufficient condition for growth and for poverty alleviation. Indeed, Castells's (2001) discussion of the digital divide in *The Internet Galaxy* makes a similar point.

At the micro-level, this formulation suggests that an individual or organization with new access to voice conductivity, enabled by mobile phones, now has more potential to connect to the networks of production and exchange (of commodities, labor, or information) that create prosperity in the informational age. Certainly, stories from the Fourth World suggest that some individuals, such as fishermen (Abraham 2006; Jensen 2007), farmers (Tobar 2004), and some microentrepreneurs (Donner 2004, 2005; Samuel, Shah, and Hadingham 2005), are able to use wireless connectivity to participate more productively and frequently in formal markets. But access does not immediately translate into participation in economic networks. Other barriers such as literacy remain (Warschauer 2003). Recent surveys suggest that rural telecommunications users are much more likely to use phones in emergencies, or to stay in touch with friends and family, than for business purposes (Souter et al. 2005).

At the community level, enabling factors need to be co-present in order for significant reductions in poverty to take place (Grace, Kenny, and Qiang 2001). Without systems for health care, education, sanitation, power generation, transport, financial transactions, and security (to name a few), stable markets cannot develop, and poverty traps (Sachs 2005) will remain. With some modifications, mobile applications such as m-government, m-learning, m-commerce, and m-health can each make a significant contribution, but it is too optimistic to think that a rollout of these applications in the Fourth World would resolve these complex environmental and infrastructure problems completely.

At an even broader level, considering regions as a whole, it is clear that connectivity alone does not guarantee participation in the flows and exchanges around information production, characterized by the milieu of innovation (Castells 1996; Porter 1990). ICTs enable local networks, which help create the patents, research, investment mechanisms, and knowledge that are the building blocks of wealth-creating industrial and informational clusters, but such clusters will not emerge simply because a population has better access to mobile telephony.

Instead, for at least the next decade, the steady growth in adoption will result in a partial diffusion of mobile communication technologies into the communities of the Fourth World. Due to access constraints, some of its citizens will have no opportunity to use a mobile; others will have access but will not use one. Of those who do use mobiles, some will use them in ways to increase their material prosperity; others will choose to stay in touch with friends and family and to improve their qualities of everyday life (Souter et al. 2005).

Changes to mobile technologies and their supporting systems, such as the introduction of calling-party-pays and prepaid cards (Oestmann 2003), the development of predictive SMS text support for multiple languages, and the revision of licensing rules already have improved significantly the accessibility, affordability, and utility of mobile technologies for the poor. Further innovations to address the accessibility and affordability challenges remain in the hands of the market, the hardware and service providers, and regulatory agencies. As Forestier, Grace, and Kenny (2002) point out, telecoms are generally "pro-poor," but they can be "sub-pro-poor" if they are concentrated in cities and among the rich, or "super-pro-poor," if they are broad-based and accessible. This argument is the legacy of universal access and provides the mandate for further technical innovations, such as voice-over IP, and business model reforms, such as limited-mobility mini-franchises (Engvall and Hesselmark 2004), that can extend access and affordability beyond what market forces alone might otherwise provide.

Discussion of Pivotal Issues

This chapter starts by reintroducing two general questions about mobiles in society, concerning mobiles' primary impact and about their diffusion, and has summarized how the answers remain different vis-à-vis the Fourth World. Mobiles provide affordable person-to-person connectivity (like landlines), yet are now more accessible in the Fourth World than landlines ever were. However, they are no panacea—no magic key for growth nor for poverty alleviation. Thus, we can expect the widespread availability of mobiles to contribute to the erosion of the Fourth World, but should not expect the former to eliminate the latter. To conclude this discussion, I highlight two additional questions left unresolved by the evidence to date, and consider some implications for the broader issues facing mobile theory.

Does Denser Equal Better?

One question is whether we can presume that the introduction of mobiles into a society is a linear good, in the sense that if 10 percent of the population have mobiles, then 20 percent would be better, and 40 percent better still. When we focus on the penetration figures to the exclusion of an understanding of the intracommunity distribution of these devices, we engage in this kind of thinking. Alternatives to the "linear good" frame come in two forms: optimistic and cautionary.

The optimistic approach would suggest that Metcalfe's law (Gilder 2000) applies: an increased density of network devices yields increased returns to all the users of the network. This theme appears in Townsend's (2000) assertions about how mobiles accelerate "the urban metabolism." My work (Donner 2006) with microentrepreneurs in Kigali supports this view. As both buyers and sellers—and husbands and wives—acquire mobiles, they are more able to use them to efficiently microcoordinate (Ling and Haddon 2003) their daily activities.

The cautionary approach points to the problems that occur when one segment of a population actively uses a technology while another does not. Whether one calls this the knowledge gap (Tichenor, Donohue, and Olien 1970) or the digital divide (Katz and Aspden 1997), it points to a process whereby those fortunate (prosperous) enough to have mobiles are best positioned to take advantage of them, further exacerbating gaps between rich and poor (Forestier, Grace, and Kenny 2002). There is evidence that the presence of even a single phone in a village leads to a reduction in price uncertainty and higher returns for a village's agricultural outputs (Bayes 2001; Eggleston, Jensen, and Zeckhauser 2002). However, one could imagine scenarios in rural areas where those with mobiles have an ability to link to streams of remittances (Tall 2004), educational services, price information, or emergency services that those without mobiles do not. Indeed, Souter et al. (2005) found that telecommunications, including mobiles, are more valued and more intensely utilized by prosperous users in rural Africa and Asia than by less prosperous users; similarly, an early assessment of the Grameen Village Phone Program in Bangladesh (Bayes 2001) found that 85 percent of users of the service were nonpoor individuals (as opposed to the poor users we might imagine would be first in line to use the service). Also, a study of Nigerian fabric weavers illustrated how unequal mobile access (restricted to the wealthier producers and traders) ended up further marginalizing those without the wherewithal to purchase a mobile (Jagun, Whalley, and Ackerman 2005). Unfortunately, these instances can be seen as the latest in a set of observations that support Castells's linkage of ICTs and the Fourth World in the first place. Thus, it is important for analyses of mobiles in the developing world to move beyond the assumption of a linear good, and to look at ways in which mobiles, like other ICTs, support and legitimize the goals and structural positions of the powerful at the expense of those of the less powerful (Thompson 2004). To do so requires research approaches that include users and nonusers. Unfortunately, the data on these potential gaps and their impacts on equity and growth remain sparse; more research is urgently needed to explore the dynamics within communities where connectivity is now present but not universal.

Connecting Where to Where?

A second open question involves the link between mediated communication, social networks, and local space. Research has demonstrated that the mobile plays a role in facilitating rural-urban links (Gamos 2003; Oestmann 2003) and remittances (Paragas 2005; Tall 2004) in the developing world. However, quite apart from enabling anytime-anywhere conversation and long-distance ties, other research suggests that mobiles enable primarily local, proximate interactions (de Gournay and Smoreda 2003; Ling and Yttri 2002). Unfortunately, the amplification of ties within a community at the margins of the main networks of information may do little to position that community for meaningful participation in the networks of production and exchange.

When one examines mobile phone behavior from this perspective, accounting for both physical and networked space, it seems more likely that mobiles are not transforming or directly challenging the structural properties of the Informational Society as proposed by Castells. Instead, mobile and wireless personal communication devices may be reinforcing those same structures, albeit at a finer, more granular, and perhaps even more amplified level. If the fiber-optic, landline, and Internet systems of ICTs were the arteries of the Informational Society circa 1998, then mobiles might have added billions of capillaries. Similarly, we might consider the spread of mobiles within communities, asymmetrically amplifying power in some places and further diminishing it in others, as a fractal—a smaller scale replication of the structure defined by Castells in his original trilogy. This too would be a fruitful direction for future research, as again our data on the intracommunity dynamics of mobile use in poorer communities remains scarce.

Implications for Mobile Theory

This chapter focuses on the implications of the spread of mobile technologies for the Fourth World. However, the exercise sheds light on a couple issues that are worth noting for thoughts about mobile theory in general. First, the emerging shared-access and limited-mobility models challenge our prevailing theories of what mobile technologies mean to users, and force us to reintegrate the importance of basic connectivity into our discussions. Indeed, the pressing economic needs of users in the developing world, and in the Fourth World in particular, underscore the importance of understanding microeconomic impacts and implications of mobile use—even when the bulk of the calling behavior might be with friends and family (Donner 2006; Samuel, Shah, and Hadingham 2005).

Second, the enormous economic disparities and complex social contexts within nations, such as the economic differences between (and linkages among) the rural areas and the urban boomtowns of India and China, illustrate possible complications with the cross-cultural approach to mobile theory. With socioeconomic status remaining a primary driver of mobile adoption, analyses of mobile communication behavior under conditions of economic scarcity should be included in general theories of mobile use.

In closing, the world continues to move from conditions of 80 percent mobile coverage and 32 percent penetration in 2005 (ITU 2006) toward broader coverage and penetration by the end of the decade. The spread of mobile and personal communication devices will provide additional opportunities for *some* individuals, households, firms, and communities to engage with the dominant networks of information production and exchange, and to exit the Fourth World. But not all people will have this opportunity, and the opportunities themselves do not guarantee a transition toward inclusion and equality. As Castells warns, the Informational Society's tendencies to create isolation and polarization are "not inexorable," but rather can be confronted by "deliberate public policies" and "conscious action (1998, p. 364). As researchers interested in the social and economic impacts of mobile and wireless technologies, and with the potential to help guide the agenda for technology deployment and regulation, we have a role to play in this conscious action.

References

Abraham, R. 2006. Mobile phones and economic development: Evidence from the fishing industry in India. Paper presented at the Conference on Information Technologies and International Development (ICTD 2006), Berkeley, CA.

Alhassan, A. 2004. *Development Communication Policy and Economic Fundamentalism in Ghana.* Tampere, Finland: University of Tampere Press.

Baines, S. 2005. The next billion mobile users. http://networks.silicon.com/mobile/0,39024665,39150979,00.htm.

Bayes, A. 2001. Infrastructure and rural development: Insights from a Grameen bank village phone initiative in Bangladesh. *Agricultural Economics* 25(2–3): 261–272.

Castells, M. 1996. *The Information Age: Economy, Society and Culture, Volume I: The Rise of the Network Society.* Malden, Mass.: Blackwell Publishing.

Castells, M. 1997. *The Information Age: Economy, Society and Culture, Volume 2: The Power of Identity.* Malden, Mass.: Blackwell Publishing.

Castells, M. 1998. *The Information Age: Economy, Society and Culture, Volume 3: The End of Millennium.* Malden, Mass.: Blackwell Publishing.

Castells, M. 2001. *The Internet Galaxy: Reflections on the Internet, Business, and Society*. Oxford, New York: Oxford University Press.

Castells, M., J. L. Qiu, M. Fernández-Ardèvol, and A. Sey. 2007. *Mobile Communication and Society: A Global Perspective*. Cambridge, Mass.: The MIT Press.

de Gournay, C., and Z. Smoreda. 2003. Communication technology and sociability: Between local ties and "global ghetto." In *Machines that Become Us: The Social Context of Personal Communication Technology*, edited by J. E. Katz. New Brunswick, N.J.: Transaction Publishers.

de Sola Pool, I., ed. 1977. *The Social Impact of the Telephone*. Cambridge, Mass: The MIT Press.

Dholakia, N., and D. Zwick. 2004. Cultural contradictions of the anytime, anywhere economy: Reframing communication technology. *Telematics and Informatics* 21(2): 123–141.

Donner, J. 2004. Microentrepreneurs and mobiles: An exploration of the uses of mobile phones by small business owners in Rwanda. *Information Technologies for International Development* 2(1): 1–21.

Donner, J. 2005. The social and economic implications of mobile telephony in Rwanda: An ownership/access typology. In *Thumb Culture: The Meaning of Mobile Phones for Society*, edited by Glotz, P., S. Bertschi, and C. Locke. Bielefeld, Germany: Transcript Verlag.

Donner, J. 2006. The use of mobile phones by microentrepreneurs in Kigali, Rwanda: Changes to social and business networks. *Information Technologies and International Development* 3(2): 3–19.

Economist. 2005. Calling an end to poverty. *The Economist* 376: 51–52.

Eggleston, K., R. T. Jensen, and R. Zeckhauser. 2002. Information and telecommunication technologies, markets, and economic development. In *The Global Information Technology Report 2001–2002: Readiness for the Networked World*, edited by G. Kirkman, P. Cornelius, J. Sachs, and K. Schwab. New York: Oxford University Press.

Engvall, A., and O. Hesselmark. 2004. Profitable universal service providers. http://www.eldis.org/fulltext/profitable.pdf.

Forestier, E., J. Grace, and C. Kenny. 2002. Can information and communication technologies be pro-poor? *Telecommunications Policy* 26(11): 623–646.

Gamos. 2003. Innovative demand models for telecommunications services. http://www.telafrica.org.

Gilder, G. 2000. *Telecosm: How Infinite Bandwidth Will Revolutionize Our World*. New York: The Free Press.

Goodman, J. 2005. Linking mobile phone ownership and use to social capital in rural South Africa and Tanzania. http://www.vodafone.com/assets/files/en/AIMP_09032005.pdf.

Grace, J., C. Kenny, and C. Qiang. 2001. Information and communication technologies and broad-based development: A partial review of the evidence. http://poverty2.forumone.com/files/10214_ict.pdf.

Hamilton, J. 2003. Are main lines and mobile phones substitutes or complements? Evidence from Africa. *Telecommunications Policy* 27(1–2): 109–133.

Hodge, J. 2005. Tariff structures and access substitution of mobile cellular for fixed line in South Africa. *Telecommunications Policy* 29(7): 493–505.

Horowitz, I. L. 1972. *Three Worlds of Development: The Theory and Practice of International Stratification* (2nd ed.). New York: Oxford University Press.

ITU. 2006. Online statistics. http://www.itu.int/ITU-D/ict/statistics/.

Jagun, A., J. Whalley, and F. Ackerman. 2005. The impact of unequal access to telephones: Case study of a Nigerian fabric weaving micro-enterprise. Presented at the ITS 16th European Regional Conference, Porto, Portugal.

Jensen, R. 2007. The digital provide: Information (technology), market performance, and welfare in the South Indian fisheries sector. *Journal of Economics* 122(3): 879–924.

Katz, J. E., and M. Aakhus. 2002. Conclusion: Making meaning of mobiles—a theory of Apparatgeist. In *Perpetual Contact: Mobile Communication, Private Talk, Public Performance*, edited by J. E. Katz and M. Aakhus. Cambridge, UK: Cambridge University Press.

Katz, J. E., and P. Aspden. 1997. Barriers to and motivations for using the Internet: Results of a national opinion survey. *Internet Research Journal: Technology, Policy & Applications* 7(3): 170–188.

Ling, R., and L. Haddon. 2003. Mobile telephony, mobility, and the coordination of everyday life. In *Machines that Become Us: The Social Context of Personal Communication Technology*, edited by J. E. Katz. New Brunswick, N.J.: Transaction Publishers.

Ling, R., and B. Yttri. 2002. Hyper-coordination via mobile phones in Norway. In *Perpetual Contact: Mobile Communication, Private Talk, Public Performance*, edited by J. E. Katz & M. Aakhus. Cambridge, UK: Cambridge University Press.

Manuel, G., and M. Posluns. 1974. *The Fourth World: An Indian Reality*. New York: The Free Press.

Molony, T. 2006. Non-developmental uses of mobile communication in Tanzania. In *A Handbook of Mobile Communication and Social Change*, edited by J. Katz. Cambridge, Mass.: The MIT Press.

Murdock, G. 2004. Past the posts: Rethinking change, retrieving critique. *European Journal of Communication* 19(1): 19–38.

O'Neill, P. D. 2003. The "poor man's mobile telephone": Access versus possession to control the information gap in India. *Contemporary South Asia* 12(1): 85–102.

Obadare, E. 2004. The great GSM boycott: Civil society, big business and the state in Nigeria. http://www.isandla.org.za/dark_roast/DR18%20Obadare.pdf.

Oestmann, S. 2003. Mobile operators: Their contribution to universal service and public access. http://rru.worldbank.org/Documents/PapersLinks/Mobile_operators.pdf.

Paragas, F. 2005. Migrant mobiles: Cellular telephony, transnational spaces, and the Filipino diaspora. In *A Sense of Place: The Global and the Local in Mobile Communication*, edited by K. Nyiri. Vienna: Passagen Verlag.

Pertierra, R., E. F. Ugarte, A. Pingol, J. Hernandez, and N. L. Dacanay. 2002. *Txt-ing Selves: Cellphones and Philippine Modernity*. Manila, Philippines: De La Salle University Press.

Porter, M. E. 1990. *The Competitive Advantage of Nations*. New York: The Free Press.

Qiu, J. L. 2005. Accidental accomplishment of little smart: Understanding the emergence of a working-class ICT. Paper presented at the USC Annenberg Research Network Workshop on Wireless Communication and Development: A Global Perspective, Marina del Rey, CA.

Rafael, V. L. 2003. The cell phone and the crowd: Messianic politics in the contemporary Philippines. *Public Culture* 15(3): 399–425.

Reck, J., and B. Wood. 2003. What works: Vodacom's community services phone shops. http://www.digitaldividend.org/pdf/vodacom.pdf.

Sachs, J. D. 2005. *The End of Poverty: Economic Possibilities for Our Time*. New York: Penguin Press.

Samuel, J., N. Shah, and W. Hadingham. 2005. Mobile communications in South Africa, Tanzania, and Egypt: Results from community and business surveys. http://www.vodafone.com/assets/files/en/AIMP_09032005.pdf.

Santos, M. 1979. *The Shared Space: The Two Circuits of the Urban Economy in Underdeveloped Countries*. New York: Methuen.

Saunders, R. J., J. J. Warford, and B. Wellenieus. 1994. *Telecommunications and Economic Development* (2nd ed.). Baltimore, MD: Johns Hopkins University Press.

Souter, D., N. Scott, C. Garforth, R. Jain, O. Mascararenhas, and K. McKerney. 2005. The economic impact of telecommunications on rural livelihoods and poverty reduction: A study or rural communities in India (Gujarat), Mozambique, and Tanzania. http://www.telafrica.org/R8347/files/pdfs/FinalReport.pdf.

Tall, S. M. 2004. Senegalese émigrés: New information and communication technologies. *Review of African Political Economy* 31(99): 31–48.

Thompson, M. 2004. ICT, power and developmental discourse: A critical analysis. *Electronic Journal of Information Systems in Developing Countries* 20(4): 1–25.

Tichenor, P. J., G. A. Donohue, and C. N. Olien. 1970. Mass media flow and differential growth in knowledge. *Public Opinion Quarterly* 34: 159–170.

Tobar, H. 2004. They can hear you now: cellphones moving the developing world into the global village. *Los Angeles Times*, A1.

Townsend, A. M. 2000. Life in the real-time city: Mobile telephones and urban metabolism. *Journal of Urban Technology* 7(2): 85–104.

Varbanov, V. 2002. Bulgaria: Mobile phones as post-communist cultural icons. In *Perpetual Contact: Mobile Communication, Private Talk, Public Performance*, edited by J. E. Katz and M. Aakhus. Cambridge, UK: Cambridge University Press.

Warschauer, M. 2003. *Technology and Social Inclusion: Rethinking the Digital Divide*. Cambridge, Mass.: The MIT Press.

Waverman, L., M. Meschi, and M. Fuss. 2005. The impact of telecoms on economic growth in developing nations. http://www.vodafone.com/assets/files/en/AIMP_09032005.pdf.

Williams, M. 2005. Mobile networks and foreign direct investment in developing countries. http://www.vodafone.com/assets/files/en/AIMP_09032005.pdf.

Wireless Intelligence. 2005. Worldwide cellular connections exceeds 2 billion. http://www.gsmworld.com/news/press_2005/press05_21.shtml.

World Bank Global ICT Department. 2005. Financing information and communication infrastructure needs in the developing world: Public and private roles, Working Paper No. 65. Washington, DC: The World Bank.

4 | Mobile Traders and Mobile Phones in Ghana

Ragnhild Overå

When Ghana deregulated its telecommunications sector in 1994, there were 0.3 landlines per 100 inhabitants—the same teledensity as in 1950 (Michelsen 2003). Ten years later, there were nearly 1.5 landlines, 8 mobile phone subscribers, and 1.8 Internet users per 100 inhabitants (ITU 2004). In 1997, there were only twenty-five pay phones nationwide but within two years they numbered five thousand (Segbefia 2000). Telecommunication technology's beneficial effects are particularly pronounced in developing countries where it has been estimated that the positive impact of mobile telephony on economic growth may be twice as large compared to developed countries (Waverman, Meschi, and Fuss 2005). Studies have shown that adoption of mobile phones reduces transportation and transaction costs, and enhances trust among members of trade networks (Overå 2006). Emily Chamlee-Wright (2005) argues that telecommunications are a crucial factor enhancing microenterprise. Village phone schemes seem to have a broad transformative potential beyond the emergence of "pockets of modernization" (Aminuzzaman, Baldersheim, and Jamil 2003). Yet the extreme urban bias in the geographical distribution of telecom services and their high costs, especially of mobile phones, limit many people's possibility of using new telecommunication technology.

This chapter examines how improved access to telephones, and mobile phones in particular, change daily life and economic opportunities for Ghana's traders in agricultural produce. It focuses on how traders change social, economic, and spatial practices when they acquire mobile phones. Mobile phones are great communication tools for rural-urban traders who move around a lot since they often have no registered business in an office with an address. They need of course to exchange information on prices, supply, and demand across long distances but also are often illiterate or semiliterate and therefore prefer to communicate verbally instead of having messages written for them. Yet the expense of service remains high, and vast rural areas are still without coverage (see figure 4.1), creating a digital divide across socioeconomic strata and by locale.

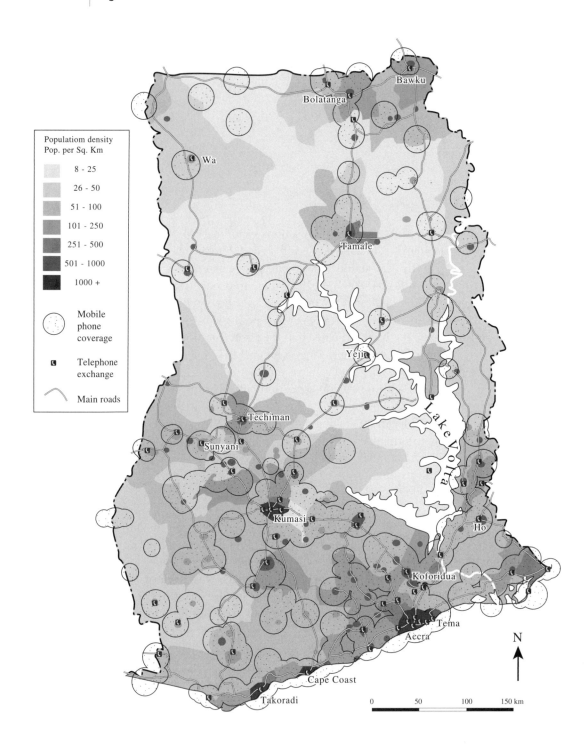

Population density
Pop. per Sq. Km

- 8 - 25
- 26 - 50
- 51 - 100
- 101 - 250
- 251 - 500
- 501 - 1000
- 1000 +

Mobile phone coverage

Telephone exchange

Main roads

Bawku

Bolatanga

Wa

Tamale

Yeji

Techiman

Sunyani

Kumasi

Lake Volta

Ho

Koforidua

Tema

Accra

Cape Coast

Takoradi

N

0   50   100   150 km

So despite its many advantages, I argue that unequal access to teleservices and especially mobile communication in Ghana marginalizes the remote and the poor relative to the urban and rich, thus eroding the positive impact of telecommunications on economic growth. To provide on-the-ground evidence for these arguments, I discuss the strategies traders employ when they start using telephones and the barriers they face in adopting the new technology. I describe how and with whom rural-urban traders communicate to organize the purchasing, transportation, and marketing of goods, and how they adapt these strategies to an improving, but still inadequate, telecommunication infrastructure.

This chapter is based on my research in Ghana before and after the telecom revolution (Overå 1993, 1998, 2005, 2006). Knowing about the "before" situation is an advantage, since once telephones are available they quickly become taken for granted (table 4.1). I, together with research assistant Charlotte Mensah from the University of Ghana, interviewed traders. She often functioned as an interpreter, mainly in the big wholesale markets and small shops in Accra, and in Tema Fishing Harbour. Through previous research we were familiar with the rural areas in which the interviewed traders purchased goods such as fish, onions, salt, tomatoes, and maize. The interviews were informal conversations at each trader's work place and lasted from a few minutes to two hours. The traders were selected through a snowball strategy.

The Rural-Urban Market Chain and Traders' Information Needs

Ghanaian marketplaces are important institutions and well organized according to gender and ethnicity with leaders for each commodity group (Robertson 1984; Clark 1994; Chamlee-Wright 1997; Overå 1998). Agricultural produce often is brought to local or regional markets by the producers themselves, but long-distance large-scale trade is mostly in the hands of itinerant wholesalers, or "travelers" (Clark 1994). Wholesalers in food stuff are for the most part women, except in some commodities like meat and onions, which are often sold by male Muslims from northern Ghana and neighboring countries. The wholesalers purchase goods from the producers in rural supply areas and organize transportation to regional wholesale markets, or to urban wholesale and retail markets. There the goods are sold to wholesale retailers, who resell to retailers or to consumers and to petty traders, who retail in even smaller quantities or hawk in the streets. The market system can thus be viewed as a predominantly female hierarchy

Figure 4.1
Telecommunication service access in Ghana, 2006. Sources: http://www.ghanatelecom.com/gh, http://www.spacefon.com, and http://www.gsmworld.com. Population data: The Center for International Earth Science Information Network (CIESIN). Map made by Kjell Helge Sjøstrøm, Department of Geography, University of Bergen.

Table 4.1
Utilization, Benefits, and Adaptation Barriers of Telecommunication

Utilization of telecommunication	Benefits of telecommunication	Barriers against adoption of telecommunication
Substitute traveling or messengers with calls	Save time and transportation costs	Infrastructure: inadequate supply of landlines and mobile phone coverage
Monitor and supervise trade partners and employees in distant locations	Wider geographical reach	High costs (landline subscriptions and mobile phone units)
Take orders from customers and inform them when supplies are available	Better security on long journeys	Long waiting lists for landlines and poor services by telecom service providers
Make orders from suppliers and receive information when consignments are ready	Coordinate activities more efficiently	Calls are cut due to congested networks (too many mobile phone subscribers per network)
Discuss prices and demand situation in distant markets	Better timing of supply and demand (higher prices and less spoilage)	Crime (e.g., mobile phones stolen, lines tapped by illegal users)
Arrange for transport and delivery of goods	Offer greater variety of goods	Many areas do not have coverage (phones are of no use to do business in many rural areas)
Receive updates, complaints about quality, delays	Be more available	Few people have phones (the number of customers and suppliers phone owners can call is limited)
Communicate with children/spouse/maid at home	Improve customer service, attract more customers	Face-to-face communication is required in negotiation of large contracts, credit requests, and exchange of sensitive information
Communicate with distant relatives and social contacts	Build a good reputation	
	Easier to combine work and family life	
	More funds and time available for expansion and diversification of business	

Source: Fieldwork, Accra, 2001 and 2003

where a small wealthy elite operate on a large scale on the top and a majority of poor petty traders struggle at the bottom of the pyramid (Robertson 1984).

Goods pass through numerous hands on their way from the producers to the consumers. The trade system can be viewed as a commodity chain in which traders act as links adding value to the product throughout each stage (Dicken 1998). It is also an "information chain" in which transactions are socially embedded. Informal institutions like the court of the market leaders (Queen Mothers or ohemma) and traders' unwritten "rules of conduct" are important providers of risk-reducing mechanisms in this "imperfect market" (i.e., North 1995). The system is based upon trust and traders' personal reputations are their most valuable assets (Fafchamps 1996; Overå 2006). Obviously, personal communication plays a pivotal role not only in practical terms but also is the glue maintaining the institutions of the market.

Before—and still for those without access to telecommunications—the main mode of communicating information apart from face-to-face contact personally or through an intermediary messenger, was—and still is—written messages, either delivered by hand (by an intermediary messenger), sent as a letter (which takes days), as a fax, or since 1995 as an e-mail. Since the majority of traders depend on others for the writing of their messages, there is always a risk that the information may be misunderstood or misused by the writer or the carrier of the message. Most traders thus spend enormous amounts of time and money traveling on bad (in rural areas) and congested (in urban areas) roads to make orders, ask for credit, collect debts, inquire about whether goods are ready, and so on. Segbefia (2000) estimates that 23 percent of all travelers on Ghanaian roads travel to exchange information.

Traders communicate often across very long distances, often while being on the road, and thus acutely experience the dilemmas of time-space temporality (see Harvey 1989). They constantly face the problem of needing information from a different place than where they are situated physically at a particular moment in time. From the rural supply end of the chain, they need information about purchase prices, the quality and quantity of goods, and the "trader-density" in particular villages or regional wholesale markets. From the urban market end of the chain, traders need information about the flow of goods influencing prices according to current demand in particular markets. The better informed a trader is about these multiple factors throughout the rural-urban market chain, the more sensible decision he or she can make with regard to where, when, how much, and at which price it will be most profitable to purchase and sell goods.

Aminuzzaman, Baldersheim, and Jamil (2003) use the notion of "information poverty" to denote a situation in which "inadequate telecommunications infrastructure leads to limitations on the choices available to individuals because high costs of telecommunications makes it too costly to seek out information about alternative courses of action" (p. 329). Hence, when traders acquire mobile phones, information asymmetries can be reduced, even when the parties do not meet physically (provided that they trust each other and are willing to share information). In the next section we see how traders reduce costs and risks and achieve advantages when they replace personal travel and intermediation with exchanging information via phone calls.

New Trading Practices: Phones and Phoning

No statistics are available on phone ownership among informal traders. A survey conducted in 2003 in Accra of one hundred informally employed men and women (at both low and high income levels) found that six had a landline at home, seventeen had a mobile phone, and two had both (Overå 2007). There was no clear correlation between gender and phone ownership, but income was decisive. In our interviews with food traders, this impression was confirmed: large-scale wholesalers more often owned phones than those operating on a medium scale, whereas none of the small-scale retailers did. With regards to income differences, the monthly incomes of maize retailers were often as low as 200,000 cedis/USD 23 while maize wholesalers' monthly incomes could be 2 million cedis/USD 233 and more. Likewise, an onion retailer could earn as little as 400,000 cedis/USD 46 a month while an onion wholesaler could earn 4 million cedis/USD 465 (Overå 2006). It is therefore not surprising that the affordability of phones is highly unequal.

Generally, the traders wish to own and use a phone but cannot afford it. Some receive mobile phones as gifts from relatives abroad, but they rarely use it for anything other than receiving incoming calls from those relatives. Even those who can afford the purchase of a mobile phone (at approximately USD 100) choose not to use it regularly. A reason for this is the considerable difference in call charges of mobile phones and landlines. Compared with the call charges of landline phones (regional 150 cedis/USD 0.016 and long distance 200 cedis/USD 0.022), a mobile phone call costs between 1600 cedis/USD 0.17 (within the same network) and 2700 cedis/USD 0.30 (to other networks) per minute. From mobile networks to landlines a minute's call costs 2100cedis/USD 0.23; and from landline to mobile networks 1800cedis/USD 0.20 (in 2005). Many therefore only receive incoming calls on their mobile phones, which is possible even if their prepaid card has run out of units, while making their own calls from a communication center. Another strategy invented to overcome high call charges is "flashing," whereby one avoids spending money on talking time by having an agreement to be called back after the ringing signal. Often, the ringing has a specific meaning, for example "I have arrived," so that calling back is unnecessary.

When mobile phone call charges are expensive and installing a landline at home is a bureaucratic process and unaffordable for most people, the "com centers" play a very important role (see Falch and Anyimadu 2003). These are private enterprises and can be large and well equipped offering a variety of services including Internet, but many com centers are simply a 2×2-meter shed inside a marketplace with one landline phone. Those working in the com centers provide an important service in delivering messages or fetching persons called. According to the estimates of interviewed com center staff, about 30 percent of the traders' calls are business related while 70 percent are family related. Obviously, privacy is not guaranteed when communicating through com centers. The possibility to control "talking space," keeping both personal and business information secret to unwanted listeners, was mentioned as an important advantage with mobile phones.

Reducing Information Poverty

Traveling to buy maize from a farmer in an area without telecommunications, a trader cannot call in advance to make orders or gather information about the current supply situation. Neither can she take the latest information about urban prices and demand into account when negotiating purchasing prices with the farmer once she has arrived in the village. The trader (and the farmer) will have to rely on information circulating by word of mouth. Alternatively, to obtain updated information from contacts in the city, the trader can travel from the village to the nearest place with mobile phone coverage or a com center with a landline. Then she must travel back and buy the goods. This is expensive, both in terms of time and travel costs. Another alternative, of course, is for the trader to quit often long-standing supply relationships with farmers in "unconnected" remote areas, where personal trust and credit relationships have been established over time, and buy goods in "connected" areas instead. Rural areas that have telephone lines installed thus get more attractive as sources of supply. For example, a salt trader buying salt in the coastal towns of Ada and Nyanyano explained that when telephone lines were installed in Ada, she could suddenly call her suppliers (via a com center) and order salt. Instead of her traveling back and forth to Ada, the suppliers could bring salt directly to her in Accra. The result was that she reduced salt supplies from Nyanyano and increased her trade with Ada. However, when Nyanyano also got a telephone line and a com center, she adapted the same strategy as in Ada.

Without access to phones, traders must rely on information from colleagues returning from market trips, which may quickly become outdated since the supply and demand situation is volatile. The alternative is to travel to places where one has had luck before and hope for the best. Auntie Gladys is a seventy-five-year-old "garden egg" (eggplant) seller outside the Makola market. She has long experience and a wide contact network. She is too old to travel herself, so her daughter travels for her to Kumasi, Techiman, Sunyani (see figure 4.1), and many other wholesale markets in the Brong-Ahafo and Ashanti regions. Gladys gets information from colleagues returning with eggplants to Accra every day, but even if she knows that her daughter is on her way to a market she just heard is saturated with eggplant buyers or to a market where the last supplies just finished, she is not able to convey these vital messages. Her daughter thus often ends up traveling around to many places to find a market with fewer traders, more garden eggs, and lower prices. This takes time, costs money, and is exhausting. Gladys says: "I always pray for her safety. I have had serious accidents myself. This work is life threatening!"

Elizabeth (35 years old) is another garden egg trader traveling to the same areas. Since she has a landline in her house (her husband is the driver of a ministry), she calls various suppliers to inquire about when supplies are ready. They can also call her directly. Based on this information, she plans her trips. She still has to travel physically, but avoids the laborious searching process, and spends less time and less money on bus fares. Better security and information flow while on the road is mentioned by many as an important advantage, especially of mobile phones. The danger of accidents and robbery is very real on Ghanaian roads where enormous amounts of people and vehicles move day and night to transport goods. Not only does the possibility to call for help in case of an emergency improve security (and calm down worried mothers like Gladys when her daughter is out on a trip), it also makes contract fulfillment more feasible. One driver explained how his customer relations had improved after he got a mobile phone. He frequently experiences punctures or motor breakdowns. After his boss in the truck company equipped him with a mobile phone, the driver could call Accra in the case of such emergencies. The employer can send a new truck to replace the broken down vehicle. The load of perishable tomatoes or plantains can be reloaded, reach the market, and be sold instead of rotting on the roadside. Importantly, the driver is able to call the customer anxiously waiting for her load to arrive, perhaps suspecting that he has driven off to sell it somewhere else, and explain the reason for the delay. And best of all: although the consignment is delayed, it is fulfilled, which makes it likely that the customer will entrust her goods to be transported by the same company again.

Coordination, Monitoring, and Timing

Large-scale traders often coordinate extensive networks of trade partners and employees across vast distances. For example, much of the onions sold in Accra originate from northern Ghana, Burkina Faso, Mali, and Niger. The wholesalers are mostly men often involved in the entire production and distribution process. Mohammed (32 years old) is from Bawku, where he has a wife and children and access to family land where they grow onions (July–November). In addition, he rents land on the outskirts of Accra where hired workers plant and harvest onions (May–August). Together with his uncle and a network of other kinsmen, Mohammed sells these onions (harvested in different locations and ready for market in suitable portions from August to December) at the Agbobloshie market in Accra. They also travel to purchase truckloads of onions in Burkina Faso and Niger.

There are both practical and social problems involved in coordinating a large number of people involved in many different activities—from planting and harvesting to transporting, selling, and extending credit to customers. Mohammed, his uncle, and some other "core persons" in their network therefore acquired mobile phones as soon as coverage was extended to Bawku in 2001. In this kind of trade, which is large-scale, profitable, involves a number of activities in different locations, and extends over more than a thousand kilometers, it goes without saying that mobile phones are very useful tools. A quite revolutionary effect is that persons located far apart can exchange information almost simultaneously and make collective decisions instantly. The timing of onion supplies into the market is especially important to achieve maximum profits. With mobile phones, the network of traders can coordinate harvesting, packing, and transportation so that the right quantity of onions arrives in Agbobloshie market at the right time (depending on current supply and demand). Monitoring of employee activities also becomes easier. If one of the hired workers in Accra steals from the harvest, this can be immediately reported to the leader of the onion network, Mohammed's uncle, even if he happens to be in Niger at the moment. Being in charge, he can decide on sanctions immediately. Reports (or even rumors) staining one's precious reputation can travel much further and faster with telecommunications, which may in certain situations prevent opportunistic behavior.

In the distributive end of the onion chain, there are also gains to be made by investing in a mobile phone. Mohammed has extended his network of regular customers considerably after getting a mobile phone. He has become much more accessible in the sense that he can be contacted at any time, when he is not physically present at the marketplace, and from a wider radius. Customers call or send text messages from beyond Accra to order onion bags to be sent by bus, or they call from the Accra suburbs about whether supplies are available and avoid making the noisy and dirty trip to the Agbobloshie market. Customers are also encouraged to call and complain if the onion quality is bad. This invokes trust in Mohammed and is good advertisement: sometimes he attracts customers at the expense of traders without mobile phones who cannot offer these services. One precondition is, of course, that the customers also have phones. As a consequence, Mohammed's customers increasingly belong to the "connected" segment of the urban population.

Reconfigurations of Power

Control over information can be decisive for access to resources and contracts. Unequal access to telecommunications can therefore reinforce unequal power relations. In the fishing town Moree near Cape Coast, women are entirely in charge of fish processing and trade, and their role as creditors—pooling profits from fish trade back into the fisheries—is essential. Some women have also invested in canoes, outboard motors, and nets and employ men to fish for them. During the canoe fishery's off-season, external sources of fish supply is vital for fish processors/traders' business. Telecommunications has played an important role in the access of the most privileged traders to the two main external supply sources: by-catch (untargeted species or fish of low quality) from trawlers and imported cartons of frozen fish from companies in Tema.

When trawlers first started delivering by-catch in Moree in the late 1970s, contact between the female traders and the crew on board the trawlers was mediated by a man, who had a car and frequently traveled to Tema (the trawlers' port) to negotiate consignments of by-catch supply. The participants in the by-catch trade were the richest traders, who were able to pay up front. Many became so rich that they invested in canoes for the purpose of fetching by-catch at high seas. There were no telephones in Moree, but after some time a man with a walkie-talkie working for the trawler by-catch suppliers arrived. He married one of the traders and began contacting the trawlers on the VHF radio to make sure his wife was favored in the supply of by-catch. He made a lot of money this way, especially considering that he had similar arrangements through wives in two other towns. Access to by-catch thus initially reinforced the richest traders' wealth and position, but when access to the new resource could be manipulated through communication technology accessible to one person only, power relations were altered again. Today the trawlers do not call at Moree anymore, but the "walkie-talkie man" and his wife still maintain their contracts with the trawlers. To avoid the social conflicts that their unacceptable strategies caused in Moree, they now land the by-catch in a nearby town.

In 1998, Moree got its first telephone when a pharmacist installed a WILL (Wireless Local Loop) phone in his shop and created the town's first com center (in 2003, there was no telephone line yet and mobile phones were not common because of the high costs and—unlike now—hardly any coverage). For the richest traders, the com center became useful in their ordering of frozen fish cartons from Tema. Instead of traveling to Tema, where they would previously often have to wait for days and nights for supplies and to negotiate prices, they now call via the com center to make inquiries. There are also com centers in Tema. This means that when the large scale traders do go to Tema they can call smaller scale traders in Moree, who do not have the capital to buy imported fish in Tema and therefore buy cartons of frozen fish on credit from the richer traders, and inform them about prices and quantities. Better information flow resulting in easier ways of accessing imported fish supply creates employment and benefits the community as a whole during the local fishery's off-season. However, since it is the richest traders who have capital to invest and therefore have more to gain on improved telecom services, their position is strengthened in relation to the poorer traders. As a consequence, the poor traders' dependence on the rich traders in terms of fish supply and credit is reinforced.

Discussion and Conclusions

Time used and transportation costs are reduced when traders substitute travels with calls, and improves the efficiency and profitability of trade phenomenally. Importantly, even if the mobile phone is the most significant technological innovation, improved access to public telecom services is a more important improvement for the poorest traders.

To own and use a mobile phone is clearly an asset for traders. By reducing information asymmetries, traders are able to reduce costs. Mobile phones not only change traders' social and economic practices but also their position in the market hierarchy. Traders with mobile phones in some instances improve their services, number of customers, and sales at the expense of traders without phones, who may lose out in competition. Telecommunications development thus reduces information poverty, but only for those with access to the new technology.

The growing differentiation between the connected and the unconnected occurs not only between individuals, but also between rural and urban areas and among regions. To illustrate, a souvenir dealer who purchases wood carvings in rural areas said: "Those illiterates in the rural areas are even more 'in the dark' now than before, and less interesting for me to deal with." Increasingly, he makes orders in areas he can call with his mobile phone, and he has even started advertising his souvenirs internationally on a Web page promoting "African art."

Despite the telephone's efficiency, cultural values and institutional constraints in the Ghanaian market can require traveling and face-to-face communication. Place-based and socially embedded face-to-face communication continues to be important in traders' screening of partners' reputation, observation of behavior, and economic transactions. Nevertheless, an enormous amount of traveling to simply exchange practical and nonsensitive information would be avoided if telecommunications was accessible and affordable to the average Ghanaian.

As the empirical examples show, geography and income largely decide whether Ghanaian traders, as well as their suppliers and customers, can benefit from the space-time compression (Harvey 1989) enabled by telecom. The new technology's benefits are often clear at the enterprise level; yet national policy to date puts those in low-density rural areas at a disadvantage, the very places where the bulk of Ghana's agricultural production occurs. This conclusion suggests that governmental resources be directed to reducing the costs of access and use of telecom services, since doing so would benefit the national economy as a whole.

References

Aminuzzaman, S., H. Baldersheim, and I. Jamil. 2003. Talking back! Empowerment and mobile phones in rural Bangladesh: A study of the village phone scheme of Grameen Bank. *Contemporary South Asia* 12(3): 327–348.

Chamlee-Wright, E. 1997. *The Cultural Foundations of Economic Development: Urban Female Entrepreneurship in Ghana.* London and New York: Routledge.

Chamlee-Wright, E. 2005. Fostering sustainable complexity in the microfinance industry: Which way forward? *Economic Affairs* 25(2): 5–12.

Clark, G. 1994. *Onions Are My Husband. Survival and Accumulation by West African Market Women.* Chicago: The University of Chicago Press.

Dicken, P. 1998. *Global Shift: Transforming the World Economy.* London: Paul Chapman Publishing.

Fafchamps, M. 1996. The enforcement of commercial contracts in Ghana. *World Development* 24(3): 427–488.

Falch, M., and A. Anyimadu. 2003. Tele-centres as a way of achieving universal access—the case of Ghana. *Telecommunications Policy* 27: 21–39.

Harvey, D. 1989. *The Condition of Postmodernity.* Oxford: Basil Blackwell.

ITU. 2004. ICT statistics by country. http://www.itu.int/ITU-D/ict/statistics/.

Michelsen, G. G. 2003. Institutional legacies at work in African telecommunications (Report No. 80). Bergen: Department of Administration and Organization Theory, University of Bergen.

North, D. C. 1995. The new institutional economics and third world development. In *The New Institutional Economics and Third World Development*, edited by J. Harriss, J. Hunter, and C. M. Lewis. London and New York: Routledge.

Overå, R. 1993. Wives and traders: Women's careers in Ghanaian Canoe Fisheries. *Maritime Anthropological Studies (MAST)* 6(1/2): 110–135.

Overå, R. 1998. Partners and competitors. Gendered entrepreneurship in Ghanaian Canoe Fisheries. Ph.D. dissertation, University of Bergen, Bergen.

Overå, R. 2006. Networks, distance and trust: Telecommunications development and changing trading practices in Ghana. *World Development* 34(7): 1301–1315.

Overå, R. 2007. When men do women's work: Structural adjustment, unemployment, and changing gender relations in the informal economy of Accra, Ghana. *Journal of Modern African Studies* 45(4): 539–563.

Robertson, C. C. 1984. *Sharing the Same Bowl: A Socio-economic History of Women and Class in Accra, Ghana.* Bloomington: Indiana University Press.

Segbefia, Y. A. 2000. The potentials of telecommunications for energy savings in transportation in Ghana: The dynamics of substituting transport of persons with telecommunications in the Greater Accra Region. Ph.D. dissertation, University of Ghana, Legon.

Waverman, L., M. Meschi, and M. Fuss. 2005. *The Impact of Telecoms on Economic Growth in Developing Countries.* Newbury: Vodafone Group.

5 | Mobile Networks: Migrant Workers in Southern China

Pui-lam Law and Yinni Peng

Migrant workers are among those belonging to the lowest income group in southern China, yet the number of them owning mobile phones has increased sharply in recent years. More interestingly, their mobile phones usually cost three to four times their monthly income. In view of this intriguing phenomenon, we have conducted research on the social consequences of the adoption of mobile phones among migrant workers in Dongguan City, which is under the jurisdiction of Guangdong Province, since 2003. We chose workers in Dongguan for the following reasons: According to the data from the Ministry of Information Industry (2005), mobile phone penetration rates in Guangdong are the highest in the country. Among the cities in Guangdong, Dongguan is considered to be a city of migrants, as migrant workers make up more than five million of the total population of seven million. These workers come from different provinces and are working in various kinds of factories. In other words, Dongguan can serve as a typical city in studying mobile telephony and migrant workers in southern China. From 2003 onward, we have conducted in-depth interviews with migrant workers in Dongguan on their use of mobile phones. By mid-2005, we had interviewed altogether 59 people, comprising 47 migrant workers, 6 factory proprietors and managers, and 6 people running odd-jobs companies. Of the migrant workers, at least 28 were interviewed more than three times; 14 were male and 14 female, ranging in age from 16 to 30. They came from different factories in the villages of Dongguan City. The size of the sample is relatively small, but the interviews inform some essential themes that are central to the study of the social consequences of the adoption of mobile phones among migrant workers.

Based upon the data collected in the past two years, this chapter presents how mobile phones have been conducive to the formation of mobile networks among the workers and how these networks empower their social lives. First, the discussion focuses on how mobile phones provide the conditions for workers to contact both their families in their home villages and kinsmen scattered far and wide in Guangdong more conveniently and frequently; it also discusses how mobiles prolong new social networks developed in their workplace. Second, it sheds light on how the expanded networks are helping migrant workers empower themselves in improving their working conditions. Third, the spotlight is placed on the odd-jobs workers, discussing how mobile phones help them freely organize their lives.

Mobile Communication and Mobile Networks

Since the implementation of economic reforms and the policy of opening up to the outside world from early 1980s onward, the internal migrations from the western and central regions of the country to the eastern coastal region have been ceaseless. The floating population, predominantly migrant workers, has increased sharply from 30 million in the early 1980s to 140 million in the early 2000s. Migrant workers in Guangdong are largely from rural areas in the central or western regions of China, such as Hunan, Sichuan, and Guangxi, where a patrilineal structure is still prevalent in the villages.

Some observations suggest that the drastic economic reforms in the countryside have had a substantial impact on the patrilineal structure. Young villagers, as wage-laborers, are economically independent and some even make a significant contribution to the household's income, giving rise to a new balance of power in the family, where the autonomy of the individual has increased considerably (Thireau 1988). The out-and-return migration processes have also constituted the change of values and goals of migrants and potential migrants (Murphy 2002). According to our study, the development of new communication technologies such as the growing popularity of television sets in the villages, the adoption of landline phones, and the recent rapid penetration of mobile phones have made the villagers' homes in the less-developed regions more open to developments in the cities of the coastal regions. In other words, the new communication technologies have been conducive to bringing closer the coastal with the central and western regions. New forms of city lives and new values and ideologies have penetrated insensibly through these new technologies to the less-developed villages. This would undoubtedly have shaken the deep-rooted tradition of patriarchy, and have had an effect on the young and on potential migrants.

Indeed, large numbers of young rural villagers are flocking to Guangdong, particularly to the Pearl River Delta, not only to hunt for jobs to improve their standard of living, but also to experience the city life. As migrant workers, these young villagers believe that when they leave their home villages they will also be free from traditional cultural and social fetters. One informant told us precisely in the interview that:

When we first came here (Guangdong), we felt that we were freer than before. In the past (in the home village), the standards of measurement, the judgments, they were all the same.

When they leave their home villages to seek their fortunes in Guangdong, these young migrants do not find life to be easy. Their labor is exploited, they often encounter hostility from local workers or villagers, and the feelings of desolation and insecurity they experience by being in a strange place have caused some to return home. But for those who are determined to stay in Guangdong, contacting their families in their places of origin and joining their kinship network in Guangdong provides important emotional and social support during their stay. Mobile phones, of course, have becomes a necessary tool in maintaining these existing networks in expanded spatiotemporal contexts (Pertierra et al. 2002).

In the past, without mobile phones, maintaining connections with families in their home villages through fixed phones was very inconvenient. Some migrant workers said that not all the rooms in the dormitories had fixed phones. Even when landline phones were installed, they were always engaged. During important festivals, when the workers call home to extend their seasonal greetings, huge crowds always form around the public phones in the factories and on the streets. Having mobile phones allows migrant workers to connect with their families more conveniently. A male worker said he had felt safer emotionally since he bought his mobile phone because he could maintain regular contact with his family in his home village. Also, a female worker said:

When I want to talk to my mother (at home village), I would use my mobile phone to call her. It is not that expensive—just 0.25 yuan . . . . I always call back home. In fact, there isn't anything important. I just want to chat (with them).

Mobile phones can provide them the immediate absent presence (Gergen 2002) of their families far away in their home village when workers are feeling lonely and thinking of them.

In the past, it was also very inconvenient for workers to maintain contact with their relatives even when they were working in factories nearby. When they wanted to meet their relatives, what a worker could do was go in person to the factories where their relatives worked and wait outside before they got off work. A mobile phone can reduce these difficulties and allow workers to contact their relatives to arrange gatherings freely and easily. If their kinsmen are scattered far and wide in Guangdong and meetings can barely be held regularly, they can connect either by making a call or by sending an SMS. There is an interesting story about a young migrant worker from Guangxi that demonstrates the strength of the connectivity of mobile phones. The worker had a job arranged for her by her father in their home village but she slipped and worked in Guangzhou City alone. Eventually she was caught because her father tracked her mobile phone and she was located by relatives. She escaped again and has now been working in Dongguan. She told us:

I don't want to have mobile phone anymore otherwise everyone can find you out easily. I just don't want to be found by my father [again].

Although a mobile phone is a personalized mobile device (Green 2003) for the migrant workers, it is indeed a very powerful tool in building mobile cyber-kinship networks in Guangdong. If a large portion of the inhabitants of a village are working in a coastal region, which is a common phenomenon, and if they have become connected via mobile phones, a strong mobile cyber-kinship network would be formed there.

Mobile phones are not only helping migrants to maintain existing kinship relationships in expanded spatiotemporal contexts, but also to prolong new social relationships developed in the workplace. Workers come to know each other when working on the same assembly lines or staying in the same dormitories. They develop friendships among workers from different villages, towns, or provinces even though they are not kinsmen and would normally consider each other to be "outsiders" (Metzger 1998). In traditional kinship relationships kinsmen are defined as "insiders" upon whom one can lay one's trust; others are "outsiders" and have to be measured in instrumental terms. Thus, industrialization in Guangdong provides a platform for developing a new kind of social relationship where people are not simply polarized into either "insiders" or "outsiders." Migrant workers are nevertheless highly mobile and most of them do not stay in one factory for a long period of time. Once they find other factories offering higher incomes and better welfare provisions, they will leave for these factories immediately. Thus in the past, friendships developed would rarely be prolonged when workers left the factories. This situation recently has been changed substantially. A worker told us that he could maintain regular contact with his good friends through mobile phone:

We send SMS to friends for maintaining contact. When we receive messages that means our friends are still thinking of us. Three days (without contact) would be okay. If I don't receive anything from them for a while, and I don't know what they are doing, I will send SMS to them. We have to keep the connection, and if we lose contact, it is just like a thread broken.

Another worker said, "When you received SMS from your friend, just one or two words, you would feel happy." He further maintained that he could develop his own circle of friends after he had gotten a mobile phone.

Mobile phones, which free individuals from temporal and physical constraints (Kopomaa 2000), serve an important function of preserving this new kind of social network among the migrant workers with different places of origins.

With expanded social relationships and new ways of connecting, migrant workers are more resourceful in obtaining information. By simply making calls to their kinsmen and friends, migrant workers can more easily learn about job opportunities in Guangdong. These networks are essential in empowering the workers when they are facing difficulties. The following section discusses how these networks could empower the workers and subsequently increase the rate of mobility of the migrant workers, both directly and indirectly.

Networks Expanded and Workers Empowered

In Dongguan, formal channels for accessing information about the job market were very limited in the past and finding a job was very inconvenient. Each factory would post vacancies on their front doors when they wanted to recruit workers, so young villagers had to go around to the factories in each village to see whether there were vacancies. If these villagers had relatives working in Guangdong, it was relatively easier for them to gain access to information about the job market; but even if their relatives found a job vacancy for them they sometimes missed their chance when they were traveling around Guangdong and could not be reached immediately if they did not have pagers or mobiles. Now they have more convenient access to job information than ever before when they have mobile phones. A worker described what he usually did when his kinsmen called him:

Usually they send me an SMS from the home village and inquire whether there are chances, and if the answer is positive they will make a voice call to find out more about the details. If they have already come to Dongguan, then I will call my friends or send SMSs and see whether there are any factories that need workers. If there are, then I will accompany them to the factory or give them my friend's mobile phone number for them to make the connection.

In the past, when migrant workers had difficulty gaining access to information about the job market, they would not readily give up their jobs because finding another one, and particularly a better one, was not an easy task. If they were fired they had to go back to their home village. Thus, when faced with poor treatment they had to be humble and submissive, or even to make abject apologies to the factory management. Otherwise, they would incur more unfair treatment. If they logged complaints against maltreatment, they would be violently assaulted by the factory guards and eventually dismissed. The assaulting of migrant workers has been a common phenomenon (Chan 2001).

Now, mobile phones have strengthened the bargaining power of migrant workers with their factory proprietors. For example, there is a story concerning a fight between two workers. One was injured and demanded that the other compensate him for his injuries. After investigating the fight, the factory manager supported the injured party and asked the other to pay 200 yuan toward medical costs or he would be fired. The worker refused to pay the compensation and left the factory without any hesitation. Before he left, he reminded the manager to call his mobile phone to tell him if he was due any wages so that he could come back and collect them. The factory manager told us that the worker was still staying in the village because he had found a job in another factory soon after being fired. In this case, the manager felt that it was more difficult to control the workers, as they are no longer afraid of being fired. He pointed out that it is precisely the penetration of mobile phones among the workers that has imperceptibly empowered them to a great extent.

Another factory proprietor, whose business is producing garments, told us that he was very afraid of the use of mobile phones among the workers in his factory. He said that, for instance, during the lunch break workers can use SMSs to share information about the salaries, benefits, promotion opportunities, and working conditions of other factories. Once they discover that any of these conditions are better at another factory, they will quit their jobs immediately. They will introduce their relatives and fellow villagers to this factory as well. Usually, there will be a chain effect. One worker resigns and goes to the other factory, and his or her relatives and fellow villagers will also resign and join that worker in that factory. The proprietor said that one time he had more than seven workers, all belonging to one family, leave his factory after receiving SMSs during the lunch break. He maintained that he really had increased the basic salary of his workers; for instance, the salary of the skilled workers already had risen to 1,000 yuan, which is a very good income for garment workers in Dongguan. Even though the workers were happy with the salary, they demanded more holidays and less overtime work. To satisfy as many of their demands as possible, his factory now needs more workers than before. He complained:

If they have less information about the job market, they will be less likely to move around.... Even if we install some kind of interference technology they can still call their kinsmen or friends after work. The only way to minimize their contacts with others is to move our factory to a place where mobile phones cannot receive any signals at all.

This proprietor believed the adoption of mobile phones among the workers has been a key factor affecting mobility rates in Dongguan, even though he knows that the shortage of labor in Guangdong has made it easier than before for workers to be mobile, as he is very close to the workers and understands their daily life very well.

As networks have expanded and connectivity has become highly efficient, the flow of job market information has also become rapid and extensive. This, in turn, has provided abundant information about job markets that is extremely easy for workers to access nowadays. Mobile phones have helped workers fight for their rights more successfully than they were able to in the past (Solinger 1999).

Odd-Jobs Workers: New Mobile Network

From our interviews with the migrant workers, we found that there were workers who preferred to do odd jobs rather than station themselves in a factory because this approach gave them more freedom. This phenomenon has emerged in the past five years, and the workers were connected via pagers when they got the job opportunities. With the shortage of skilled labor and the increasing popularity of mobile phones among the workers, the number of odd jobs workers recently has increased substantially in Guangdong. The head of a group of odd-jobs workers informed us that in an industrial zone of a village in a township in Dongguan there are about six thousand transient odd-jobs workers. According to a relatively conservative estimate made by factory proprietors and odd-jobs workers, there are approximately one hundred thousand odd-jobs workers in the whole township, which is comprised of twenty-eight industrialized villages and three street committees, a figure close to the permanent population of that township.

We have interviewed people who were running an odd-jobs company. They were skilled garment production workers and had themselves become odd-jobs workers after working in factories for years. They have run their odd-jobs company for more than four years and the workers they have on their lists are predominantly skilled garment production workers as well. When they started their business, they promoted themselves by faxing the details of their companies to factories in Dongguan, and a pager was the essential tool for maintaining contact with the factory proprietors, managers, and the workers as well. Now, the mobile phone has replaced the pager. When we interviewed the owners, their mobile phones rang incessantly, and the interviews were interrupted numerous times when the owners answered their calls.

Although these people are based in Dongguan, they are highly mobile and they have sent hundreds of workers to the cities of Shenzhen and Zhongshan. The largest business that one of the companies we interviewed had was providing six hundred workers for a factory. One of them told us they could call up to three thousand workers, and one of them said he had already saved in his mobile phone the mobile phone numbers of more than five hundred workers. When they have business they never send SMSs but make voice calls because they are direct and fast.

We also interviewed odd-jobs workers. They are largely skilled workers and have very good experience in their respective job areas. Yet they prefer taking on odd jobs to the stability of being a factory worker, as they prize their autonomy and dislike being bound by the rules of a factory. Indeed it is true that even a very small factory with only fifty workers has rules that its workers are expected to comply with. One of the odd-jobs workers said:

As an odd-jobs worker, I have freedom. They would not require you to go to the factory on time. The wages are double. You work half month, and you can play for another half a month.

Another remarked:

We have the same amount of money (as compared with those stationed in a factory), and we have freedom. In any case, we can't earn all the money in the world. It is okay as long as you find that your life is happy....

If you are stationed in a factory, you have to get up at 8:00 a.m. and have your work card punched. But if I don't have an odd job to work on, I can get up late. In a factory, even if I don't have work to do, I would have to apply for release. I would have to ask this or that supervisor to sign for me, then go to the security office to have the application stamped. Applying for sick leave is also very complicated, particularly if you work in a large factory. You have to fill out a number of forms. But now as an odd-jobs worker, I can do whatever I want to do.

In fact, in addition to rules, workers seldom have the chance to organize their lives freely if they are working in a factory. Particularly during the busy season, workers are required to do overtime work every day until midnight. Even though they get extra pay for the overtime work, they hardly have any leisure time, which the new generation of workers treasures. It seems that this group of workers demands more freedom than those who consider the stability of having a factory job to be more important.

The odd-jobs workers usually register with an odd-jobs company. Mobile phones are necessary for them, since when jobs are available they can be connected instantly. They would not consider whether those running the companies are their kinsmen or not. What is important is whether the company can provide jobs for them and negotiate good wages for them. When they receive a call for work, they will be informed on the phone, either by SMS or by voice call, the time they need to gather, the name of the factory, and its address if it is close to their home. If it is far away and they have to stay there for a while, they will also be advised to bring with them items of daily use, and the head of the company will hire a van to bring them there. Usually, they will not know their fellow workmates until they all meet in the factory.

Although the emergence of odd-jobs workers or the odd-jobs companies should not be ascribed to the introduction of the mobile phone alone, the mobile phone makes it possible for these workers who demand more autonomy to be connected instantly and live their lives as odd-jobs workers in Guangdong. In addition, with the emergence of odd-jobs companies that link up workers, the workers have more opportunities to meet with other odd-jobs workers from different places of origin, and to build a larger mobile network.

Conclusion

In recent years, young migrant workers have started to put more emphasis on their individual freedom and enjoyment. Some tend not to contact their fellow kinsmen or villagers for fear of being subjected to traditional norms and values. If they can live alone, they can act as freely as they desire. Yet, as sojourners, the unfavorable social conditions and the feeling of loneliness in the host society have pushed them back into the bosom of tradition. The mobile phone, a personalized mobile device freeing the highly mobile migrant workers from spatiotemporal constraints, which makes possible the physically absent present, has made these workers re-establish connections both with their families in their home village and with their kinsmen or fellow villagers working in Guangdong, or even in other provinces. Traditional kinship networks have been weaved in the form of a mobile kinship network. Mobile phones also provide the conditions for expanding networks other than traditional ones, by linking ex-workmates through cyber connectivity.

In addition, the odd-jobs networks are also conducive to the formation of new mobile networks. When workers are stationed in a factory, they usually expand their kinship ties by introducing their fellow kinsmen to the factory. But as an odd-jobs worker, each contract is short and transient and it is therefore difficult to work together with their fellow kinsmen or villagers each time. Instead, the workers are always meeting new people, resulting in new mobile networks and leading to a more vagrant way of life than that led by the average factory worker. This might dilute the influence of or even replace the mobile kinship network.

With the emergence of both the traditional and new cyber networks, the workers can easily get emotional support and readily obtain more job market information from their mobile networks. They have been empowered in fighting for their worker rights, and their working conditions have been improved despite the fact that the room for changes remains limited. They have also been capable of living a more autonomous life in Guangdong. The extensive penetration of mobile phones among the migrant workers has significantly changed their social lives. We believe that the rapidly evolving mobile phone will continue changing the social lives of the migrant workers in the future, both directly and indirectly.

References

Chan, A. 2001. *China's Workers under Assault: Exploitation and Abuse in a Globalizing Economy.* Armonk, New York: M. E. Sharpe.

Gergen, K. J. 2002. The challenge of absent presence. In *Perpetual Contact: Mobile Communication, Private Talk, Public Performance,* edited by J. E. Katz & M. Aakhus. Cambridge: Cambridge University Press.

Green, N. 2003. Community redefined: Privacy and accountability. In *Mobile Communication: Essays on Cognition and Community,* edited by K. Nyiri. Vienna: Passagen Verlag.

Kopomaa, T. 2000. *The City in Your Pocket: Birth of the Mobile Information Society.* Helsinki: Gaudeamus.

Metzger, T. 1998. The Western concept of the civil society in the context of Chinese history (Hoover Essays No. 21). Stanford: Hoover Institution on War, Revolution and Peace, Standard University.

Ministry of Information Industry. 2005. Penetration rates of mobile phones. http://www.mii.gov.cn.

Murphy, R. 2002. *How Migrant Labor Is Changing Rural China*. Cambridge: Cambridge University Press.

Pertierra, R., E. F. Ugarte, A. Pingol, J. Hernandez, and N. L. Dacanay. 2002. *Txt-ing-selves: Cellphones and Philippe Modernity*. Manila: De La Salle University Press.

Solinger, D. 1999. *Contesting Citizenship in Urban China: Peasant Migrants, the State, and the Logic of the Market*. Berkeley: University of California Press.

Thireau, I. 1988. Recent changes in a Guangdong village. *The Australian Journal of Chinese Affairs* 19/20: 289–310.

6 | Mobile Communication in Mexico: Policy and Popular Dimensions

Judith Mariscal and Carla Marisa Bonina

Mexico specifically and Latin America generally have followed many of the mobile communication trends seen elsewhere in the world. It has seen unexpectedly rapid growth, and access to mobiles is outstripping fixed access. During the mid-1980s, mobiles were considered a rich person's device, but mobile telephones are today proliferating among the poor, often providing them with their only source of telecommunication access. This chapter examines major contours of the situation of mobiles in Mexico, and reports on a snapshot survey on mobile usage among youth undertaken specifically for this volume.

Development of the Mexican Mobile Industry

In Latin America, from 1995 to 2005, the number of mobile subscribers increased nearly 57 times, from 4 million to 227 million in 2005. This increase is yet more dramatic when one considers that until 1997, mobile telephony was a secondary business for incumbent companies. Fixed teledensity surpassed mobile penetration and investment in fixed telephony, being relatively sheltered from competition and operating within a relatively weak regulatory environment, seemed to promise a major source of income. Mobile telephony firms, on the other hand, were subject to intense competition. Therefore, as the mobile companies were facing serious difficulties in generating profits, the firms in the fixed sector owning mobile sister companies did not consider this branch of their business as very promising (Mariscal and Rivera 2005). After 1998, while fixed teledensity tended to stagnate in most countries, mobile telephony began to grow at double-digit rates. This dramatic growth changed the access to voice communications; what initially appeared as a means of communications restricted to high income groups was transformed into the main means of telecommunications access to the poorer groups of the region.

This same pattern of growth was experienced in Mexico. Mexico initiated in 1990 a process of major reforms in its telecommunications sector, with the aim of modernizing the network on the one hand, and opening the country to international trade and investment on the other. The second phase of reforms began in 1994 when national and international long distance services were opened to competition. Although in 2001 Mexico had nine competitors in the mobile arena, due to consolidation and business strategies, after a half-decade less than half remained in the market. Telcel dominates with 75 percent of the market, with concomitant impact on prices. Among the other market participants, Telefónica Movistar is in second place with 14 percent, and Iusacell, Unefón, and Nextel have a small residual. Like the rest of Latin American, even though the mobile communication segment is open to competition, the market has become a duopoly.

The Pattern of Growth in the Mexican Mobile Market

Similar to most Latin American countries, Mexico's growth in mobile telephony has been extraordinary. While in 1990 Mexico had 64,000 subscribers, mostly limited to the higher echelons of society, by 2005, the number had increased to 44 million. Prepaid mobile systems introduced in 1995 and "calling party pays" modality (CPP) introduced in 1999 have resulted in impressive growth and penetration rates, as can be seen in figure 6.1. Growth in mobile telephony far surpasses that of fixed telephony. Figure 6.1 depicts the evolution in the penetration of fixed and mobile telephony in Mexico.

A significant characteristic in the dynamic growth of the Mexican mobile sector is the predominance in prepaid subscribers as a proportion of total subscribers in the country. As can be seen in the figure 6.2, Mexico has the highest rate of prepaid subscribers (93 percent) in Latin America.

This phenomenon may be due to the fact that a prepaid modality was introduced right after Mexico's December 1994 economic crisis and prepaid services were pro-

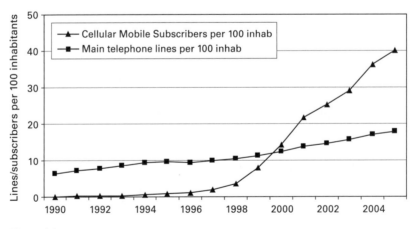

Figure 6.1
Fixed versus mobile telephony in Mexico, 1990–2005.

moted aggressively by Telcel to avoid any credit risks associated with telephone credit. The second reason has to do with the lower costs in the chains of distribution; Telcel is a member of the Carso Group, which is a conglomerate made up of not only telecommunication but also financial and other companies as well. Hence, the costs of distributing prepaid cards were very low given the large number of sales points available within the company.

In sum, as in other developing countries, the rapid diffusion of mobiles in Mexico has had a stronger impact on obtaining the policy goal of universal access than had traditional policies aimed at this goal. Among the policies for mobiles that have fueled the dramatic growth are prepaid cards and "calling party pays." Together, these policies have helped millions overcome barriers that low income people have traditionally faced when seeking to gain mobile service.

Mobile Usage in Mexico: Gender, Age, and Socioeconomic Levels

This section identifies the usage patterns in different groups in Mexico, particularly in terms of gender, age, and economic status. It draws on a Telefónica Movistar de México (TEMM) nationwide survey conducted in May 2005. This nationwide survey was of about four thousand people above the age of fourteen.

Gender

In Mexico, there are relatively more men than women among current mobile phone users (55 percent men versus 47 percent women). Among former users, 14 percent were men versus 9 percent women. Interestingly, the surprisingly high percentage of

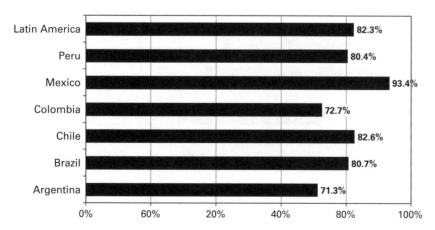

Figure 6.2
Prepaid subscribers as a percentage of total subscribers in Latin American countries, 2004. Source: Authors.

mobile phone "dropouts" fits with the discovery of a comparable phenomenon in the United States by Katz and Rice (2002). In Mexico, more than nine out of ten users have prepaid plans. While men have a slightly higher proportion of postpaid plans (10 percent) compared to women (8 percent), the difference is probably not very consequential.

According to another nationwide survey carried out in June 2003, the two main reasons for females to get a cell phone were to be easily reachable (30 percent) and for security or an emergency (23 percent). On the other hand, men's main reason for getting a mobile was to make personal calls (26 percent), followed by making job-related calls (22 percent). Noteworthy was that only 10 percent of females considered work purposes as the main reason for purchasing a mobile, compared to 22 percent of males. Prices or costs were not important determinants of cell phone acquisition (but of course those without cell phones were not included in the survey).

Age
There are differences in usage of mobile telephony by age. As can be seen from figure 6.3, in 2005 young adults age twenty-five to thirty-four show the highest adoption rate. Perhaps surprising is that teenagers and youth, fifteen to twenty-four years old, also widely adopt the cell phone, especially when compared to those in the next older (and presumably richer) category. This high level, though, is understandable when one considers that the younger generation is a target of mobile operators in Mexico, who address their new products and publicity campaigns to this generation. For instance,

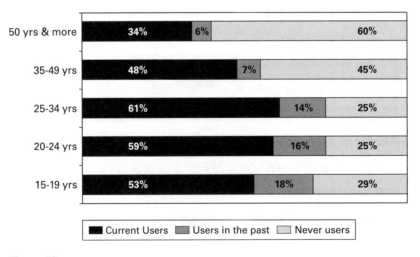

Figure 6.3
Mobile phone penetration by age group in Mexico. Source: Authors, derived from TEMM 2005.

Telcel—the most important player in the industry—has focused on the teenage and even the children's market by launching new phones based on popular cartoons and television characters. The fourth section of this chapter explores in more detail the youth and teenagers market.

The fifty-and-older age group has the lowest penetration rate, which is typical of the pattern worldwide. The distribution suggests that adoption is heavily a function of social location and not income. It may also be due to the typical resistance that older people often show to new technology, as seen in figure 6.3.

Preferences regarding payment options show, again, that the prepaid modality is preferred by every age group in more than 85 percent of cases (figure 6.4). People age thirty-five to forty-nine show the highest usage rate of postpaid plans. This might be due to their higher participation in the labor market and the resulting higher average income.

Socioeconomic Levels

The concept of "socioeconomic levels" (SEL), an industry standard defined by the Mexican Association of Market Research and Public Opinion Agencies (AMAI), can be used to analyze the growing use of cellular telephones by low income groups. The SEL are divided into five groups—A/B, C+, C, D+, D, and E—where the A/B group encompasses the highest income ranges of the population while the E group covers those with the lowest income level and quality of life.

Drawing on data generated by two Telefónica Movistar of México surveys, we can analyze the use of mobile telephony by low income groups. Table 6.1 provides

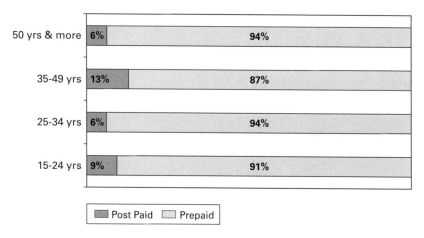

Figure 6.4
Age-based usage by payment options. Source: Authors, derived from TEMM 2005.

Table 6.1

Mobile Penetration by Socioeconomic Level, 2003

	Socioeconomic level (in percent)		
	A/B, and C+	C and D+	D and E
Overall distribution of population	10.8	32.9	56.3
Postpaid subscribers	19	8	6
Prepaid subscribers	81	92	92
Mobile penetration within level	85	43	9

Source: Telecom CIDE (2006) derived from a 2003 nationwide TEMM survey of approximately 5,000 people

indicators for 2003. In 2003 the use of mobile telephones was predominately in the higher income sectors of the population (85 percent of all people in the highest SES category). On the other hand, in 2003, the lowest income group also included users of mobile telephony—one in every eleven had a mobile telephone!

Nevertheless, penetration in the past two years tells a different story. According to recent data provided by TEMM, the mobile telephone has become a common tool among the lower income sectors. While in 2003 only 9 percent of the individuals classified within the D and E socioeconomic levels were users of mobile telephony, by 2005 the number had tripled and reached 27 percent of the population in those brackets.

In the higher income sectors, on the other hand, the number has not changed significantly. This could be expected since the percentage of the population using mobile telephones in that income bracket was already high. (It also suggests that there may be barriers to virtually 100 percent penetration, even when costs are of little real consequence.) Still, the middle class, associated with SEL C and D+, has also shown a growing use of mobile telephones as evidenced by an increase from 43 percent in 2003 to 51 percent in 2005. This is seen in table 6.2.

The increasing use of mobile telephones by the low income groups is mostly due to the lower access and usage costs enabled by the prepayment system and the "calling party pays" (CPP) arrangement. (Under the prepaid system, users have the advantage of controlling their telephone expenses, eliminating the risk of escalating debts. Users have no fixed monthly charges and can determine their level of expense and usage. Together with the CPP modality, even if the telephone no longer has credit, the user can continue receiving calls, allowing for continuing connectivity.) When analyzing the segment of prepayment specifically, using tables 6.1 and 6.2, both in 2003 and 2005, the groups most intensively using this modality are those falling within SELs D and E. This provides lower income people with increased autonomy from other alter-

Table 6.2
Mobile Penetration by Socioeconomic Level, 2005

Subscription status	Socioeconomic level (in percent)				
	A and B	C+	C	D+	D and E
Postpaid subscribers	28	12	6	6	4
Prepaid subscribers	72	88	94	94	96
Mobile penetration (per group)	89	75	67	42	27

Source: Telecom CIDE (2006) based on nationwide survey by TEMM

natives such as community centers, where there are often restrictions to receiving calls. It is important as well to have the means to be located in order to get jobs, since among the lower income groups temporary employment is predominate. So the main reasons mentioned by the mobile users of socioeconomic level D for purchasing a cellular telephone include needing to be located, making personal calls, and making job-related calls.

In sum, prepaid services were preferred by every group. Together with the introduction of prepayment in mobile telephony, the overall adoption of the "calling party pays"—where the user does not have to finance incoming calls—has translated into a major increase in demand and contributed to a major growth in coverage in Mexico.

Mobiles and Youth in Mexico City: Findings from a Small Survey

Youth and teenagers are the most enthusiastic users of mobile telephony in many countries around the world. Mobile phones have become not only a status symbol and a fashionable good for young people but also a new mode of socializing, particularly in developed countries but elsewhere as well (Katz 2003). The ITU has even claimed that "many teenagers don't recognize the difference between speaking on their mobile phone and meeting face-to-face" (ITU 2004a, p. 12).

In Mexico, young people are increasing their use of mobile services, thereby transforming the way they interact and creating new social innovations. In this section, we explore how teenagers and youth are using cellular phones in Mexico, building our own research upon prior studies by the ITU (2004b) and MACRO (2004). The ITU study was designed to explore mobile usage patterns and trends of young students from the United States, and the MACRO report replicates it in the Indian context. We sought to use the same variables in our Mexican study, hoping in part to build upon previous findings in other countries. However, our survey was not aimed to be a rigorous scientific study but rather to give a first look at the current situation; certainly the topic is worthy of more detailed study, which we hope to do in the future, and given

Table 6.3

Respondents by Age and Gender

Age Category	Gender (in percent)*		Category subtotal
	Females	Males	
15–19	44	29	N=28
20–24	25	43	N=26
25–29	31	29	N=23
Subtotal in percent	100	100	N=77

Source: Author's survey, 2005

*May not total to 100 due to rounding

the dearth of studies on the subject in Mexico we were happy to get at least a small project underway.

Our nonrandom sample was drawn from young people (age fifteen to twenty-nine) in high school, college, or graduate school in the west area of Mexico City; questionnaires were distributed principally at the Centro de Investigación y Docencia Económicas (Teaching and Research in the Social Sciences Center) (CIDE) and a private high school. Most of the respondents were full-time students though some were also working. While the number of observations in the sample is small, it still provides an initial overview of how youth use mobile services in Mexico. Out of seventy-seven respondents, 53 percent were female and 47 percent were male. Table 6.3 shows age and gender distribution of the sample.

About 90 percent of the overall sample of students had mobile phones, so only eight respondents did not. Of the eight respondents not owning a cell phone, three plan to buy one in the near future. Their main reason for believing they would be getting a cell phone was in anticipation of it being needed for work. Regarding gender and age groups, those who reported not owning a cell phone were male between twenty and twenty-four years old. Respondents from the youngest age group showed the highest rate of users, which is due to two main factors: many of them belong to a high income group and, as it was already mentioned, cellular phones have been spreading rapidly among teenagers during the past few years. On the other hand, 65 percent of the cases in the 25–29 age category were working and many of them cited they own a cellular telephone because of that.

There were at least four main reasons for those who do not own a cell phone; high costs represent the most important barrier. Interestingly, in other countries an important reason for not having a cell phone has to do with not being allowed to; this was not an important factor to the respondents of this survey. In fact, no one chose that answer. (For U.S. students and in an Indian study, "not being allowed to" was the second most frequently given reason for not owning a cell phone (ITU 2004b and

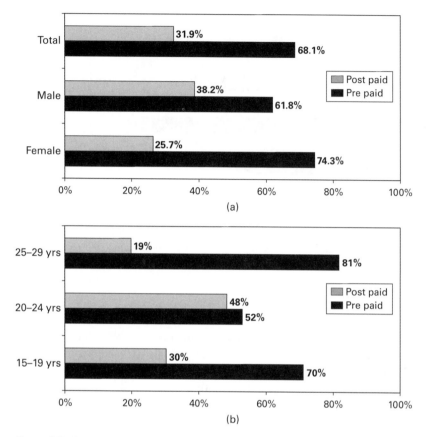

Figure 6.5a–b
Payment options by gender and age groups. Source: Authors' survey 2005.

MACRO 2004).) Having lost their mobiles was the other reason the respondents cited as to why they did not have a cell phone.

Modalities of Payment

As stated earlier, prepaid services are preferred in every category of analysis. From the total users, 68 percent were prepaid users while 32 percent had monthly rate services. When analyzing by gender, females showed a higher proportion of being on prepaid payment modality than men. Assuming that Mexico's reality is consistent with other surveys that find that females talk more on their cellular phones than males (MACRO 2004, pg. 18), the possibility of budgeting telephone expenses using prepaid services can explain this gender inclination toward this modality.

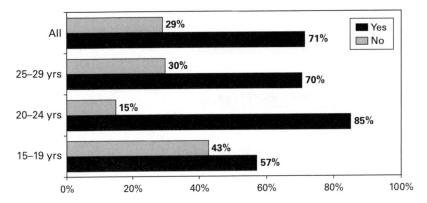

Figure 6.6
Perception of mobile phone versus face-to-face interaction. Answers to the question: "Do you think there is any difference between speaking on your cell phone and meeting face to face?" Source: Authors' survey 2005.

Calling and Usage Patterns

When asking if there is any difference between speaking on your cell phone and meet-ing face-to-face, almost a third of total respondents gave a negative answer. Interest-ingly, the amount of respondents in the 15–19 age group that stated not recognizing any difference between these two ways of communicating where the highest in the sample (43 percent). This pattern reaffirms what was pointed out at the beginning of this section regarding the blending of mediated with face-to-face communication.

In terms of mobile functionality, sending and receiving text messages are the most common activity among teenagers. As table 6.4 depicts, text messaging is the most common activity. Making local calls is important as well, but doing so is less fre-quent than using short message services (SMS). This trend was also found in other countries such as India and the UK where young people may prefer text to voice. This accords with ITU studies: in the UK more than eight out of ten people under the age of twenty-five are more likely to send someone a text message than to call (ITU 2004a, pg. 13). In the study of mobile phone usage in Mumbai, India, making local calls and text (SMS)-ing were reported as the most common activities as well (Macro 2004, pg. 22). On the other hand, activities such as downloading ringtones, playing games, and send-ing photos were reported as the less common actions. In the case of playing games on a cellular phone, the frequency proportion found among the respondents is surprisingly small compared with other countries. As found in the MACRO report (2004), industry reports indicate that mobile gaming can be considered the "next big thing" after SMS and ringtones (MACRO 2004, pg. 22). According to Mexican operators, while gaming

Table 6.4

Mobile Functionality

Functionality	Self-described frequency of use (in percent)		
	Always/Often	Sometimes/Rarely	Never
Sending text messages	93	15	2
Local calls	65	19	6
Long distance calls	18	53	29
Playing games	8	44	48
Sending photos	2	23	75
Downloading ringtones	2	21	77

Source: Author's survey 2005
N=69 for all functions

Table 6.5

Importance of Mobile Phone. Question: "Could you live without your cell phone?"

	Female	Male
Yes	51%	74%
No	49%	26%
	100%	100%
	n=41	n=36

Source: Author's survey, 2005
Female n=41; male n=36

still represents an incipient service, it will experience major advances in quality and use in the near future. The three main reasons explaining this are high costs, lack of habit of using mobile services other than voice, and the incipient penetration of modern devices capable of supporting gaming and other mobile services.

Another interesting question was related to the possibility of living without a cell phone. Surprisingly, 38 percent of our sample says that they would not be able to spend even a day without their cell phones. Many of the respondents said that they were quite used to having their cellular phones every day and needed to be constantly accessible to their friends, family, or colleagues. Moreover, some declared they get very nervous and anxious if they forget their mobiles at home, in their cars, or at the office (table 6.5).

Regarding gender analysis, females report being more prone to mobile dependence than were men. As table 6.5 shows, 49% of females said they could not live without a cell phone while males were at about a quarter only. Girls have less independence from their families than do males and that may explain this situation; indeed, women are more likely to call home than men. Thirty-four percent of females stated they direct one out of every two calls to their families versus 28 percent of males; another 51 percent of females said they call home one of every three calls while the proportion of men in the same category was 40 percent.

Regulation and Usage Dimensions

Due to its popularity and rapid growth, the cell phone has for Mexico constituted the most effective tool yet for advancing universal access to communication. The country is enjoying its concomitant benefits. Regulatory changes in the industry and pricing strategies such as "calling party pays" and prepay systems have contributed to and support this development. The result is dramatically improved access to voice and text communication. A technology that initially appeared as a means of communication for only the highest income groups has been transformed into the principal means of telecommunication for the poorer groups of the region.

Analysis of usage patterns reveals that Mexicans by far prefer prepay modality. This is independent of gender, age, and income variables and is due mainly to the benefits offered by the prepayment system both to the operators as well as to the users. To the companies, advantages include reducing fraud, monthly bills, and collection costs. To consumers, advantages include control of telephone costs and ease of acquisition. Regarding gender, men show a higher proportion rate of usage than women. This difference may be explained by the higher participation of males in the labor market. In particular, mobile technology was found to be very helpful for lower income groups to find employment.

Mexico shows a robust and sustainable growth mobile market. The youth market is becoming the focus for mobile operators, equipment manufacturers, and other service providers. In terms of social impact, quickly diffusing mobile usage among the young is changing their behavioral routines and social interactions. We point to the data showing that teenagers consider no difference between speaking on the cellular phone and meeting face-to-face. But as yet, the consequences of these changes in mobile communication–driven behavior are but little understood, especially in terms of full participation in the various aspects of life in the information society.

References

International Telecommunications Union (ITU). 2004a. Social and human considerations for a more mobile world. ITU Workshop On Shaping The Future Mobile Information Society, SMIS/04, Geneva. http://www.itu.int/osg/spu/ni/futuremobile/socialaspects/.

International Telecommunications Union (ITU). 2004b. Mobile phones and youth: A look at the U.S. student market. ITU. Geneva. http://www.itu.int/osg/spu/ni/futuremobile/socialaspects/.

International Telecommunications Union (ITU). 2003. *World Telecommunications Indicators 2003*. Geneva.

Katz, J., ed. 2003. *Machines that Become Us: The Social Context of Personal Communication Technology*. New Brunswick, N.J.: Transaction Publishers.

Katz, J., and R. Rice. 2002. *Social Consequences of Internet Use: Access, Involvement, and Interaction*. Cambridge, Mass.: The MIT Press.

Mariscal, J., and E. Rivera. 2005. New trends in mobile communications in Latin America. Prepared for the Annenberg Research Network on International Communication Workshop, Marina del Rey, Calif.

Market Analysis and Consumer Research Organization (MACRO). 2004. Study of mobile phone usage among the teenagers and youth in Mumbai. http://www.itu.int/osg/spu/ni/futuremobile/socialaspects/IndiaMacroMobileYouthStudy04.pdf.

Telecom CIDE. 2006. Contribuciones sociales y económicas de la telefonía móvil en México (Social and economic contributions of mobile communications in Mexico). Prepared for Telefónica Movistar México, Mexico City.

Telefónica Movistar de México. 2003. Datos sobre el Mercado de usuarios de telefonía celular en México (Market data on mobile users in Mexico). Mexico City.

Telefónica Movistar de México. 2005. Datos sobre el Mercado de usuarios de telefonía celular en México (Market data on mobile users in Mexico). Mexico City.

7 Reducing Illiteracy as a Barrier to Mobile Communication

Jan Chipchase

The mobile phone enables personal and convenient synchronous and asynchronous communication—in essence, it allows its users' communications to transcend time and space in the context of their choosing. It is unsurprising, therefore, that with these characteristics many people consider their mobile phones to be one of the essential objects to carry when leaving home (Chipchase et al. 2005).

As traditional markets for mobile phones such as Sweden, the UK, and Singapore reach saturation, handset manufacturers seek growth in "emerging markets" such as India, China, Vietnam, Brazil, and Indonesia whose populations number many hundreds of millions. Targeting products and services at new markets generally requires addressing the needs of potential customers.

Yet vast portions of people in these new markets have limited formal education and consequently lower levels of literacy and numeracy. The United Nations estimates the total number of illiterate adults to be about 800 million worldwide, 270 million of whom are located in India alone, and defines illiteracy as a "person who cannot with understanding both read and write a short simple statement on their everyday life" (UNESCO 2004). I use instead the term *textually nonliterate* to reflect that they are many ways to define literacy (for example the ability to complete a task or understand a problem), as well as to emphasize that illiteracy is often a result of lack of opportunity rather than of ability.

The key question for mobile phone manufacturers who wish to address the communication needs of this potential customer base is how textual illiteracy affects the ability of mobile phone users to make effective use of mobile phones. Based on the answer to this question, manufacturers can consider how communication tools could be designed that draw on the knowledge and experiences of textually nonliterate users so that they can more effectively use the mobile device.

This chapter draws on studies by researchers in Nokia research centers in Tokyo, Beijing, and Helsinki that were launched to understand and improve the communication experiences of nonliterate people. (This research does not imply the development by Nokia or its partners of products and services proposed in this chapter.)

Framework

If a mobile phone's sole purpose is a status symbol, then holding it up to one's ear and pretending to speak to a remote other may be sufficient to the user and no textual literacy is required. However, the primary benefits of the mobile phone as a tool for personal and convenient synchronous and asynchronous communication, and secondary benefits such as contact management, time keeping, and time planning, can be extremely challenging to access for someone with limited mastery of words and numbers and their meanings.

There are many ways to learn how to use a device or complete a task. A useful distinction is to think about structured and unstructured means. *Unstructured learning* includes visual feedback—how it looks, observation—how it behaves and how other people interact with it in the world around them, tactile—how it feels, and aural—how it sounds. For example, a person may never have picked up a mobile phone, but based on advertising and television alone would be able to ascertain the right way to orient the device to the face. Product, industrial, and user interface designers try to use as many of these cues as possible to make the mobile easy and logical to use. Textual and numerical literacy is typically learned through *structured learning*, or schooling. Since the mobile phone interface includes both numbers and letters it is understandable that a degree of textual and numerical literacy is required to use many of the features on the phone. This is a problem if, as in India, structured learning and consequently levels of literacy and numeracy are low.

There are a number of ways to define literacy. While the U.N.'s definition for a literate person is someone who can "with understanding, both read and write a short simple statement in his or her everyday life," the Chinese government, on the other hand, uses criteria that better account for what is needed in that user's context—a literate person is someone from the countryside who can read fifteen hundred Chinese characters, or employees in corporations or citizens in towns and cities who can read two thousand Chinese characters. The definition also covers a person's ability to carry out simple accounting (Chinese Government 1993). There are many reasons for being nonliterate, including the need to forego schooling to enter the workforce to financially support the family or even lack of educational infrastructure. Our assumption is that just about everyone has the potential to become textually literate; people just do not all have the opportunity.

While definitions for illiteracy can help frame the discussion, the more relevant question is where does textual and numerical illiteracy become a barrier to device competency.

In emerging markets, a user's experience will be affected by other factors. The user may be literate or semiliterate in a language that the phone user interface does not support. Or the device itself may well have been bought used and is mechanically unreli-

Figure 7.1
Learning to use a new interface is especially challenging when the keypad is badly worn (Hang-zhou, China). Source: Author.

able, perhaps continuously repaired by one of the many street-repair services. Buttons may be worn out, as is the case with the phone presented in the accompanying photo-graph (figure 7.1). Alternatively, if the network coverage is weak and oversubscribed to, multiple attempts to call must be made before a connection is made. Calls may be dropped frequently. Though each of these factors may not present an insurmountable inconvenience by itself, the difficulty in learning to use the device is compounded when the situated learning experience itself is unpredictable.

Icons: The Quick Fix?

Icon-driven menus are often proposed initially as the solution to the illiteracy problem—after all, it must be that everyone can understand the meaning of a few pic-tures. Why not create an icon-based mobile interface?

While richer iconic support could assist a textually nonliterate user, this assistance is conceptually far distant from designing a mobile phone relying totally on an iconic interface. Icons by themselves are not the answer. To begin, the meaning and subsequent use of icons are best understood when initially accompanied by textual descriptions (Wiedenbeck 1999). Understanding can be improved by successfully completing tasks, which implies an understanding of the textual annotated steps that make up a task, a degree of prior device understanding, and exploration. Many tasks, like configuring General Packet Radio Service (GPRS) settings, are so abstract from the user's real-world knowledge that it is implausible that even the most talented icon designers could successfully solve the problem. Lastly, were icons to be designed for every phone feature they would need to be comprehensively tested with each diverse user group, perhaps requiring hundreds or even thousands of icon variations.

Field Research to Find Answers

To explore these issues, initial research was conducted in India in 2004 with follow-up studies in 2005 in India, China, and Nepal as part of Nokia's exploration of future user interface requirements. Study locations were selected because of a mixture of a high level of textually nonliterate participants and research partners with suitable available skills.

Local research partners assisted with data collection, cultural interpretation, and synthesis. The studies included eleven nonliterate participants who were engaged in a variety of manual trades, for example, as a cook, a cleaner, a gas station attendant, and a caretaker. Data from these participants were collected using qualitative techniques: shadowing, observations, and contextual interviews including screening criteria for literacy and numeracy. We looked at what devices our nonliterate participants currently used, studied how they managed to maintain contact information, and documented their strategies for coping with written material.

Our aim was to understand the world from the perspective of a nonliterate person—how they survived (or even thrived) in a world of words and numbers, and the bottlenecks in their desire to communicate.

General Observations

Our first observation is that our textually nonliterate participants generally lead more predictable lives than more literate counterparts from other studies. There are multiple possible explanations for this, one being that because some of our textually nonliterate participants had limited disposable income—since they were largely able to obtain only entry-level manual work, which paid relatively little. Disposable income provides options increasing the range of what is on offer. The second reason can be explained by thinking about the acceptable amount of effort required to complete a given task. Choosing a dish from a restaurant menu requires asking the restaurant staff or literate fellow diners what is on offer. Sometimes this is fine, but multiply this task for every time literacy is a barrier and it soon begins to grate on the person to the point where it is easier to narrow down choices to what is already known.

Our second observation is that textually nonliterate users can complete tasks requiring a degree of textual literacy, but these tasks typically take considerably longer to complete. Being asked to fill in a form at work may take a literate person five minutes—whereas for a textually nonliterate person it becomes an overnight task involving the availability of a literate relative or friendly neighbor. This is sometimes called "proximate literacy"—the ability to rely on others who either are sufficiently competent in using the device, or are literate and can take the user through the steps requiring textual understanding. For example, one participant in India sent text messages on her mobile phone via her literate daughter and required her daughter to read the responses. Families or even whole villages may share the use of a single mobile phone. The primary reason for this is the cost of ownership and use, but also because in societies where fixed line penetration is limited the mobile phone is the first phone available to them. (See chapters on the Philippines and Bulgaria in Katz and Aakhus (1999).)

Our third observation is that there is a "parallel universe" of cues that are visible if only you know how to see it. Bank notes are a good example in that they are required to be usable by all members of society, and provide multiple cues to their authenticity and value. While you might be thinking of print quality and watermarks, our nonliterate participants picked up on texture (China) and scent (India, for 500 rupee notes) of the notes. Additional cues can be built into the product design that does not impinge on use of literate users as well.

Fourth, with sufficient application of intellect and memory, rote learning can be used to memorize the steps needed to carry out most tasks. However, rote learning is not understanding, and when things go wrong understanding is often required to solve the problem. Remember that the used/shared mobile phone and network may be less reliable, and problems are more likely to arise.

Lastly, most of our participants worked very long hours with little or no holiday time. The research team was left wondering who has the greater need for personal and convenient synchronous and asynchronous communication—someone working nine to five, five days per week, or someone working five to nine, seven days per week. While the option of whether to purchase a mobile phone may be constrained by income levels, based on observations our assumption is that the synchronous and asynchronous communication has the potential to benefit everyone.

How Nonliterate Users Get By

The simple answer to how do nonliterate users get by is that nonliterate mobile phone users can call, but cannot text message or use the address book. The subtleties are more interesting than this.

Two basic tasks were easy for almost all our participants to complete, namely turning on the phone and answering an incoming call. Beyond this there were various degrees of success. Dialing a local phone number is relatively easy, but problems can occur when there are variations such as dialing a national or international number, or using IP telephone prefixes. Dialing an incorrect number may require starting from the beginning of the task since the cancel button is not always understood.

Our hypothesis is that once the nonliterate user has learned how to make and receive phone calls to their close circle of contacts, their primary reason for owning a mobile phone largely has been met. There is, therefore, less motivation to spend additional time rote learning other features on the phone, unless someone can proactively demonstrate the worth of the features, and spend the time to teach them the steps required to complete the task.

Phone features that require text editing such as creating a contact, saving a text message, and creating a text message present too great a barrier to use.

Information is often relayed as part of a phone call, but taking a verbal message during a phone call requires the user to remember the message details since this cannot be written down. This increases the likelihood that the message will simply be that a person called, rather than the content of the message itself. It may or may not be possible to write down numbers, and names if written are often annotated with rudimentary markings understood only to the writer. The call log serves as an ad hoc address book, albeit one in which the user needs to remember the number of calls since the person they wish to communicate with last called.

Several of our participants kept paper phone books. Typically, contact information was written and updated by a literate family member, and sometimes annotated by the textually nonliterate user as an aid to remembering who was who. Specific contact information was remembered based on a number of criteria including on what page in the address book it was written, what color pen was used to write the number and what position on the page it was in. The ability to put contact information in the most appropriate format significantly supports the user's ability to gather it in one convenient place.

We noted that textually nonliterate users of public call offices often took a scrap of paper with a phone number scrawled on it to the owner and asked them to dial the number. This system is open to errors caused by inaccuracy, either because the number was not clearly transcribed, or simply because the paper on which the number was written was worn and faded from being carried.

User interface designers often talk about the user's mental model of a system, and how it maps to the reality of how a device actually functions. It is typical for designers to use metaphors such as the "desktop" or "soft keys" to support the building of an accurate model. Textually nonliterate users will not have access to textual cues so their mental model may well be poor. While a poor mental model is not a problem within a limited range of (rote learned) tasks, if and when errors occur users may adopt the wrong strategies to correct the problem. Designers use a myriad of audio, visual, and textual cues to support the user's understanding of how the mobile phone works. Literate persons are able to quickly absorb (and subsequently ignore) this textual information and apply the knowledge in practice. A positive outcome reinforces their understanding of how the system works and helps build an accurate mental model. Textually nonliterate people are required to make assumptions for the textual prompts based on how the device responds to their actions. A plausibly positive result is sufficient to believe that that is how the system works regardless of how well it maps to the actual system.

One method of learning how a device works is through trial and error. Our hypothesis is that the user's willingness to explore the user interface boils down to perceived risk versus perceived consequences. (Here permit an aside: We did not study this in a systematic way, but there were comments by participants that tasks were not tried because there was a risk of breaking the device. It's a tricky issue because it is very context-dependent, including whether or not there is a technologically literate person in proximity to advise or fix if things go wrong).

As with many tools, once people have achieved their primary goal, for example, to be able to make or receive a call, their motivation to learn beyond that is also reduced. However, based on anecdotal evidence I have heard, people in emerging economies (including nonliterate people) generally try to use most of the potential of any device they do have.

A nonliterate user's willingness to explore features on a mobile phone requires weighing the perceived risk of factors such as changing settings so that things no longer work, past experiences of things going wrong, deleting data that cannot be recovered, becoming lost and not being able to retrace steps, or physically breaking the phone. Perceived risk is not the same as actual risk. Where there are three menu options to choose from and one of them *might* delete the call log entries, how likely are you to use trial and error? There are individual and cultural differences in attitudes to risk and a person's perceptions of risk will change according to circumstance.

One way of thinking about the issues of context and exploration is to consider figure 7.2. Assume you are wandering around a market looking for a toilet. If you understand the signs written in one of the two languages in the picture, Hindi or English, you can easily interpret the purpose of this building. If you are textually nonliterate, there is other information you can rely on: the pictures/icons of the man; asking a stranger; experimenting by following men through the door if you are male; taking a step back and observing females going into a similarly pictured entrance next door. You could rely on your sense of smell, or maybe you've used this building, or one like it before.

Now consider the issue of risk. How sure are you that this is indeed a toilet? What would be the costs of making the wrong choice? Embarrassment perhaps if it turns

Figure 7.2
The Public Call Office is another communication options (Ahmedabad, India). Source: Lokesh Bitra.

out to be a hairdresser's or the headquarters of a local political candidate—whose mustachioed mural adorns its walls. But what if the cost is greater, for example, a month's wages, or in the context of exploring a mobile phone the cost of an expensive call or breaking the phone—the valuable lifeline to your loved ones.

Our challenge is that many mobile phone features rely on some degree of textual understanding, the tasks are much more abstract, and rich context is missing.

Our research team also explored alternative communication channels available for our participants, for example, Public Call Offices (PCOs) in India (figure 7.3). We interviewed more than twenty owners and users in an effort to understand how other communications infrastructures were used.

We identified three areas to improve the user mobile experience and in effect bring personal and convenient synchronous and asynchronous communication within the reach of textually nonliterate users: on the phone, in the communications ecosystem, and on the carrier network.

Figure 7.3
Much of the rich context that helps nonliterate individuals understand their surroundings is missing from the phone's user interface (Bangalore, India). Source: Author.

Improving the Device

A simple mobile phone with a minimal feature set is the short answer. In practice this means supporting incoming and outgoing calls, with a call log adapted for use as an address book. Contact management and text messaging features could be a setting that the user has to activate before they appear in the menus (a task that would require a literate person to complete). Menus could have additional iconic support, and hardware buttons other than soft keys as much as possible should be reserved to one button for one task. A two-way rocker button can confuse and may be perceived as one button.

Wherever possible phone settings should be automated to avoid the need for editing—for example, by default setting the time and date on the phone from the network.

Successful outcomes can be reinforced with audio feedback including, for example, playing back the number that was dialed prior to calling. Another option is spoken menus, though again this is a nontrivial undertaking given the scale of languages and dialects to support. One radical approach could be to replace the digital contact management tool with a physical/digital hybrid that the user could annotate by pen or pencil.

A mobile phone equipped with a sufficiently high quality camera and display would enable the capturing and location shifting of written text, for example, taking a photo of a hazardous materials sign at work and showing it to a literate relative at home. However, cost issues currently make this an unlikely mass-market solution on the lowest-end phone models.

Different ways of bringing the benefits of asynchronous communication to nonliterate users are through services such as Short Audio Messaging (Nokia 2003), or simply leaving a message on an answering machine. For all these solutions, however, accessing in-coming communication is unlikely to be a problem, compared to the complexities of saving, editing, deleting, and replying.

To avoid the potential social stigma associated with textual illiteracy the phone should not be noticeably different to other products on the market. This comment is drawing on observation and related research. The observation is that interactions with nonliterate people (who were not part of the study) sometimes required textual understanding, but the issue of literacy was never discussed openly—it was side-stepped by the nonliterate person. The related research I am drawing on is nonpublished Nokia market research into preferences for seniors, which include having products optimized for this market, but that do not stand out as being for seniors.

Improving the Ecosystem

The best possible solution may be one that raises the population's general level of literacy and numeracy, and the mobile phone may have a role to play in this regard. Beyond this, providing classes on how to use the phone and creating an environment for risk-free exploration can also raise device competency levels. Low tech solutions can suffice—for example, a poster showing the flow and outcomes of key tasks may familiarize users with the user interface so they feel comfortable to explore beyond what they already know.

It may also be possible to nurture commercial services that overcome textual barriers such as one for entering contacts into an address book and assigning photos to entries. Solutions such as this can build upon the rich social face-to-face interaction that already exists.

We note that IP telephony kiosks in China and PCO in India already contain a simple printer for providing receipts, and it may be possible to modify this infrastructure to create accurate and uniformly designed contact information for textually nonliterate customers.

Improving the Infrastructure

Why require text entry at all? A simpler alternative to managing contacts is to press a button and speak to an operator who connects you to whomever you want to speak with. The same principle applies with messaging and managing personal information.

Since caller ID is already used as an ad hoc relational contact management tool, why not extend the information that is sent with caller ID, including a photo, and auto-build the address book? Although it would be the target of spammers and advertisers, it may be possible to autogenerate a phone's address book entries.

Conclusion

Personal and convenient synchronous and asynchronous communication has the potential to benefit everyone. Two features of mobile phones that many users take for granted—text messaging and contact management—present significant but not insurmountable hurdles for textually nonliterate users. The market for nonliterate users is enormous and potentially profitable. Serving this market effectively could greatly improve the lives of the nonliterate, and perhaps even provide a springboard for literacy. Moreover, it is likely that mobile phones will be increasingly important to accessing governmental, commercial, and social services as these services become more digitally based. As these services expand, there also is likely to be a concomitant demand for better and more elaborate handsets. To avoid widening the already enormous digital divide that nonliterate people face, it is important to provide meaningful solutions to them. Areas in which to seek such solutions to support these users have been proposed in the realm of the phone itself, in the communications ecosystem, and on the carrier network, and Nokia research in this area is ongoing.

References

Chinese Government. 1993. http://www.law-lib.com/lawhtm/1993/9704.htm. (Chinese language.)

Chipchase, J., P. Persson, M. Aarras, P. Piippo, and T. Yamamoto. 2005. Mobile essentials: Field study and concepting. Paper presented at the Designing the User Experience DUX 05, San Francisco. http://www.dux2005.org.

Katz, J. E., and M. Aakhus. 1999. *Perpetual Contact*. New York: Cambridge University Press.

Nokia. 2003. Short audio messaging—New low cost voice communication service. Nokia Datasheet. http://www.nokia.com/downloads/operators/downloadable/datasheets/samdatasheet_net .pdf.

UNESCO. 2004. Youth and adult literacy rates by country and by gender for 2000–2004. Montreal: UNESCO Institute for Statistics (UIS). http://www.uis.unesco.org/ev.php?URL_ID=5204& URL_DO=DO_TOPIC&URL_SECTION=201.

Wiedenbeck, S. 1999. The use of icons and labels in an end user application program: An empirical study of learning and retention. *Behaviour & Information Technology* 18(2): 68–82.

8 Health Services and Mobiles: A Case from Egypt

Patricia Mechael

Mobile phones are a rapidly growing aspect of health and health service delivery (Agar 2003; Vodafone 2006), and yet are frequently overlooked in both developed and developing countries as a strategic opportunity for the health sector to maximize increased access to the technology for meeting health objectives (World Health Organization 2006). While this appears to be the situation in both developed and developing countries, it is particularly the case that the greatest leverage is likely to be obtained in developing countries where the majority of a country's population had extremely limited access to telecommunication technologies of any kind. Until recently, little research or documentation existed about how mobile communication would contribute to health, especially in the developing world (Rice and Katz 2000). The focus of many "digital divide" initiatives has erroneously been on the use of the Internet in developing countries for improved access to information (*Economist* 2005). In the following quote, Anthony Townsend attributes this to a general preference for the study of the Internet.

[Unfortunately] the advent of inexpensive mass-produced mobile communications in particular, has avoided scholarly attention, perhaps because it seems pedestrian compared to the nebulous depths of cyberspace. Yet the cellular telephone, merely the first wave of an imminent invasion of portable digital communications tools to come, will undoubtedly lead to fundamental transformations in individuals' perceptions of self and the world, and consequently the way they collectively construct that world. (Townsend 2000)

Mobile phones have the potential to enable communication in places where it was not possible in the past in addition to instantaneous dialogue and information transfer without dependence on literacy, solving some issues but problematizing other issues in the delivery of health care. For instance, patients can more rapidly access advice from physicians; however, this raises questions regarding the sorts of situations that can be addressed without a physical examination through verbal descriptions.

There are several areas that have been generally cited as aspects of health service delivery for which the basic voice and text capabilities of the mobile phone can provide support in both developed and developing countries (Vodafone 2006). The primary departure between developed and developing countries within health care settings is a focus on chronic diseases such as cancer and heart disease in developed countries and infectious diseases in developing countries. This yields differential applications of mobile phones toward addressing health needs. Health care problems in developing countries are compounded by the magnitude of child and adult morbidity and mortality as well as the absence of qualified health care personnel, particularly outside of major urban centers.

A well-documented aspect of mobile phone use for health service delivery is the increased efficiency of direct contact between health service providers and patients. Studies in the United Kingdom have documented significant cost and time savings of text message reminders for medical appointments (*Economist* 2006; Vodafone 2006). Studies in the United States and Australia have also begun to highlight the role that mobile phones are serving in saving time and increasing demand for emergency services in relation to road traffic accidents (Chapman and Schofield 1998; Horan and Schooley 2002). It is projected that in 2020, the third leading contribution to the global burden of disease will be road traffic accidents (Global Burden of Disease 1993). Much of this disease burden is due to the increasing number of drivers and automobiles as well as the increasing cost of treatment and long-term care for injury victims. In addition, it is a concern that affects countries at all income levels, particularly lower and middle income countries, which sustain 85 percent of deaths and close to 90 percent of the disability (World Health Organization 1999; Nantulya and Reich 2001). Interestingly, mobiles both help this problem by bringing emergency services to bear more quickly on the accident, but may also cause the accident in the first place. In one study conducted by Harvard School of Public Health in 2000, key findings initially highlighted how the perceived benefits of having a mobile phone while traveling by automobile outweigh the risk of automobile accidents; however, this was refuted in a follow-up study conducted in 2002 (Harvard Center for Risk Analysis 2002).

The second aspect of health care that is being supported by increased access to mobile phones is the improved capacity for chronically ill patients, particularly those with diabetes, to self-monitor their conditions and minimize complications (Vodafone 2006). For developing countries the disease burden is primarily associated with diseases such as tuberculosis (TB), malaria, and HIV/AIDS; however, with demographic shifts toward an aging population many health systems are also experiencing more chronic illnesses among patients. Although there are limited numbers of case studies on the use of mobile phones for health in developing countries, the main focus has been on the related use of text messaging for patient monitoring and treatment compliance for the detection and treatment of tuberculosis (Hedberg personal communication 2002) and for HIV/AIDS in sub-Saharan Africa (Shields et al. 2005). In the following example, provided by Carl Hedberg in a personal communication, mobile phones are being used to support mobile sputum sampling, analysis of sputum in a central laboratory, and communication of results between the central laboratory and satellite facilities. The result has been saved time and improved detection rates for TB in remote settings.

[One] significant undertaking regarding cell phone technology and health delivery I'm familiar with (beyond general cell phone use by managers and staff) is a very successful pilot project in the Libode/Port St. Johns area in the Eastern Cape, South Africa.

Passive detection of tuberculosis was abysmal in this area because of the extreme turnaround time for lab results (up to 4 weeks, and often no feedback at all). The term *passive* is used because you don't actively seek out patients through home visits, et cetera—you just ensure that all patients coming to the facility with TB-like symptoms provide sputum samples for smear microscopy and/or for culture testing and/or they are x-rayed and/or you analyze the Road-to-Health card for children to detect signs of Primary TB.

They started a pilot project about two years ago using young guys on motorbikes to pick up the sputum samples and bring them to the nearest hospital lab for analysis. Results are sent back to the nurses using SMS messages (some pagers were donated by one of the mobile network providers).

Turnaround times dropped to around twenty-four hours, and detection rates of TB went up more than 300 percent. Generally regarded as highly successful, despite logistical and financial problems getting fuel for the motorbikes, et cetera (in that area—droves of doctors have left the last year because they don't get paid). (Hedberg personal communication 2002)

The third aspect of health care that is benefiting from the use of mobile phones is the increased capability of isolated or remote groups to access health services, enhancing privacy and confidentiality in health-related discussions. In developed countries this is largely presented in relation to the use of hotlines by young people without the knowledge of family members (Vodafone 2006). In the context of developing countries mobile phones are an invaluable tool for individuals and health care providers living in rural areas. Rural denizens previously had to travel in person, with all that entails, to seek guidance from health professionals, access health information, or schedule a medical appointment.

Increasingly, mobile phones are carried throughout the world as part of an individual's desire to preserve and maintain safety and security, which has become a part of the social image of the technology (Agar 2003; Katz 2006; Ling 2004). To better understand how mobile phones are being applied to support public health objectives, it would be useful to descend several levels of analysis from the global- and society-wide concerns to that of a particularly situated locale. In this way the larger forces affecting healthcare can be dissected more meaningfully and in a fuller context. To provide this detail, I present evidence from a qualitative study I conducted in Minia Governorate, Egypt, in 2002 and 2003. The original goal of my research was to use in-depth interviews and observation to determine what changes in access to and the delivery of health services were associated with the introduction of mobile phones. This study can also serve as a more generalizable template for the way in which mobile communication can both solve many problems but also introduce new complications.

My ethnographic study included interviews with sixty-six health care professionals and members of the general population living in rural and urban settings. I present my data by moving from macro-level syntheses to micro-level details of the perceptions and experiences of users in situ. It is very much the case that Egypt's situation mirrors and refracts some problems both of developing and developed countries. These include demographic changes and shifts in health priorities from infectious diseases to more chronic health conditions that come from improvements in health care as well as an aging population (Mehanna and Winch 1998).

Mobile Phones in Egypt

Mobile phones were first introduced to Egypt in 1997. According to the World Bank, Egypt moved from 51.1 fixed-line and mobile telephones per 1000 people in 1996 to 107.7 in 2000 (World Bank 2001). At the time of my study, there were more than 3.3 million mobile phone subscribers in Egypt in a population of 65 million people. Since then the number of subscribers has climbed to 14 million, representing 20 percent of the country's population (World IT Report 2006).

Before mobile phones, most Egyptians outside of the major cities had to go outside the home to make phone calls. Their options included not making phone calls (and "leaving things to God") or using telegrams, local telephone stations known as *centrales* mostly located in urban and periurban communities, pay phone booths, and private landlines owned by wealthier neighbors willing to share their fixed-line telephones in emergency situations. With improved access to mobile phones a number of new developments have emerged within the health sector. The first is the distinctive use of mobile phones versus fixed-line telephones by health service providers. The second is a threshold effect whereby although a relatively small percent of the population has direct access to mobile phones, many more people are able to benefit from them. Third is the emergence of informal networks for problem solving and remote service delivery. Each of these dynamics poses critical questions to health care policy makers and managers as they strive to maximize benefits and minimize potential harm.

Combined Use of Mobile Phones and Its Predecessor

In Minia, there is a complex web of elements working simultaneously to advance and impede the use of mobile phones as a tool for health. In developing the study design and preparing for field work, I underestimated the role of fixed-line telephones in combination with mobile phones. It was in the combination of telecommunication technologies that maximal health benefits were attained by health professionals, as mobile phones were not perceived as a replacement for fixed-line telephones. Fixed-line telephones are generally more cost-effective than mobile phones, making them a first choice for both health care facilities and households.

Coordinating Personal Health Care Processes

Wireline telephones are preferred because they tend to be more reliable, do not require recharging, do not interfere with medical equipment, and have the capacity to enable Internet access (access in Egypt is provided free of charge by the government). They were deemed most useful for health professionals that work in fixed locations.

Where fixed-line telephones are not available in health facilities, mobile phones are considered beneficial, particularly in periurban and rural health facilities where health care workers often feel isolated. In addition, mobile phones were perceived as beneficial for predominantly mobile health care workers, including emergency health care professionals, specialists that are on call, and health care administrators, especially those who are responsible for staff and facility management. They reportedly enable coordination and altered action as needed, while fixed-line telephones were used mostly for consultations, which tend to involve longer periods of use.

When exploring the combined use of telecommunications technologies, failures with fixed-line telephone services in health facilities impede and discourage potential contacts. Negative perceptions regarding health service quality and reliability of fixed-line telephone services in health facilities support observations indicating that simply increasing the number of mobile phones among health professionals and the general population is not sufficient to address fundamental deficiencies within the health sector. It also underscores the sentiment that mobile phones and other technologies are tools that can support health, but that caution should be taken that health systems are not further burdened by their integration (Shields et al. 2005; World Health Organization 2005).

Associated with the combined use of mobile phones and fixed-line telephones is the use of *intermediaries* to access health services, transportation, and information particularly in relation to addressing emergency situations. Intermediaries use mobile phones to access services and information on behalf of others. Health professionals often consult other health professionals as intermediaries on behalf of a patient for more specialized information to determine the most appropriate course of action. This informal decision support consultation process broadly falls within the field of *telemedicine*. In the general population, individuals use mobile phones to mobilize support on behalf of a relative or stranger (altruism) in an effort to overcome ineffective emergency call numbers.

During the early stages of my data collection in Minia, it became clear that there was much confusion in terms of what number ought to be used to contact an ambulance. At the time there was one fixed-line telephone code that was meant to be used for ambulances, "123," which was known by most respondents. However, all the calls that respondents reported making to this number from a mobile phone were recounted with frustration because respondents had been routed through Cairo and provided access only to information about services in Cairo after a significant waiting period. This service was not particularly beneficial for citizens in Minia living four hundred kilometers south of the Egyptian capital. Instead of using "123," many mobile phone users in Minia call their relatives or friends who have fixed-line telephones to coordinate assistance or the local police, "122," in emergencies. One respondent recounted,

My cousin had an accident in Assiut and it was so difficult to call from his mobile any governmental service center, so he called his friend to send him an ambulance. (Male physician—natural group discussion; urban; user)

Families, friends, as well as the local police (as illustrated in the quote above) maintain a coordination function as intermediaries mostly because the cost in time and money of trying to coordinate emergency ambulance support from a mobile phone is high.

The prominent use of intermediaries means that the mobile phone user category can be conceptualized as wider than actual owners. The critical mass of mobile phone users extends benefits to others beyond the owners themselves. In Minia, two such illustrations of altruism were expressed by key informants. One respondent shared that his brother who frequently travels on the highway as part of his transportation business contacted him to coordinate emergency support for someone who had an accident on the highway. Another example, provided by a Ministry of Health representative, illustrates how his son contacted his mother to mobilize similar support for a stranger he witnessed have an accident:

One day my son saw an accident on the detour route, so he took the mobile of the injured to call his mother and one of his friends, and my son stayed with him until the ambulance came, also his mother and also the friend that he called. (male; age 58; urban; user)

People involved in or witnesses to emergency situations oftentimes use mobile phones to coordinate responses with people having access to fixed-line telephones, particularly family and friends. Due to cost, the most common usage is to make short calls with a mobile phone to a family member or friend to mobilize the necessary support on behalf of callers acting in an altruistic capacity or to address their own needs. The notion that the calling party pays lends itself to shorter outgoing calls from a mobile phone (Donner 2005). This combined with the much cheaper calling rates of fixed-line telephones compared to mobile phones provides the basis for such calling patterns in Egypt.

Telecommunications "Herd Immunity"

In Minia, there was sufficient access to mobile phones at the time of the study in urban settings that respondents provided accounts in which they either borrowed or lent a phone free of charge to address an emergency situation. The value of connecting to a network depends on the number of others connected to it (Haddon 2004). As I observed in Minia, each increase in the number of doctors and other health service providers as well as lay users that have mobile phones and/or fixed-line telephones improves their utility.

In the public health *lingua franca* I characterize this as a *telecommunications herd immunity* or threshold effect for health professionals and the general public. In vaccination programs, the notion of *herd immunity* states that if a specific critical mass of individuals within a population have been immunized against a disease that the entire population benefits and is protected from the disease (Gordis 1996). In my study in Egypt, I found that the increase in the number of mobile and fixed-line telephones is benefiting many more than individual subscribers. It is benefiting the general population as the overall teleaccessibility (number of lines per one hundred households) increases.

In social networks such as health professionals in Egypt, trends for the adoption of mobile phones include the pre-existence of a critical mass of others within the network that have access to the technology. Leslie Haddon (2004) cites two reasons for similar trends in uptake among various social networks. The first is that with more users there is more help available to gain familiarity with potential uses. Second, fewer users within a network likely yield a more limited range of use (Haddon 2004). In the case of telemedicine and improving access to health services, the more health professionals that have access to telecommunications the more accessible they become to each other as well as to their patients. This also provides a broader environment for learning new ways of applying the technology as functions become more familiar to larger groups of people. The following physician's quote illustrates how the critical mass of mobile phones is increasing informal collaboration among health professionals, potentially resulting in improved health service delivery, and gives some warnings of the dangers of overdependence.

If you are only one doctor with a mobile phone, then it does not serve a purpose. But if there are many with the phone then you can get somewhere.

Probe: So what is it good for?

If I need advice or someone needs advice from me, if I need to get in contact with someone directly, if I need consent or permission to do something it is much easier with a mobile. I can get the right people at the right time to the right place. This is not to say that the whole system is great. There are still many places in large cities, especially Cairo, where there are dead spots. In the heart of large buildings this is a major problem. This creates problems. Now with advances in telephones we can recognize important calls (caller ID is a popular service in Egypt now for ground lines).

Probe: So how would you quantify the health impact?

You can't do this in a percentage.... There is an impression, yes, but no statistics. But on the other hand people who should be in hospitals all the time leave their duties, saying, "I have a mobile and you can contact me any time." This is not right [because of] failures with the mobile phone networks and traffic. In spite of its usefulness, it encourages people to do wrong things. (Emergency response specialist from Cairo)

The Evolution of Informal Networks for Health Service Delivery

As indicated in the preceding quotation, health care professionals are availing themselves of one another's guidance and services. This is manifesting itself in the form of remote patient monitoring, informal phone consultations for guidance on complicated cases, and improved coordination of disease prevention and control. When asked about their perceived benefits of mobile phones within the health sector, the majority of key informants described the primary benefit as the ability of doctors and patients to initiate contact and be reached at any time and in any location to address health problems. Among health professionals, the domestication of mobile phones into their everyday work life has resulted in increased mobility, facilitated work in several places, and increased remote patient monitoring.

Contacting other health professionals, as described by respondents, is usually either consultative or associated with organizing health service delivery. Eight respondents mentioned the use of mobile phones for coordination purposes in response to emergencies and in requesting guidance from physicians with particular specializations. Thirteen respondents discussed contacting other health professionals to access second line staff, obtain consent or permission for action, and to receive and discuss lab test results. Each of these interactions contributed to the mobilization of remote patient care. In many ways this ability of health professionals to instantaneously transfer information or directives off- or on-site in support of treatment and care-related decision making is new and unique to the introduction of mobile phones into Egyptian society due to the limited access to fixed-line telephones.

Mobile phones enable health professionals and others to multitask. They create free time for physicians to address other professional and personal obligations. Figure 8.1 illustrates such interactions between patients and health professionals, whereby physicians based in health facilities (on-site) can verify patient information and proposed actions to be taken or seek direction from the primary health professional (off-site) responsible for a particular patient in his or her absence.

The essential benefit derived is the minimization of unnecessary travel for both health professionals as well as patients, potentially resulting in the saving of time and

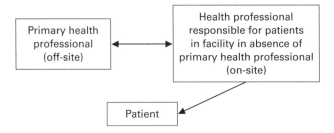

Figure 8.1
In-patient care coordination by a junior health professional.

money. The question this raises, however, is *how effective are phone consultations in contrast to a physical encounter and examination?* As acknowledged by the earlier-quoted physician, caution must be asserted that improved capacity to receive decision support is not a substitute for seeing patients that need to be seen.

For patients requiring treatment guidance from health professionals, mobile phones offer instantaneous answers as well as reassurance. The following quotation provides an introduction to the nature of remote mobilization of treatment support by patients themselves through the use of mobile phones. The nurse describes two situations in which patients that were uncomfortable with the treatment regimen or desired to be released circumvented health facility staff to directly contact their primary physicians to gain information or catalyze action.

I am working in a hospital, and we were treating a patient by giving her a dose of a particular drug. The problem was that the patient was aware of everything about her condition and the dose she was taking, and she was very worried when someone of us was giving her the dose, and if another doctor gave her the dose other than her doctor she would be very worried. And, once the assistant gave her the dose and he added twelve drops, so she called her doctor and told him that they had increased her dose, so he told her that there was no problem and he assured her, and it would be a big problem if the doctor wouldn't assure her.

Also, once there was a patient who needed to take permission to leave the hospital, and his doctor was very busy, so the patient called him on his mobile, so in his turn the doctor called his assistant and told him to make the necessary arrangements to let him leave, and after that the doctor would take care of him at his clinic. (nurse; age 32; female; urban; nonuser)

Providing reassurance to patients via mobile phones was recounted as an increasingly normal part of their work. Patients are more involved in their situations and can access the reassurance they need from their physicians as well as mobilize action as necessary in the absence of their physician (figure 8.2).

Mobile phones increase access to information from sources preselected and trusted by the patient, increasing the sense of involvement they have in their treatment. Related to this, using the phraseology, *following up on patients,* eight out of the twenty-four

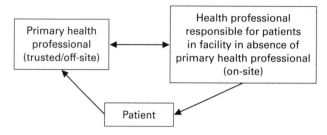

Figure 8.2
Personal health care coordination by a patient.

health professionals interviewed mentioned that mobile phones help them to be reached by patients mostly with questions about prescriptions and by other health professionals on behalf of in-patient ward occupants.

For many physicians, this increased remote contact with patients was described as a positive movement toward enhanced patient care. However, it raises questions regarding liability of facility-based staff for off-site guidance and the increasing capacity of patients to play an active role in discussing treatment with health care professionals.

Directly related to coordination of health service delivery, access to advice and insight was readily discussed by most of the health professionals whom I interviewed. The mobile phone was described as helpful because a physician can more easily reach a colleague in Minia or outside of Minia to ask about a specific case. This exchange of experience and information is critical to ensure optimal care for patients. For most health professionals the term *telemedicine* often elicits grandiose notions of hi-tech equipment although it is simply the process of healing (medicine) from a distance (tele). Formal telemedicine generally refers to pre-established institutional relationships between two or more health facilities. Informal telemedicine or *teleconsultation* ranges from friends calling each other for social purposes and asking for an opinion on a case or contacting a specialist to ask for guidance without a pre-established institutional arrangement. As for more formalized telemedicine, most of the health professionals did not seem ready to embrace this in Minia particularly between *dakatra sagheer*, "small" or less experienced, and *dakatra kubar*, "large" or expert, doctors, but it is happening in informal ways among close colleagues at the same level. Some of this may be a reluctance to acknowledge limitations in one's own skills, experience, and knowledge to someone who is not familiar or trusted.

Apart from the use of mobile phones for the transfer of information for the treatment of patients, several respondents also discussed in detail health applications related to disease outbreaks, containment, and prevention. Of more recent public interest, due to the global media coverage regarding Avian Influenza, was an interview that

I conducted with a mobile phone user in one of the study villages who was an animal health worker with a chicken farming and distribution business. He described his work-related use of mobile phones in the identification of disease, treatment, and immunization of chickens.

> For example someone can call me to tell me that the chickens have diarrhea. I can advise him with what to do and which medication to use in the fodder or in the water and he can ask when to give it to them. I can tell him during sunset this is because if chickens were exposed to the sun after they took the medications.... So, the mobile is useful for the worker to find me at any time. Sometimes I cannot specify the case of the chickens so I have to go to check by myself, to check their abdomens, their livers to see what is wrong with them and to give them the medication which is suitable for them and I cure them. (Animal health worker; age 34; male; rural; user)

Conclusion

In many developed and developing countries, the role of mobile phones and an overall increase in access to telecommunications in health is largely undocumented. For developing countries, efforts to address chronic and infectious diseases simultaneously pose a significant burden to an already overworked health care system and human resource base. Mobile phones within the general population and among health professionals are creating new paradigms for improving access to emergency and general health services and improving coordination and collaboration among users in most countries. The potential result is increased efficiency and effectiveness of health service delivery. Using case studies developed through a systematic analysis of qualitative data collected in Egypt, I discovered the importance of mobile phones as one of a number of telecommunications devices with its own benefits and drawbacks. I also learned that a critical mass of users is needed to maximize benefits for various familial, social, and professional networks to access health services and information, extending benefits beyond owners. And finally, within the mobile phone interactions of health professionals there are a number of areas that require future research, particularly in the areas of liability for information and directives provided by phone as well as the development and testing of standard protocols for remote patient monitoring, telemedicine/consultation, and infectious disease prevention and control.

References

Agar, J. 2003. *Constant Touch: A Global History of the Mobile Phone.* Cambridge: Icon Books Ltd.

Chapman, S., and W. N. Schofield. 1998. *Lifesavers and Cellular Samaritans: Emergency Use of Cellular (Mobile) Phones in Australia.* Sydney: Department of Public Health and Community Medicine, University of Sydney.

Donner, J. 2005. The rules of beeping: exchanging messages using missed calls on mobile phones in sub-Saharan Africa. Presented at the 55th Annual Conference of the International Communication Association, New York, N.Y.

The Economist. 2005. The real digital divide. *The Economist* 374(8417): 11.

The Economist. 2006. A text a day. *The Economist* 374(8470): 84–85.

Global Burden of Disease. 1993. *The Global Burden of Disease: A Comprehensive Assessment of Mortality and Disability from Diseases, Injuries, and Risk Factors in 1990 and Projected to 2020.* Cambridge, Mass.: Harvard University Press.

Gordis, L. 1996. *Epidemiology.* London: W. B. Saunders Company.

Haddon, L. 2004. *Information and Communication Technologies in Everyday Life: A Concise Introduction and Research Guide.* Oxford: Berg.

Harvard Center for Risk Analysis. 2002. Updated study shows higher risk fatality from cell phones while driving. http://www.hcra.harvard.edu/cellphones.html.

Horan, T. A., and B. L. Schooley. 2002. Interorganizational Emergency Medical Services: Case Study of Rural Wireless Deployment and Management. *Information Systems Frontiers* 7(2): 155–173.

Katz, J. E. 2006. *Magic in the Air: Mobile Communication and the Transformation of Social Life.* New Brunswick, N.J.: Transaction Publishers.

Ling, R. 2004. *The Mobile Connection: The Cell Phone's Impact on Society.* London: Morgan Kaufmann.

Mehanna, S., and P. Winch. 1998. Health Units in Rural Egypt: At the Forefront of Health Improvement or Anachronisms? In *Directions of Change in Rural Egypt*, edited by N. S. Hopkins and K. Westergaard. Cairo: The American University in Cairo Press.

Nantulya, V. M., and M. R. Reich. 2001. *Road Traffic Injuries in Developing Countries: Strategies for Prevention and Control.* Cambridge, Mass.: Harvard School of Public Health.

Rice, R. E., and J. E. Katz, eds. 2000. *Internet and Health Communication: Experience and Expectations.* Thousand Oaks, Calif.: Sage.

Shields, T., A. Chetley, N. Hopkins, and K. Westergaard. 2005. ICT in the health sector: Summary of the online consultation. *infoDev.*

Townsend, A. M. 2000. Life in the real-time city: Mobile telephones and urban metabolism. *Journal of Urban Technology* 7(2): 85–104.

Vodafone. 2006. *The Role of Mobile Phones in Increasing Accessibility and Efficiency in Healthcare.* United Kingdom: Vodafone.

World Bank. 2001. Egypt, Arab Rep. Data Profile. http://devdata.worldbank.org/external/CPProfile.asp?CCODE-EGY&PTYPE-CP.

World Health Organization. 1999. *World Health Report 1999*. Geneva: WHO.

World Health Organization. 2005. *eHealth Tools and Services: Needs of Member States*. Geneva: WHO.

World Health Organization. 2006. *Connecting for Health: Global Vision, Local Insight*. Geneva: WHO.

World IT Report. 2006. Egypt: Number of mobile subscribers hit 14 million. http://www .worlditreport.com/Africa/96752-Egypt__Number_of_mobile_subscribers_hit_14_million.htm.

9 How the Urban Poor Acquire and Give Meaning to the Mobile Phone

Lourdes M. Portus

While there have been many studies of affluent teens and other groups' usage of the mobile phone, especially in the industrialized world, little attention has been paid to how the urban poor acquire, use, and ascribe meaning to the mobile phone (though exceptions are presented in other chapters of this volume). Given the potential role of the mobile phone in overcoming social and economic obstacles, and concerns about digital divides, it is important to understand poor people's perceptions of the mobile phone. The Philippines provides a useful case in this regard: it has a high poverty rate (34 percent according to the National Statistics Coordination Board) and a large expatriate community as well as significant terror attacks. It also has high uptake rates for mobiles: about 30 percent of the Philippines' eighty-five million people are owners or users. Sending an average of about two hundred million SMS text messages daily (Gurango 2005), the highly literate population (94 percent) has gained the reputation as the "text-ing capital of the world" (Pertierra et al. 2003).

The Philippines is significant for other reasons, too: it is the home of the famed "People Power I and II" revolutions of 1986 and 2001, respectively, in which the mobile phone and texting played an important role (see chapter by Rheingold).

To learn about the role of mobile communication among the urban poor, I met with thirty-five people in four focus groups, each with seven to ten participants, in January 2005. The groups were: 1) all adult women, 2) all adult men, 3) mixed adult men and women, and 4) mixed young men and women. The male informants' ages ranged from forty-four to sixty-one years old, while the women were from twenty-five to fifty-three years old. Most youth informants were between the ages of thirteen and twenty-six.

Five strategic informants were also interviewed: four were officers of urban poor organizations, and one was a manager and community organizer under the government's National Housing Authority (NHA). Focus group participants were recruited by an NHA community organizer. Participants had to travel to the College of Mass Communication to attend the discussions and were given a small token of appreciation for their efforts. Most participants had reached either the high school or college levels, reflecting a surprising disconnect between educational achievement and economic status within the group. The males had irregular or seasonal jobs, such as plumbing, carpentry, and electrical works, while all the women were unsalaried homemakers. The housewives supplemented the meager income of their husbands through a home-based buy-and-sell enterprise, tending *sari-sari* (variety) stores, or accepting sewing and laundry jobs at their homes. For the women, the mobile phone served as the main tool in their at-home businesses, facilitating ordering and closing deals with customers.

Some women simply stayed home to look after their young children. Their perceived maternal duties forced them to become "homebodies," and made them financially dependent on their husbands. Hence, they wanted a device such as a television or mobile phone that would connect them to the outside world, more particularly, to their husbands, who were at work, and to their children, who were in school.

Most of my youth-informants were high school students. The older youths, having dropped out of school because their parents could not afford to pay for their studies, worked as crew members of fast-food chains.

The informants live in Sitio San Roque, a squatter community of about eight thousand families. San Roque consists of fifty-three hectares of prime land located at the heart of Quezon City and surrounded by a mall, the city hall, a hospital, and schools including the University of the Philippines in Diliman.

Sitio San Roque itself started with a single makeshift house, though soon grew into a vital community teeming with people, availing itself of government and private institutions' services and facilities, inevitably catching the attention and interest of politicians, nongovernment agencies, and other institutions. Stores sprouted, supplying needed household goods and services. Dress shops, barber shops, beauty salons, and electronic and mobile phone repair shops also popped up. Almost without exception, the stores displayed poster advertisements of their merchandise, which included mobile phones and accessories. This helped alert residents to mobile phones and stimulated a desire to acquire them.

Mobile Phone Mania

Despite being an urban poor community, Sitio San Roque managed to keep pace with the times—setting up the necessary infrastructure and support services for the mobile phone users among its residents. More than 50 percent of the local establishments sold not only foodstuff and household items but also mobile phone accessories and prepaid cards. They also provided technical services such as mobile phone repairs and electronic loading or "e-loading."

The informants' initiation to the mobile phone resulted from either personal or mediated communication, or from both. They became aware of the mobile phone through the media, particularly the TV. They often saw their favorite soap opera characters, actors, and actresses holding a mobile phone. They also saw numerous advertisements of different brands of mobile phones during the commercial breaks in TV programs.

Almost half the informants saw the mobile phone gadgets in the hands of friends or acquaintances who were members of rich families. They also witnessed rich persons in malls brandishing mobile phones. As a result, some of the informants found themselves yearning for the same gadgets.

Significantly, the informants interacted with people from the upper economic and social classes. Urban poor in-school youth intermingled with wealthier peers. The women entrepreneurs visited affluent families to sell their wares and services, such as laundry, sewing, manicure/pedicure, and house cleaning. Men who worked in construction or repairs almost invariably had rich people among their customers. The rich, who bought expensive gadgets, were described by one focus group discussion (FGD) participant as "angry with their money. They simply don't know what to do with, or how to get rid of, their money."

The Urban Poor's Acquisition of Mobile Phones

The high rate of mobile phone acquisition in Sitio San Roque that I found demonstrates that one's economic situation does not necessarily pre-empt one's ability to be technologically equipped. All the FGD participants owned and used mobile phones. Apart from the mobile phone, they also managed to acquire a range of popular technologies such as radios, TVs, karaoke machines, video compact discs (VCDs), DVDs, computers, and the Internet. It had taken them some time to acquire the hardware, but the informants managed somehow to keep abreast with their rich counterparts.

Urban poor informants got their mobile phones through either installment or cash payments. Regardless of the mobile phone's condition (new, used, upgraded, or reconditioned), by far the most common path was by installment payments. Sometimes units were purchased from an enterprising friend, who, for a 20 percent profit margin, would be paid in four equal installments. The installment system was known locally as *paiyakan* ("shedding of tears" due to the difficulty in collecting payments).

The installment method seems suited to the coping mechanisms of the urban poor, who rely mainly on daily wages for their subsistence. While costly, this mode enables them to satisfy a real or imagined need. Not surprisingly, therefore, the urban poor accumulate debts from all their possible sources of credit.

Cash was usually used to purchase second-hand units. Buyers found these relatively cheap (at USD 28 to 38 per unit) but failed to consider, or simply ignored, the aspects of the phone's quality or longevity. The few who managed to buy brand new mobile phones with cash actually did so because of financial support from overseas foreign workers (OFW), usually parents or rich relatives. Otherwise, they scrimped on basic needs to save money for a mobile phone instead of saving the money or using it for basic household needs.

One avenue by which mobile phone ownership is pursued is by lotteries or raffles. Feeling sure that they would never acquire a mobile phone if they relied merely on their meager wages, some informants tried their luck with lottery. One of them won a new mobile phone at a Christmas party raffle. Although it was an isolated incident, its impact on the raffle winner and his neighbors, who merely learned about it, lingered on. For several months, the superstitious among them continued to bet in lotteries or buy raffle tickets, hoping that Lady Luck would also favor them.

Finally, pawn and trade are used in the acquisition process. This practice suggests the economic usefulness of the mobile phone when used as collateral for a sum of money in times of emergency. Failure to repay it gives the pawnee the right to own the mobile phone. Some of the informants regarded the mobile phone as an investment, not unlike jewelry, that could be pawned for cash. A youth informant would pawn his mobile phone whenever he needed emergency funds for school enrollment fees: "Of course, since we live a hand-to-mouth existence, I wouldn't have money for matriculation. But the cell phone has a value because I can pawn it. There was an instance when I couldn't redeem it due to lack of funds, so I simply sold it."

Motivation for Acquiring a Mobile Phone

For the male urban poor informants, the mobile phone was a necessity for emergencies; unlike their wealthier counterparts, the informants tended to have many untoward incidents in their neighborhood: gang rumbles, robberies, and fires. Since they were often away from home, the mobile phone was important for staying in touch. Still, it seems the greatest consideration of the urban poor was accessibility of the husbands to their wives and children (Portus 2005). Thus, almost always, the husband would obtain the first mobile phone in the family. This process also indicated the importance accorded to the husband—more particularly, his role as protector of the family.

Meanwhile, the women informants often stated that they had bought their mobile phones because they wanted to imitate the rich and to seemingly belong to the "high society." They aspired for the latest gadgets and would invariably think they would buy the top-of-the-line mobile phones, if only they had the money. One said, "We want to keep up with the rich. We want to belong to the high society. You can call this envy. That cannot be avoided."

Mrs. Josefina, one of my FGD participants, agreed: "Of course if you had the money, why not? If you can afford it, buy the latest model, so that you can keep up with people in high society, just like the rich." Reflective of this attitude, some informants would go on a buying spree when the opportunity presented itself, and suffer lack of necessities later.

For the urban poor women informants, the mobile phone also served as an extension of parental authority over their children and allowed them to monitor their children's location, activities, companions, and schedules. It afforded mothers the opportunity to exercise their role even outside the home and lessened their stress because they knew where their children were.

Text messages to their husbands revealed the wives' anxiety regarding the husbands' safety and well-being. In the past, this expression of care and concern was vague and inconsistent, but now it has become more regular and unequivocal. Whenever their husbands came home late, or worked overnight, or stayed out with friends, using the mobile phone allayed the wives' apprehension. One informant said: "We would ask them if they had already taken their meal or whether they were already bound for home or whether they were in good condition or safe. But we send more text messages to our children than to our husbands."

Competency of Mobile Phone Use

Most members of the family were adept in operating the mobile phone. However, it was usually the young who were the more dexterous and adventurous in trying the various functions or applications in the mobile phone, such as alarm clock, calendar, radio, organizer, calculator, ringtones, and other features.

Using the mobile phone for the first time was a struggle for the informants. They learned from their children, neighbors, or friends how to send text messages. An adult informant recalled, "At first, I did not know what to press. I had to ask my children where the space and the period were. I would get irritated because I was very slow and always made mistakes. I had to ask my child to do the texting for me."

An informant, in describing his feelings when he first used the mobile phone, likened himself to someone in authority. He described his feelings:

When I started owning a cell phone, I felt like I was a mayor, an important person, a person in authority. I know the feeling is not right. It should be that I got one because I want to be effective in my work. In fact, I was rather late in years in acquiring one. There are schoolchildren who have already been given cell phones by their parents. My cell phone is already outmoded.

The most popular activity among all the informants was sending text messages and, on very rare occasions, making calls that were deliberately shortened to cut costs. As one informant explained:

It is expedient to use mobile phones mainly for sending text messages, rather than for calls, because it is cheaper at one peso (USD 0.02) per transmission. But, at times, I am constrained to make a call for an important matter or if I don't get an immediate reply to my text message. My message might not have been received. At any rate, I make it a point to end the call as fast as possible.

Whenever the informants were short of cash, they would "miscall" a person. This would alert the recipient of the call when his own mobile phone rang only once or twice and induce him to reciprocate the call, and thus, assume the call charges instead. Some respondents revealed that they never made a call to anyone to avoid spending P8 (USD 0.15) for a minute or less of conversation.

In the beginning, when families had only one mobile phone, sharing it among parents and children was common. One parent said that at home the mobile phone was for the use of every member of the family:

This is how it is: When I'm at home, this [mobile phone] is for all of us—anyone in the house can use it. This is why, sometimes, I'm surprised when I start to use it to find a message there that wasn't intended for me, but for my child. Oh no! One time, I found out a woman was inviting him out. Ha! Ha! I thought I was the one being asked for a date.

Having a top-branded phone model connotes status, especially for women and youth. Concomitantly, there is a degree of embarrassment about displaying older models in public. Said one youth:

I was with my friends, most of whom have cell phones, but my cell phone was the weakest or oldest model. My phone rang while kept inside my pocket. I didn't look at the message even though my friends were already telling me that my phone was the one ringing. I said that it wasn't mine because I was embarrassed to show it to them, because my phone belongs to the "Jurassic Age" and they might laugh at me once they see it. When they were not looking, I secretly turned off my phone. It was so embarrassing!

Adult men were often indifferent. None said he was aware that it was embarrassing to show the old or obsolete models of mobile phones. However, they said that it was a symbol of status in their role as head of the family.

Mobile Phone Service Provider and Mode of Payment

All informants used prepaid cards rather than subscriptions. One reason was that they had found subscription rates more costly. Another reason was they could not meet the mobile phone companies' subscription requirements, that is, regular income, income tax returns, proof of employment and salary, credit investigation, and billing address. As it was, finding a squatter-resident's address would surely be a daunting task for any bill collector.

Choosing the prepaid card option matched with the informants' irregular occupational profile. After all, some of my informants had never even filed an income tax return.

Engendering the Mobile Phones

Gendered differences, women and men informants said, appeared in the way they actually used the mobile phone:

1. Women were talkers and big spenders, while men used the mobile phone for important things and provided the funds for the phone and cell card purchases. Some male informants claimed that women's nature to gossip resulted in big expenses on prepaid cards. They said, "Women are fond of visiting and talking with neighbors, since there is not much to do at home. They always send text messages. That is why they spend a lot on cell cards. Anyway, it is better that they simply stay at home rather than go to the neighbor's house."

In addition, the men credited the relatively big expenditure on mobile phones to the accessories and the constant replacement of screen logos, ringtones, backlights, cases or pouches, colorful straps, ornamental holders, and even SIM cards. Unlike the women, the men did not have adornments on their mobile phones and simply put them in their pockets.

Another reason cited by the men, in jest, was that women "have two mouths," that is, that the women were fond of idle talk. The women, however, regarded the frequent use of the mobile phone as socializing and news updating as opposed to rumor-mongering. One of them countered:

The thing is, whenever we talk with someone, they would immediately conclude that we were engaged in gossip-mongering. For us women, that is not gossiping but rather small talk—a form of enhancing friendship. We're only exchanging or sharing news about what's happening in our place, or with our relatives or other friends in the province or abroad. We're also greeting one another whenever one is celebrating his/her birthday, or when it's Christmas time, the New Year, or Valentine's Day.

One of the FGD participants discovered an important use of the cell phone—counseling someone who was emotionally unstable. Through the cell phone, she said, she was able to provide comfort and advice to one such person. The informant would regularly send her inspirational and religious messages. She claimed that the person was now closer spiritually to the Lord. She proudly related, "I would keep sending inspirational messages to her about the Lord. We used the cell phone so that I could advise her. Otherwise, something might already have happened to her."

The male informants attributed the women's overly solicitous text messages to lack of trust. The women informants did not deny this allegation because some of them had had sad experiences of their husbands' infidelity.

2. The women informants were mainly responsible for their children, while the men, as authority figures at home, remained aloof. Some male informants believed that women were "created" to be nurturers of their children and husbands. Moreover, the

men said that it was fitting and proper for the women, being the mothers, to take care of the children as well as contact them through text messaging. This was one reason why, at times, they could own a mobile phone ahead of their husbands.

3. Women were home mainstays, while the men were engrossed in community concerns and their job searches. Because of this belief, women needed the mobile phone to have someone to talk and consult with. The belief cut the women from the outside world and implied the lack of exposure to events and world-views. This evokes the view that women's own limited experiences imprison their consciousness and lessen their world-making activity. Information media, particularly the mobile phone, can provide a solution to this problem. The mobile phone's capacity to allow women to form their own world-views enables them to expand their network and sustain social or, at times, even romantic relationships.

Reminiscent of Moores's study of TV, the women use the mobile phone as their "lifeline contact with the world" (Moores 1993, p. 36). Like the TV, the mobile phone in its turn has become "...an integral part of their day" (Hobson in Moores 1993, p. 36) and combats the loneliness felt by the housewives as a result of their confinement to domestic space (Moores 1993, p. 36).

On the other hand, the male informants contended that as breadwinners they had better things to do. Having no steady jobs, they were engrossed in perpetual job hunting, and were away from home most of the time, thus leaving the women to take care of the house and the children.

4. Women had peripheral concerns while the men dealt with essential issues such as community and family welfare. Some of the FGD discussants claimed that the mobile phone should be used only for important matters, referring to their own roles and functions as decision-makers and breadwinners, for example, when they discussed issues with important personalities. They even criticized the women for using the mobile phone for unimportant or frivolous matters, referring to their idle talk and their husband-monitoring activities. The men informants claimed: "Women's use of the cell phone or messages sent is not that important—these are used mostly for idle talk or to monitor the whereabouts of their husbands. This is especially true for the jealous type among them."

The above shows the men's lack of appreciation for the wives' sociability and for the concern they show for their husbands. It also indicates that the husbands regard the wives' sociability and concern primarily as engaging in *tsismis* (gossips), and as distrusting their husbands.

5. The women informants seemed insecure and the men were constrained to reassure them of their love through text messaging. The women manifested this insecurity by sending text messages to their husbands more frequently than the husbands did to them and their children. Thus, the husbands would send texts messages to their wives to persuade them against feeling neglected and jealous.

Unless necessary, the husbands would hardly send text messages to their children, but they regularly sent text messages to their wives because most of them felt obligated to do so, especially when their wives had sent a message telling them to *Tx bak*, Text a reply, or *Sagot ka*, Answer this. Jokingly, some of them would interpret the message as *Sagot ka* to mean *Lagot ka*, You'd better answer this, or else.... A husband informant divulged, "But with my wife, there's no way that I won't text her. I'm constrained to do so. Otherwise, she might think that I'm disregarding her and she might get jealous."

Improving communication or relationships with their wives was not the husbands' primary concern for acquiring the mobile phone. But husbands did claim that using the mobile phone for emergency or business purposes actually demonstrated genuine concern for their wives and children. One of them said, "Of course, I love my wife and child, and they are the reasons why I go to work. And that's why I bought a cell phone—so that I can go home quickly whenever I'm needed there. It's important to maintain in contact with them for their own protection."

There also seemed to be a sense of insecurity on the part of the women informants. Perhaps, this insecurity—or feeling left out and lonely—is because they were not formally employed and simply spent most of their time at home. When boredom crept in, the wives would turn to the mobile phone to communicate with their husbands, children, relatives, and friends (Portus 2005). This partly explains why the mobile phone has become very important to them.

Women said the reason for regularly sending text messages to their husbands was due to jealousy or the suspicion that their husbands were fooling around with other women. Some of the informants nonchalantly admitted: "Surely, this couldn't be avoided. After all, these are men. Sometimes, the women would seduce them. Well, you know, they are not gays or homosexuals."

Indicating a persistent double-standard morality, the women expressed tolerance and even rationalized the right of the men to be unfaithful. The wives said that they—"because they are women"—should stick to only one partner, thus tacitly underscoring their notion that men are polygamous by nature and thus enhanced their husbands' chauvinism.

Economic Class and the Mobile Phone

The mobile phone became an indicator of class status and a site of class discourse, as exemplified by the following discussion points.

Lagging Behind the Rich

Although the urban poor are constantly upgrading their phones, they still lag far behind their rich counterparts. As one male FGD discussant said, "No matter what we do, it is always the rich who'll be served first and have the first crack on everything. Mobile phone brands become cheap when they [rich folks] no longer buy these. It's only then that the poor are able to buy the brand."

Mobile Phone Type vis-à-vis Class Matching

An informant averred that the type of mobile phone that most urban poor possessed was appropriate to their economic class and status. It would not look good if they had expensive mobile phones while their stomachs were experiencing hunger pangs.

Thus, owning an expensive mobile phone with camera would be regarded as scandalous since they could not even enjoy the basic necessities of life. It would be morally wrong on their part, being squatters, to be sporting top-of-the-line mobile phones. One of them contended:

> It is only natural and proper that, for us who are poor, our equipment should be appropriate to our poor economic condition. It would not look good if we had expensive possessions, with our stomachs experiencing hunger spasms. It is shameful and immoral to have top-of-the-line cell phones, but then to be residing in a squatter area.

The informant emphasized that: "You should have a partner or a counterpart with the same cell phone features. Otherwise, where would you send the picture? However, you won't be able to identify a partner because no one here owns that kind of cell phone. Why would you buy such a phone, then?"

Snatching Top-of-the-Line Mobile Phones

There was an interesting reverse snobbery when poor people somehow had possession of a top-line mobile phone. One male youth informant expressed his resentment at the way the rich considered the urban poor who possessed expensive mobile phones as being a "snatcher," and as having gotten the phones from illegal sources. An informant explained his cause of irritation: "Our world is rather small. Indeed, we are poor, yet the rich would still look down on us. For instance, if an urban poor person were to have an expensive high-tech cell phone, the rich would have doubts and think that the poor guy had snatched it from his victim."

Negotiations and Resistance: Effects of the Mobile Phone on the Urban Poor

Some of the informants were keenly aware of the various issues surrounding the mobile phone that generally carried with it a value system of consumerism and materialism. These issues are as follows.

Mobile Phone as Financial Burden

The mobile phone was adversely affecting the already inadequate budget of the family. Hooked on the gadget, the family would give up some of their more important household needs in favor of a P30 (USD 0.56) "e-load." (E-loads are locally initiated electronic transfers of call credit and, while inexpensive compared to regular prepaid cards, expire after only three days.)

Mobile phones required a consistent flow of money to sustain their use. As a result, the owners would run out of money. According to some male FGD discussants: "The cell phone implies new and additional expenses. This is like a vice that you have to pay for, like smoking or drinking, and also like sending a child to school. Our money is all going to the cell phone. It is necessary that one should have a permanent job to continue using the cell phone."

For the urban poor, the problem was less that of acquiring a mobile phone than of financing its continued use. Some contemplated returning or selling their mobile phones, while others actually stopped buying prepaid cards and decided to wait for better times, that is, when they would have a regular income. Unfortunately, they were not aware that the mobile phone's SIM card would expire if not used or loaded within a certain period, leading initially to shock but then a lesson well learned.

Mobile Phone Companies Seen as Exploiting the Poor

As part of our research project, we interviewed four officers of poverty action organizations. These were the president and secretary of Kasama, Inc., an umbrella organization that advocates on behalf of the landless urban poor, and the president and village leader of the biggest squatter colony, Zone One Tondo Organization (ZOTO). The four officers of urban poor organizations speculated that mobile phone companies were making excessively great profits by selling mobile phones, accessories, prepaid cards, and services to the urban poor. The male FGD informants in particular pointed out that

1. The mobile phone companies strategically shifted their focus from the rich to the poor consumers.

2. The mobile phone companies made prepaid cards, or loads, available to the poor at the cheapest price possible in order to tap the large number of poor people.

3. In a not so subtle approach, mobile phone companies would automatically issue reminders whenever there were only a few pesos left in one's prepaid account. The informants found this a rather *makulit* (irritating) market promotion strategy since they were being pressured to procure a load as fast as possible to avoid the inconvenience of a service disconnection.

An informant felt that this was the outcome or a logical system of a commercialized world, saying, "This is becoming too commercialized. Although my load has not yet been fully used, there is already a reminder for me to buy a fresh load."

4. The cell card was *not* a consumable item that could be saved for several months. Since it had an expiration date—usually two months—one would have to use it within the specified period. Afterward, one would need to buy a fresh card to continue using the mobile phone. The situation was even worse for e-loads, of course.

Mang Roger, a lead informant, was conscious of the additional expense that accompanied the ownership and use of a mobile phone. His experience with a mobile phone prepaid card made him feel short-changed, particularly because of the given expiration date. Once, he tried his best to minimize texting in order to prolong the capacity of his P300 (USD 5.66) load. He had only used P60 (USD 1.13) worth of text messages when, suddenly, his load balance fell down to zero. He didn't know that he had to consume his load within two months.

When told that a phone text or voice message (in the English language) might have reminded him about this at the time that he procured a load, he confessed that he was not familiar with the English language. He said, "I did not realize that you cannot be frugal as far as your load is concerned. These companies will find a way to oblige you to spend your money."

5. The *pasa* (to transfer or share) load rendered a service to mobile phone users, who had a zero load balance in their cards. They could ask acquaintances or friends among the mobile phone users to share part of their load at a minimal cost of only one peso (approximately USD 0.02) per transfer.

The informants perceived this to be another business tactic that intended to induce mobile phone users to spend more money. After all, the *pasa* or shared load has a limit of only P15 (approximately USD 0.28), which has to be expended within 24 hours.

6. The so-called free texts given by mobile phone companies served only to entice or entrap customers. When finally "hooked" on the mobile phone, the users would continue using it with normal loads, even when the companies had ceased granting the freebies. The informants realized that "they tricked us. In the beginning, there were one hundred free texts, later there were only sixty. And, still, the number continued to go down. Now there are almost none!"

7. Promotion gimmicks lured some informants to participate in texting their answers to promo questions. Later, they realized that they had fallen into a trap, since the questions continued endlessly. Therefore, it became virtually impossible to get the promised prize.

An informant reported that he tried joining the contests and his fresh load was used up in no time. The question-and-answer portion—including a promised prize—turned out to be money-making gimmicks.

Awareness of Ownership and Control of the Mobile Phone

An informant explained that the Smart and Globe companies, which have mobile phone businesses, were competing to gain more profits. According to him, the owners of Globe and Smart have become billionaires because of the mobile phone.

This is so because Smart is owned by the Philippine Long Distance Telephone Company (PLDT) and Globe is owned by the Ayala Corporation. They are the same—conglomerates—the elites of the Philippine society. As such, they are pursuing their class interests rather than necessarily working against each other.

Mobile Phones Give Rise to Criminality

The informants sometimes regarded the mobile phone as a magnet for robbers, thus posing a threat to their security. They also referred to it metaphorically as a possible *mitsa ng buhay*, a dynamite's fuse to one's life, that may likely lead to one's demise.

In Sitio San Roque, its environs, and elsewhere in Metro Manila, the mobile phone has become a favorite target among snatchers and robbers because it is easy to hide and highly saleable since it is society's current obsession.

Many of the mobile phone owners and users hesitated when it came to parting with their units. Despite the risk to life and limb, they would resist or refuse surrendering their mobile phones. In fact, many robberies have resulted in injury and death to victims.

Some of the urban poor informants became victims of these robberies despite their owning old or obsolete mobile phone models. Seemingly unmindful of the age or condition of their loot, the lawless elements could still get something out of a sale. No matter how cheap the mobile phone, it was still worth something in cash.

Threat to Relationship

The mobile phone was blamed for marital rifts between husbands and wives. As reference to these problems, informants were quick to cite cases of celebrity quarrels. One such case involved two popular Filipino movie actors, Kris Aquino, the daughter of a former president, and Joey Marquez, a former city mayor, that started with the text message of a woman, allegedly Marquez's lover.

This proves that the powerful effect of the mobile phone on relationships is real, not only among the urban poor, but also, across other social classes. Romantic messages or any seemingly innocent message could cause one jealous partner to be suspicious and to start nagging the other.

Some of the FGD participants attested to such problems that threatened their relationships with their spouses. A husband informant confirmed that his wife believed in gossips, which later led to quarrels between them.

The so-called gossips could be transmitted through the mobile phone. Although text messages might come from anonymous sources, they could still arouse suspicion in one of the spouses and spark violent confrontations. The resulting shouting bouts and harangues could eventually lead to the spouses' separation.

A couple of the informants actually did break up as a result of text messages about the husband's womanizing. A wife informant said that some people had really spread the gossips to stir up a quarrel between spouses. She said: "A strange woman may pretend to call one's husband to sow intrigue, which may cause a family's break-up. There may be truth in some, untruth in others."

It is hard to attribute marital problems to the mobile phone itself, yet the device does appear to open new opportunities for problems to arise or be exacerbated.

Conclusions

All of the informants have mobile phones, most of which are second-hand units acquired through credit. They use prepaid cards, with Smart Communications, Inc. as their primary service provider.

The study suggests that the mobile phone allows the urban poor to pursue their gender-defined functions more effectively. Husbands are more accessible for home emergencies and also better equipped to perform their role as protector. Wives are able to monitor remotely their children while staying home and in the interim sustain social relationships and home-based activities.

Possession of the mobile phone is also a source of desire and envy, most apparently among the women. They want to emulate the rich and successful. Yet, the data on class discourse underline the resentment of the urban poor toward the rich, who allegedly look down on the poor. It thus becomes ironic but also understandable that some of the urban poor informants despise those whom they want to imitate.

Despite the popularity of the mobile phone, there are those among the urban poor who are critical about the effects of the mobile phone on their household budget, buying behavior, and consumerist values. They were also critical about the profit motive of mobile phone companies and the impact of the ubiquitous advertisement on the consciousness of the consumers. Their unhappiness with the mobile phone seems to stem from a more general critical view of modern society and the role the mobile phone plays in being a handmaiden to contemporary social values rather than on the many benefits that most see with the technology. These critics' views notwithstanding, for the poor area I studied, the mobile phone is a highly sought-after and increasingly vital part of their lives.

References

Gurango, J. 2005. Offshore development: RP software developers eye 2005 gains. www.itnetcentral .com/print.asp?id=14376&icontent.

Moores, S. 1993. *Interpreting Audiences: The Ethnography of Media Consumption*. London: Sage Publications, Ltd.

Pertierra, R., E. Ugarte, A. Pingol, J. Hernanadez, and N. Dacanay. 2003. *Text-ing Selves: Cellphone and Philippine Modernity*. Manila: De La Salle University Press.

Portus, L. M. 2005. World-making: Mobile phone discourses among selected urban poor married couples. *Plaridel Journal* 2(1): 108–109, 112–116.

Sociality and Co-presence

10   Always-On/Always-On-You: The Tethered Self

Sherry Turkle

In the mid-1990s, a group of young researchers at the MIT Media Lab carried computers and radio transmitters in their backpacks, keyboards in their pockets, and digital displays embedded in their eyeglass frames. Always on the Internet, they called themselves "cyborgs." The cyborgs seemed at a remove from their bodies. When their burdensome technology cut into their skin, causing lesions and then scar tissue, they were indifferent. When their encumbrances led them to be taken for the physically disabled, they patiently provided explanations. They were learning to walk and talk as new creatures, learning to inhabit their own bodies all over again, and yet in a way they were fading away, bleeding out onto the Net. Their experiment was both a re-embodiment—a prosthetic consummation—and a disembodiment: a disappearance of their bodies into still-nascent computational spaces.

Within a few years, the cyborgs had a new identity as the Media Lab's "Wearable Computing Group," harbingers of embedded technologies while the rest of us clumsily juggled cell phones, laptops, and PDAs. But the legacy of the MIT cyborgs goes beyond the idea that communications technologies might be wearable (or totable). Core elements of their experience have become generalized in global culture: the experience of living on the Net, newly free in some ways, newly yoked in others.

Today, the near-ubiquity of handheld and palm-size computing and cellular technologies that enable voice communication, text messaging, e-mail, and Web access have made connectivity commonplace. When digital technologies first came onto the consumer market in the form of personal computers they were objects for psychological projection. Computers—programmable and customizable—came to be experienced as a "second self" (Turkle 2005a). In the early twenty-first century, such language does not go far enough; our new intimacy with communications devices compels us to speak of a new state of the self, itself.

A New State of the Self, Itself

For the most part, our everyday language for talking about technology's effects assumes a life both on and off the screen; it assumes the existence of separate worlds, plugged and unplugged. But some of today's locutions suggest a new placement of the subject, such as when we say "I'll be on my cell," by which we mean "You can reach me; my cell phone will be on, and I am wired into (social) existence through it." *On* my cell, *on*line, *on* the Web, *on* instant messaging—these phrases suggest a *tethered* self.

We are tethered to our "always-on/always-on-you" communications devices and the people and things we reach through them: people, Web pages, voice mail, games, artificial intelligences (nonplayer game characters, interactive online "bots"). These very different objects achieve a certain sameness because of the way we reach them. Animate and inanimate, they live for us through our tethering devices, always ready-to-mind and hand. The self, attached to its devices, occupies a liminal space between the physical real and its digital lives on multiple screens (Turner 1969). I once described the rapid movements from physical to a multiplicity of digital selves through the metaphor of "cycling-through." With cell technology, rapid cycling stabilizes into a sense of continual co-presence (Turkle 1995).

For example, in the past, I did not usually perform my role as mother in the presence of my professional colleagues. Now a call from my sixteen-year-old daughter brings me forth in this role. The presence of the cell phone, which has a special ring if my daughter calls, keeps me on the alert all day. Wherever I am, whatever I am doing, I am psychologically tuned to the connections that matter.

The Connections that Matter

We are witnessing a new form of sociality in which the connectedness that "matters" is determined by our distance from working communications technology. Increasingly, what people want out of public spaces is that they offer a place to be private with tethering technologies. A neighborhood walk reveals a world of madmen and women, talking to themselves, sometimes shouting to themselves, little concerned with what is around them, happy to have intimate conversations in public spaces. In fact, neighborhood spaces themselves become liminal, not entirely public, not entirely private (Katz 2006, chapters 1 and 2).

A train station is no longer a communal space, but a place of social collection: tethered selves come together, but do not speak to each other. Each person at the station is more likely to be having an encounter with someone miles away than with the person in the next chair. Each inhabits a private media bubble. Indeed, the presence of our tethering media signal that we do not want to be disturbed by conventional sociality with physically proximate individuals.

When people have personal cell phone conversations in public spaces, what sustains their sense of intimacy is the presumption that those around them treat them not only as anonymous, but as close to disembodied. When individuals hold cell phones (or "speak into the air," indicating the presence of cells with earphone microphones), they are marked with a certain absence. They are transported to the space of a new ether, virtualized. This "transport" can be signaled in other ways: when people look down at their laps during meals or meetings, the change of gaze has come to signify attention to their BlackBerries or other small communications devices. They are focused on elsewhere.

The director of a program that places American students in Greek universities complains that students are not "experiencing Greece" because they spend too much time online, talking with their friends from home. I am sympathetic as she speaks, thinking of the hours I spent walking with my daughter on a visit to Paris as she "texted" her friends at home on her cell phone. I worry that she is missing something I cherished in my youth, the experience of an undiluted Paris that came with the thrill of disconnection from where I was from. But she is happy and tells me that keeping in touch is "comforting" and that beyond this, her text mails to home constitute a diary. She can look back at her texts and remember her state of mind at different points of her trip. Her notes back to friends, translated from instant message shorthand include "Saw Pont D'Avignon," "Saw World Cup Soccer in Paris," and "Went to Bordeaux." It is hard to get in too many words on the phone keyboard and there is no cultural incentive to do so. A friend calls my daughter as we prepare for dinner at our Paris hotel and asks her to lunch in Boston. My daughter says, quite simply: "Not possible, but how about Friday." Her friend has no idea that her call was transatlantic. Emotionally and socially, my daughter has not left home.

Of course, balancing one's physical and electronic connections is not limited to those on holiday. Contemporary professional life is rich in examples of people ignoring those they are physically "with" to give priority to online others. Certain settings in which this occurs have become iconic: sessions at international conferences where experts from all over the world come together but do their e-mail; the communications channels that are set up by audience members at conferences to comment on speakers' presentations during the presentations themselves (these conversations are as much about jockeying for professional position among the audience as they are about what is being said at the podium). Here, the public presentation becomes a portal to discussions that take people away from it, discussions that tend to take place in hierarchical tiers—only certain people are invited to participate in certain discussions. As a member of the audience, one develops a certain anxiety: have I been invited to chat in the inner circle?

Observing e-mail and electronic messaging during conferences at exotic locations compels our attention because it is easy to measure the time and money it takes to get everyone physically together at such meetings. Other scenes have become so mundane that we scarcely notice them: students do e-mail during classes; business people do e-mail during meetings; parents do e-mail while playing with their children; couples do e-mail at dinner; people talk on the phone and do their e-mail at the same time. Once done surreptitiously, the habit of electronic co-presence is no longer something people feel they need to hide. Indeed, being "elsewhere" than where you might be has become something of a marker of one's sense of self-importance.

Phoning It In

The expression "phoning it in" used to be pejorative. It implied a lack of appropriate attention to what might be novel about a task at hand. Now, as pure description, it provides a metric for status; it suggests that you are important enough to deliver your work remotely. The location of the high-status body is significant, but with connectivity comes multiple patterns for its deployment. In one pattern, the high-status body is in intensive contact with others, but spreads itself around the world, traveling. In another pattern, the high-status body is in retreat, traveling to face-to-face contact in order to maximize privacy and creativity. However the traveling body chooses to use its time, it is always tethered, kept in touch through technical means. Advertisements for wireless technology routinely feature a handsome man or beautiful woman on a beach. The ad copy makes it clear that he or she is important and working. The new disembodiment does not ask you to deny your body its pleasures, but on the contrary, to love your body, to put it somewhere appealing while "you" work.

Our devices become a badge of our networks, a sign that we have them, that we are wanted by those we *know*, the people on our "contact lists" and by the potential, as yet *unknown* friends who wait for us in virtual places (such as Facebook, MySpace, or Friendster). It is not surprising that we project the possibility of love, surprise, amusement, and warmth onto our communications devices. Through them we live with a heightened sense of potential relationships, or at least of new connections. Whether or not our devices are in use, without them we feel adrift—adrift not only from our current realities but from our wishes for the future.

A call to a friend is a call to a known (if evolving) relationship. Going online to a social networking site offers a place to dream, sometimes fostering a sense that old relationships are dispensable. People describe feeling more attached to the site than to any particular acquaintances they have on them. In psychodynamic terms, the site becomes a transference object: the place where friendships come from. "I toss people," says Maura, thirty-one, an architect, describing how she treats acquaintances on Second Life, an elaborate online social environment. Second Life offers the possibility of an online parallel life (including a virtual body, wardrobe, real estate, and paying job). "I know it gives me something of a reputation, but there are always new people. I don't stay in relationships long." Maura continues: "There is always someone else to talk to, someone else to meet. I don't feel a commitment." People who have deployed avatars on Second Life stress that the virtual world gives them a feeling of everyday renewal. "I never know who I'll meet," says a thirty-seven-year-old housewife from the Boston suburbs, and contrasts this pleasurable feeling with the routine of her life at home with two toddlers.

From the early 1990s, game environments known as MUDs (for multiuser domains) and then MMRPGs (massively-multiplayer-role-playing-games) presented their users with the possibility of creating characters and living out multiple aspects of self. Although the games often took the forms of medieval quests, these virtual environments owed their "holding power" to the opportunities that they offered for exploring identity. (Turkle 1995). People used their lives on the screen to work through unresolved or partly resolved issues, often related to sexuality or intimacy. For many who enjoy online life, it is easier to express intimacy in the virtual world than in "RL" or real life. For those who are lonely yet fearful of intimacy, online life provides environments where one can be a loner yet not alone, environments where one can have the illusion of companionship without the demands of sustained, intimate friendship. Online life emerged as an "identity workshop" (Bruckman 1992).

Throughout our lives, transitions (career change, divorce, retirement, children leaving home) provide new impetus for rethinking identity. We never "graduate" from working on identity; we simply work on it with the materials we have at hand at a particular stage of life. Online social worlds provide new materials. The plain may represent themselves as glamorous; the introverted can try out being bold. People build the dreamhouses in the virtual that they cannot afford in the real. They plant virtual gardens. They take online jobs of great responsibility. They often have relationships, partners, and what they term "marriages" of great emotional importance. In the virtual, the crippled can walk without crutches and the shy can improve their chances as seducers.

It is not exact to think of people as tethered to their *devices*. People are tethered to the gratifications offered by their online selves. These include the promise of affection, conversation, a sense of new beginnings. And, there is vanity: building a new body in a game like Second Life allows you to put aside an imperfect physical self and reinvent yourself as a wonder of virtual fitness. Everyone on Second Life can have their own "look"; the game enables a high level of customization, but everyone looks good, wearing designer clothes that appear most elegant on sleek virtual bodies. With virtual beauty comes possibilities for sexual encounters that may not be available in the physical real.

Thus, more than the sum of their instrumental functions, tethering devices help to constitute new subjectivities. Powerful evocative objects for adults, they are even more intense and compelling for adolescents, at that point in development when identity play is at the center of life.

The Tethered Teen

The job of adolescence is centered around experimentation—with ideas, with people, with notions of self. When adolescents play an online role playing game they often use it to recast their lives. They may begin by building their own home, furnishing it to their taste, not that of their parents, and then getting on with the business of reworking in the virtual world what has not worked so well in the real. Trish, a thirteen-year-old who has been physically abused by her father, creates an abusive family on Sims Online—but in the game her character, also thirteen, is physically and emotionally strong. In simulation, she plays and replays the experience of fighting off her aggressor. Rhonda, a sexually experienced girl of sixteen, creates an online innocent. "I want to have a rest," she tells me and goes on to recall the movie *Pleasantville* in which the female lead character, a high school teenager, "gets to go to a town that only exists from a TV show where she starts to be slutty like she is at home, but then she changes her mind and starts to turn boys down and starts a new life. She practices being a different kind of person. That's what Sims Online is for me. Practice."

Rhonda "practices" on the game at breakfast, during school recess, and after dinner. She says she feels comforted by her virtual life. The game does not connect her to other people. The game responds to her desire to connect to herself.

ST: Are you doing anything different in everyday life [since playing Sims Online]? Rhonda: Not really. Not very. But I'm thinking about breaking up with my boyfriend. I don't want to have sex anymore but I would like to have a boyfriend. My character [in Sims Online] has boyfriends but doesn't have sex. They help her with her job. I think to start fresh I would have to break up with my boyfriend.

Rhonda is emotionally tethered to the world of the Sims. Technology gives her access to a medium in which she can see her life through a new filter, and possibly begin to work through problems in a new way (Turkle 1995).

Adolescents create online personae in many ways: when they deploy a game avatar, design a Web page, or write a profile for a social-networking site such as Facebook. Even creating a playlist of music becomes a way of capturing one's persona at a moment in time. Multiple playlists reflect aspects of self. And once you have collected your own music, you can make connections to people all over the world to whom you send your songs.

Today's adolescents provide our first view of tethering in developmental terms. The adolescent wants both to be part of the group and to assert individual identity, experiencing peers as both sustaining and constraining. The mores of tethering support group demands: among urban teens, it is common for friends to expect that their peers will stay available by cell or instant message. In this social contract, one needs good cause to claim time offline. The pressure to be always-on can be a burden. So, for example, teenagers who need uninterrupted time for schoolwork resort to using their parents' Internet accounts to hide out from friends. Other effects of the always-on/always-on-you communications culture may be less easily managed and perhaps more enduring.

Mark Twain mythologized the process of separation during which adolescents work out their identities as the Huck Finn experience, the on-the-Mississippi time of escape from the adult world. The time on the river portrays an ongoing rite of passage during which children separate from parents to become young adults, a process now transformed by technology. Traditionally, children have internalized the adults in their world before (or just as, or shortly after) the threshold of independence is crossed. In the technologically tethered variant, parents can be brought along in an intermediate space, for example, the space created by the cell phone where everyone is on speed dial. In this sense, the generations sail down the river together.

When children receive cell phones from their parents, the gift usually comes with a proviso: children are to answer their parents' calls. This arrangement gives children permission to do things—take trips to see friends, attend movies, go to the beach—that would not be permitted without the phone-tethering to parents. Yet the tethered child does not have the experience of being alone with only him or herself to count on. There used to be a point for an urban child, usually between the ages of eleven and fourteen, when there was a "first time" to navigate the city alone. It was a rite of passage that communicated "You are on your own and responsible. If you are frightened, you have to experience those feelings." The cell phone buffers this moment; the parent is "on tap." With the on-tap parent, tethered children think differently about their own responsibilities and capacities. These remain potential, not proven.

New Forms of Validation

I think of the *inner history* of technology as the relationships people form with their artifacts, relationships that can forge new sensibilities. Tethering technologies have their own inner histories. For example, a mobile phone gives us the potential to communicate whenever we have a feeling, enabling a new coupling of "I have a feeling/ Get me a friend." This formulation has the emotional corollary, "I want to have a feeling/Get me a friend." In either case, what is *not* being cultivated is the ability to be alone, to reflect on and contain one's emotions. The anxiety that teens report when they are without their cell phones or their link to the Internet may not speak so much to missing the easy sociability with others but of missing the self that is constituted in these relationships.

When David Riesman remarked on the American turn from an inner- to an other-directed sense of self by 1950 (Riesman 1950), he could not foresee how technology could raise other-directedness to a new level. It does this by making it possible for each of us to develop new patterns of reliance on others. And we develop transference relationships that make others available to us at literally a moment's notice. Some people experienced this kind of transference to the traditional (landline) telephone.

The telephone was a medium through which to receive validation, and sometimes the feelings associated with that validation were transferred to the telephone itself. The cell phone takes this effect to a higher power because the device is always available and there is a high probability that one will be able to reach a source of validation through it. It is understood that the validating cell conversation may be brief, just a "check-in," but more is not necessarily desired.

The cell phone check-in enables the new other-directness. At the moment of having a thought or feeling, one can have it validated. Or, one may *need* to have it validated. And further down a continuum of dependency, as a thought or feeling is being formed, it may *need validation to become established*. The technology does not cause a new style of relating, but enables it. As we become accustomed to cell calls, e-mail, and social Web sites, certain styles of relating self to other feel more natural. The validation (of a feeling already felt) and enabling (of a feeling that cannot be felt without outside validation) are becoming commonplace rather than marked as childlike or pathological.

The psychoanalyst Heinz Kohut writes about narcissism and describes how some people, in their fragility, turn other persons into "self-objects" to shore up their fragile sense of self (Ornstein 1978). In the role of self-object, the other is experienced as part of the self, thus in perfect tune with the fragile individual's inner state. They are there for validation, mirroring. Technology increases one's options. One fifteen-year-old girl explains: "I have a lot of people on my contact list. If one friend doesn't get it, I call another." In Kohutian terms, this young woman's contact or buddy list has become a list of spare parts for her fragile adolescent self.

Just as always-on/always-on-you connectivity enables teens to postpone independently managing their emotions, it can also make it difficult to assess children's level of maturity, conventionally defined in terms of autonomy and responsibility. Tethered children know that they have backup. The "check-in" call has evolved into a new kind of contact between parents and children. It is a call that says "I am fine. You are there. We are connected."

In general, the telegraphic text message quickly communicates a state, rather than opens a dialogue about complexity of feeling. Although the culture that grows up around the cell is a talk culture (in shopping malls, supermarkets, city streets, cafés, playgrounds, and parks, cells are out and people are talking into them), it is not necessarily a culture in which talk contributes to self-reflection. Today's adolescents have no less need than previous generations to learn empathic skills, to manage and express feelings, and to handle being alone. But when the interchanges to develop empathy are reduced to the shorthand of emoticon emotions, questions such as "Who am I?" and "Who are you?" are reformatted for the small screen, and are flattened in the process. High technology, with all its potential range and richness, has been put at the service of telegraphic speed and brevity.

Leaving the Time to Take Our Time

Always-on/always-on-you communications devices are seductive for many reasons, among them, they give the sense that one can do more, be in more places, and control more aspects of life. Those who are attached to BlackBerry technology speak about the fascination of watching their lives "scroll by," of watching their lives as though watching a movie. One develops a new view of self when one considers the many thousands of people to whom one may be connected. Yet just as teenagers may suffer from a media environment that invites them to greater dependency, adults, too, may suffer from being overly tethered, too connected. Adults are stressed by new responsibilities to keep up with email, the nagging sense of always being behind, the inability to take a vacation without bringing the office with them, and the feeling that they are being asked to respond immediately to situations at work, even when a wise response requires taking time for reflection, a time that is no longer available.

We are becoming accustomed to a communications style in which we receive a hasty message to which we give a rapid response. Are we leaving enough time to take our time?

Adults use tethering technologies during what most of us think of as down time, the time we might have daydreamed during a cab ride, waiting in line, or walking to work. This may be time that we physiologically and emotionally need to maintain or restore our ability to focus (Herzog et al. 1997; Kaplan 1995). Tethering takes time from other activities (particularly those that demand undivided attention), it adds new tasks that take up time (keeping up with e-mail and messages), and adds a new kind of time to the day, the time of attention sharing, sometimes referred to as *continuous partial attention* (Stone 2006). In all of this, we make our attention into our rarest resource, creating increasingly stiff competition for its deployment, *but we undervalue it as well*. We deny the importance of giving it to one thing and one thing only.

Continuous partial attention affects the quality of thought we give to each of our tasks, now done with less *mind share*. From the perspective of this essay with its focus on identity, continuous partial attention affects how people think about their lives and priorities. The phrases "doing my e-mail" and "doing my messages" imply performance rather than reflection. These are the performances of a self that can be split into constituent parts.

When media does not stand waiting in the background but is always there, waiting to be wanted, the self can lose a sense of conscious choosing to communicate. The sophisticated consumer of tethering devices finds ways to integrate always-on/always-on-you technology into the everyday gestures of the body. One BlackBerry user says: "I glance at my watch to sense the time; I glance at my BlackBerry to get a sense of my life." The term *addiction* has been used to describe this state, but this way of thinking is limited in its usefulness. More useful is thinking about a new state of self, one that is extended in a communications artifact. The BlackBerry movie of one's life takes on a life of its own—with more in it than can be processed. People develop the sense that they cannot keep up with their own lives. They become alienated from their own experience and anxious about watching a version of their lives moving along, scrolling along, faster than they can handle. It is the unedited version of their lives; they are not able to keep up with it, but they are responsible for it (Mazmanian 2005).

Michel Foucault wrote about Jeremy Bentham's panopticon as emblematic of the situation of the individual in modern, "disciplinary" society (Foucault 1979). The Panopticon is a wheel-like structure with an observer (in the case of a prison, a prison guard) at its hub. The architecture of the Panopticon creates a sense of being always watched whether or not the guard is actually present. For Foucault, the task of the modern state is to construct citizens who do not need to be watched, who mind the rules and themselves. Always-on/always-on-you technology takes the job of self-monitoring to a new level. We try to keep up with our lives as they are presented to us by a new disciplining technology. We try, in sum, to have a self that keeps up with our e-mail.

Boundaries

A new complaint in family and business life is that it is hard to know when one has the attention of a BlackBerry user. A parent, partner, or child can be lost for a few seconds or a few minutes to an alternate reality. The shift of attention can be subtle; friends and family are sometimes not aware of the loss until the person has "returned." Indeed, BlackBerry users may not even know where their attention lies. They report that their sense of self has merged with their prosthetic extensions and some see this as a new "high." But this exhilaration may be denying the costs of multitasking. Sociologists who study the boundaries between work and the rest of life suggest that it is helpful when people demarcate role shifts between the two. Their work suggests that being able to use a BlackBerry to blur the line is problematic rather than a skill to be celebrated. (Clark 2000; Desrochers and Sargent 2003; Shumate and Fulk 2004). And celebrating the integration of remote communications into the flow of life may be underestimating the importance of face-to-face connections (Mazmanian 2005).

Attention-sharing creates work environments fraught with new tensions over the lack of primacy given to physical proximity. Face-to-face conversations are routinely interrupted by cell phone calls and e-mail reading. Fifteen years ago, if a colleague read mail in your presence, it was considered rude. These days, turning away from a person in front of you to answer a cell phone has become the norm. Additionally, for generations, business people have grown accustomed to relying on time in taxis, airports, trains, and limousines to get to know each other and to discuss substantive matters. The waiting time in client outer offices was precious time for work and the exchange of news that created social bonds among professional colleagues. Now, things have changed: professionals spend taxi time on their cell phones or doing e-mail on their PDAs. In the precious moments before client presentations, one sees consulting teams moving around the periphery of waiting rooms, looking for the best place for cell reception so that they can make calls. "My colleagues go to the ether when we wait for our clients," says one advertising executive. "I think our presentations have suffered." We live and work with people whose commitment to our presence feels increasingly tenuous because they are tethered to more important virtual others.

Human beings are skilled at creating rituals for demarcating the boundaries between the world of work and the world of family, play, and relaxation. There are special times (the Sabbath), special meals (the family dinner), special attire (the "armor" for a day's labor comes off at home, whether it is the businessperson's suit or the laborer's overalls), and special places (the dining room, the parlor, the bedroom, the beach). Now always-on/always-on-you technology accompanies people to all these places, undermining the traditional rituals of separation.

There is a certain push back. Just as teenagers hide from friends by using their parents' online accounts to do homework, adults, too, find ways to escape from the demands of tethering: BlackBerries are left at the office on weekends or they are left in locked desk drawers to free up time for family or leisure (Gant and Kiesler 2001). "It used to be my home was a haven; but now my home is a media center," says an architect whose clients reach him on his Internet-enabled cell. No longer a safe space or refuge, people need to find places to hide. There are technically none except long plane rides where there is no cell or Internet access, and this, too, may be changing.

A Self Shaped by Rapid Response

Our technology reflects and shapes our values. If we think of a telephone call as a quick-response system enabled by always-on/always-on-you technology, we can forget there is a difference between a scheduled call and the call you make in reaction to a fleeting emotion, because something crossed your mind, or because someone left you a message. The self that is shaped by this world of rapid response measures success by calls made, e-mails answered, and contacts reached. This self is calibrated on the basis of what the technology proposes, by what it makes possible, and by what it makes easy. But in the buzz of activity, there are losses that we are perhaps not ready to sustain.

One is the technology-induced pressure for speed, even when we are considering matters over which we should take our time. We insist that our world is increasingly complex, yet we have created a communications culture that has decreased the time available for us to sit and think uninterrupted. BlackBerry users describe that sense of encroachment of the device on their time. One says, "I don't have enough time alone with my mind." Other phrases come up: "I have to struggle to make time to think." "I artificially make time to think." "I block out time to think." In all of these statements is the implicit formulation of an "I" that is separate from technology, that can put it aside and needs time to think on its own. This formulation contrasts with a growing reality of our lives lived in the continual presence of communications devices. This reality has us, like the early MIT "cyborg" group, learning to see ourselves not as separate but as at one with the machines that tether us to each other and to the information culture. To put it most starkly: to make more "time" in the old-fashioned sense means turning off our devices, disengaging from the always-on culture. But this is not a simple proposition since our devices have become more closely coupled to our sense of our bodies and increasingly feel like extensions of our minds.

In the 1990s, as the Internet became part of everyday life, people began to create multiple online avatars and used them to shift gender, age, race, and class. The effort was to create richly rendered virtual selves through which one could experiment with identity by playing out parallel lives in constructed worlds. The world of avatars and games continues, but now, alongside its pleasures, we use always-on/always-on-you technology to play ourselves. Today's communications technology provides a social and psychological GPS, a navigation system for tethered selves. One television producer, accustomed to being linked to the world via her cell and Palm device, revealed that for her, the Palm's inner spaces were where her self resides: "When my Palm crashed it was like a death. It was more than I could handle. I felt as though I had lost my mind."

Tethered: To Whom and to What?

Acknowledging our tethered state raises the question of to whom or to what we are connected (Katz 2003). Traditional telephones tied us to friends, family, colleagues from school and work, and commercial or philanthropic solicitations. Things are no longer so simple. These days we respond to humans and to objects that represent them: answering machines, Web sites, and personal pages on social-networking sites. Sometimes we engage with avatars that anonymously "stand in" for others, enabling us to express ourselves in intimate ways to strangers, in part because we and they are able to veil who we "really are." And sometimes we listen to disembodied voices—recorded announcements and messages—or interact with synthetic voice recognition protocols that simulate real people as they try to assist us with technical and administrative problems. We no longer demand that as a person we have another person as an interlocutor. On the Internet, we interact with bots, anthropomorphic programs that are able to converse with us, and in online games we are partnered with nonplayer characters, artificial intelligences that are not linked to human players. The games require that we put our trust in these characters. Sometimes it is only these nonplayer characters who can save our "lives" in the game.

This wide range of entities—human and not—is available to us wherever we are. I live in Boston. I write this chapter in Paris. As I travel, my access to my favorite avatars, nonplayer characters, and social networking sites stays constant. There is a degree of emotional security in a good hotel on the other side of the world, but for many, it cannot compare to the constancy of a stable technological environment and the interactive objects within it. Some of these objects are engaged on the Internet. Some are interactive digital companions that can travel with you, now including robots that are built for relationships.

Consider this moment: an older woman, seventy-two, in a nursing home outside of Boston is sad. Her son has broken off his relationship with her. Her nursing home is part of a study I am conducting on robotics for the elderly. I am recording her reactions as she sits with the robot Paro, a seal-like creature, advertised as the first "therapeutic robot" for its ostensibly positive effects on the ill, the elderly, and the emotionally troubled. Paro is able to make eye contact through sensing the direction of a human voice, is sensitive to touch, and has "states of mind" that are affected by how it is treated—for example, it can sense if it is being stroked gently or with some aggression. In this session with Paro, the woman, depressed because of her son's abandonment, comes to believe that the robot is depressed as well. She turns to Paro, strokes him, and says: "Yes, you're sad, aren't you. It's tough out there. Yes, it's hard." And then she pets the robot once again, attempting to provide it with comfort. And in so doing, she tries to comfort herself.

Psychoanalytically trained, I believe that this kind of moment, if it happens between people, has profound therapeutic potential. What are we to make of this transaction as it unfolds between a depressed woman and a robot? The woman's sense of being understood is based on the ability of computational objects like Paro to convince their users that they are in a relationship. I call these creatures (some virtual, some physical robots) "relational artifacts" (Turkle 1999; 2003a; 2003b; 2004a; 2004b; 2004c; 2005b; 2005c; 2006b; Turkle et al. 2006a). Their ability to inspire a relationship is not based on their intelligence or consciousness but on their ability to push certain "Darwinian" buttons in people (making eye contact, for example) that cause people to respond *as though* they were in a relationship.

Do plans to provide relational robots to children and the elderly make us less likely to look for other solutions for their care? If our experience with relational artifacts is based on a fundamentally deceitful interchange (artifacts' ability to persuade us that they know and care about our existence), can it be good for us? Or might it be good for us in the "feel good" sense, but bad for us in our lives as moral beings? The answers to such questions are not dependent on what computers can do today or what they are likely to be able to do in the future. These questions ask what *we* will be like, what kind of people we are becoming, as we develop increasingly intimate relationships with machines.

In *Computer Power and Human Reason*, Joseph Weizenbaum wrote about his experiences with his invention, ELIZA, a computer program that engaged people in a dialogue similar to that of a Rogerian psychotherapist (Weizenbaum 1976). It mirrored one's thoughts; it was always supportive. To the comment "My mother is making me angry," the program might respond "Tell me more about your mother," or "Why do you feel so negatively about your mother?" Weizenbaum was disturbed that his students, knowing they were talking with a computer program, wanted to chat with it, indeed, wanted to be alone with it. Weizenbaum was my colleague at MIT; we taught courses together on computers and society. At the time his book came out, I felt moved to reassure him about his concerns. ELIZA seemed to me like a Rorschach; users did become involved with the program, but in a spirit of "as if." The gap between program and person was vast. People bridged it with attribution and desire. They thought: "I will talk to this program 'as if' it were a person"; "I will vent, I will rage, I will get things off my chest." At the time, ELIZA seemed to me no more threatening than an interactive diary. Now, thirty years later, I ask myself if I underestimated the quality of the connection. Now, computational creatures have been designed that evoke a sense of mutual relating. The people who meet relational artifacts are drawn in by a desire to nurture them. And with nurturance comes the fantasy of reciprocation. People want the creatures to care about them in return. Very little about these relationships seems to be experienced "as if."

Relational artifacts are the latest chapter in the trajectory of the tethered self. We move from technologies that tether us to people to those that are able to tether us to the Web sites and avatars that represent people. Relational artifacts represent their programmers but are given autonomy and primitive psychologies; they are designed to stand on their own as creatures to be loved. They are potent objects-to-think-with for asking the questions, posed by all of the machines that tether us to new socialities: "What is an authentic relationship with a machine?" "What are machines doing to our relationships with people?" And ultimately, "What is a relationship?"

Methodology Note

I have studied relational artifacts in the lives of children and the elderly since 1997, beginning with the simple Tamagotchis that were available at every toy store to Kismet and Cog, advanced robots at the MIT Artificial Intelligence Laboratory, and Paro, a seal-like creature designed specifically for therapeutic purposes. Along the way there have been Furbies, AIBOS, and My Real Babies, the latter a baby doll that like the Paro has changing inner states that respond to the quality of its human care. More than two hundred and fifty subjects have been involved in these studies. My investigations of computer-mediated communication date from the mid-1980s and have followed the media from e-mail, primitive virtual communities, and Web-based chat to cell technology, instant messaging, and social networking. More than four hundred subjects have been involved in these studies. My work was done in Boston and Cambridge and their surrounding suburbs. The work on robotics investigated children and seniors from a range of ethnicities and social classes. This was possible because in every case I was providing robots and other relational artifacts to my informants. In the case of the work on communications technology, I spoke to people, children, adolescents, and adults, who already had computers, Web access, mobile phones, BlackBerries, et cetera. This necessarily makes my claims about their lives in the always-on/always-on-you culture not equally generalizable outside of the social class currently wealthy enough to afford such things.

References

Bruckman, A. 1992. Identity workshop: Emergent social and psychological phenomena in text-based virtual reality. Unpublished paper written in partial completion of a doctoral degree at the Media Lab, Massachusetts Institute of Technology. http://www-static.cc.gatech.edu/~asb/papers/old-papers.html.

Clark, S. Campbell. 2000. Work/family border theory: A new theory of work/family balance. *Human Relations* 53(6): 747–770.

Desrochers, S., and L. D. Sargent. 2003. Work-family boundary ambiguity, gender and stress in dual-earner couples. Paper presented at the Conference "From 9-to-5 to 24/7: How Workplace Changes Impact Families, Work, and Communities," 2003 BPW/Brandeis University Conference, Orlando, Fla.

Foucault, M. 1979. *Discipline and Punish: The Birth of the Prison.* New York: Vintage Books.

Gant, D. B., and S. Kiesler. 2001. Blurring the boundaries: Cell phones, mobility and the line between work and personal life. In *Wireless World: Social and Interactional Aspects of the Mobile Age,* edited by N. G. R. H. Barry Brown. New York: Springer.

Herzog, T. R., A. M. Black, K. A. Fountaine, and D. J. Knotts. 1997. Reflection and attentional recovery as distinctive benefits of restorative environments. *Journal of Environmental Psychology* 17: 165–170.

Jones, C. A. 2006. Tethered. In *Sensorium: Embodied Experience, Technology, and Contemporary Art,* edited by C. A. Jones. Cambridge, Mass.: List Visual Art Center and MIT Press.

Kaplan, S. 1995. The restorative benefits of nature: Toward an integrative framework. *Journal of Environmental Psychology* 15: 169–182.

Katz, J. E. 2006. *Magic in the Air: Mobile Communication and the Transformation of Social Life.* New Brunswick, N.J.: Transaction.

Katz, J. E., ed. 2003. *Machines that Become Us: The Social Context of Personal Communication Technology*. New Brunswick, N.J.: Transaction.

Mazmanian, M. 2005. Some thoughts on blackberries. In Memo.

Ornstein, P. H., ed. 1978. *The Search for the Self: Selected Writings of Heinz Kohut: 1950–1978*: 2. New York: International Universities Press, Inc.

Riesman, D., R. Denney, and N. Glazer. 1950. *The Lonely Crowd: A Study of the Changing American Character*. New Haven: Yale University Press.

Shumate, M., and J. Fulk. 2004. Boundaries and role conflict when work and family are colocated: A communication network and symbolic interaction approach. *Human Relations* 57(1): 55–74.

Stone, L. 2006. Linda Stone's thoughts on attention, and specifically, continual partial attention. http://www.lindastone.net.

Turkle, S. 1995. *Life on the Screen: Identity in the Age of the Internet*. New York: Simon and Schuster.

Turkle, S. 1999. Toys to change our minds. In *Predictions*, edited by S. Griffiths. Oxford: Oxford University Press.

Turkle, S. 2003a. Sociable technologies: Enhancing human performance when the computer is not a tool but a companion. In *Converging Technologies for Improving Human Performance*, edited by M. C. Roco and W. S. Bainbridge. The Netherlands: Kluwer Academic Publishers.

Turkle, S. 2003b. Technology and human vulnerability. *Harvard Business Review*.

Turkle, S. 2004a. *NSF Report: Relational Artifacts*. National Science Foundation. (NSF Grant SES-01115668).

Turkle, S. 2004b. Spinning technology. In *Technological Visions*, edited by M. Sturken, D. Thomas, and S. Ball-Rokeach. Philadelphia: Temple University Press.

Turkle, S. 2004c. Whither psychoanalysis in computer culture. *Psychoanalytic Psychology: Journal of the Division of Psychoanalysis* 21(1): 16–30.

Turkle, S. 2005a. *The Second Self: Computers and the Human Spirit* (20th anniversary ed.). Cambridge, Mass.: MIT Press [1984].

Turkle, S. 2005b. Computer games as evocative objects: From projective screens to relational artifacts. In *Handbook of Computer Games Studies*, edited by J. Raessens and J. Goldstein. Cambridge, Mass.: MIT Press.

Turkle, S. 2005c. Relational artifacts/children/elders: The complexities of cybercompanions. IEEE Workshop on Android Science, Stresa, Italy.

Turkle, S., C. Breazeal, O. Dasté, and B. Scassellat. 2006a. First encounters with kismet and cog: Children's relationship with humanoid robots. In *Digital Media: Transfer in Human Communication*, edited by P. Messaris and L. Humphreys. New York: Peter Lang Publishing.

Turkle, S. 2006b. Tamagotchi diary. *The London Review of Books*, April 20.

Turkle, S. 2006c. Tethering. In *Sensorium: Embodied Experience, Technology, and Contemporary Art*, edited by C. A. Jones. Cambridge, Mass.: List Visual Art Center and MIT Press.

Turner, V. 1969. *The Ritual Process: Structure and Anti-structure*. Chicago: Aldine.

Weizenbaum, J. 1976. *Computer Power and Human Reason: From Judgment to Calculation*. San Francisco: W. H. Freeman.

11 | The Mobile Phone's Ring

Christian Licoppe

Introduction

The shrill ringing of a phone is a forceful event. Although the ubiquity of phones in our environments legitimizes in principle the unexpected occurrence of incoming calls, the ring is a threat to participants' "face." Though legitimate, the presence of phones binds parties in a given situation to the possibility of their personal territory being invaded, and their activities under way being disturbed. However, parties display shared normative expectations about the fact that a ringing phone should be answered, and they draw inferences when it is not answered, answered too soon, or too late (Schegloff 2002). The repetitive ring that characterized most phones until the advent of mobile phones in the 1990s is a paradigmatic example of a particular interactional sequence: that of the "summons" (Schegloff 2002).

Many other authors have commented in a more literary way on the fact that a telephone call creates an expectation—an emergency that explains why one feels compelled to answer a ringing phone, even if the call is for someone else. As Sadie Plant explains, "A telephone that rings calls for an answer. Public uses of the mobile distribute this tension to everyone within hearing, even if they are unable to answer. Only the person called is engaged" (Plant 2000). This artifact has therefore been a real test for everything that creates and maintains the face governing social order in the public sphere, to the point of the regulation of uses becoming an issue in media debate and a research subject (Katz and Aakhus 2002; Ling 2004).

There is always the risk of an incoming call to come at an improper moment and threaten the "negative face" of users (Brown and Levinson 1987). This threat and the following remedial interchanges are a regular feature of the history of the telephone, even if the configuration has varied from one period to the next. For instance, in early-twentieth-century France, the master of a household, not wanting to appear to respond to orders from "just anyone," would ask a servant to answer the telephone. An aristocrat during that period commented: "We never ran to the phone...it was a servant's duty to answer, to ask what the caller wanted and to fetch the person concerned" (Bertho-Lavenir 1984). This substitution can typically be interpreted as a measure designed to preserve the parents' negative face from this threat, as was, more generally, the instruction given to servants to answer the telephone. Yet the telephone ring also expresses and enhances the importance of the answerer of the call, thus bringing their "positive face" into play.

In the original model of telephone communication, the telephone was used as a makeshift solution, a possibility of communicating at a distance when it was difficult to meet the other person. The answerer of the call believed the caller had thought about the most suitable moment to phone. Today that model has been combined with the "connected presence" model in which the interpersonal link is maintained through constant contact, the quantity of which is as important as the quality (Licoppe and Smoreda 2006). In the latter case the participants accept the fact that the caller may have phoned on impulse, without prior thought to the relevance of the call. Configurations of interpersonal communication based on connected presence weigh particularly heavily on answerers' availability. In this context the need emerges to facilitate the answerer's evaluations and to possess new means for managing telephone rings as a summons. The moral anthropology to which "connected presence" is relevant is one in which the will to display one's availability to extend or maintain social networks, and to act upon it, may become a collective good, detached from the specifics of a given relationship (Boltanski and Chiapello 1999).

Expectations concerning the way in which the telephone ringtone invites an answer, and the inferences drawn from a particular treatment of it as a summons, are different in each case. In the former model, the answerer takes into account the fact that the caller (whose identity is usually unknown) has probably weighed his or her decision to call. In the latter, the answerer has to evaluate the relevance of answering by bearing in mind the fact that the caller may have acted impulsively. In both models the answerer has good reasons to answer, to prove to be available, but they are not based on the same evaluations, and the different resulting interactional sequences are not intelligible and accountable in the same way.

Significantly enough, many artifacts have been developed whose use is relevant to the way we perceive and treat phone rings: answering machines (which retranslate the invitation to answer, in the form of an insistent sound demanding an immediate answer, into a prerecorded message proposing delayed treatment) or caller-identification services (which provide a useful resource for recognizing the caller or the place from which they are calling, and for facilitating evaluation of the legitimacy and relevance of the call, in the situated perspective of the answerer). Early on, mobile phones have allowed their user to customize the ring (albeit within a limited set of alternatives) by setting the volume and choosing the ringing mode (e.g., vibrate or a single beep instead of the standard repetitive beating). Such possibilities are intended to enable users to adjust the interruptive nature of their telephone ring (still "mechanical" in the way it sounded) to suit the situation. It also allows them, where relevant, to clearly mark the fact that they have taken into consideration the disturbance likely to be caused around them.

A second feature, which mobile phones inherited from pagers and which is currently being extended virtually to any type of phone-like device, is their screen. Mobile phone screens have supported the development of caller-identification services, in which a glance at the ringing phone is usually enough for the answerer to identify the caller before answering—unlike the situation in the past where mutual recognition was accomplished during the first conversational turns (Schegloff 2002). Although this possibility of prior identification hardly seems to alter the rules governing the order of opening sequences of telephone conversations (Hutchby and Barnett 2005), participants nevertheless display an orientation to it (Relieu 2002). This alters the relevance of different possible opening sequences of a telephone conversation (Arminen 2005).

Musical ringtones allow the answerer to choose or design sophisticated sounds and tunes for their ringtone (extending the range of possibilities to almost any type of sound) and to assign distinctive ringtones to specific callers or group of callers. For the sake of brevity, we discuss here only the first point and leave the second to a more extended discussion. More specifically, what is a musical ringtone? Is it still designed so as to project its treatment as a summons? May it be made into more than that, according to the kind of responses a ring may elicit and their meanings? May it be retrospectively made into different kinds of summons? To what transformations of mediated sociability in private and public spaces does the success of musical ringtones point to?

Methodology

Our survey consisted of two phases: first, an online semistructured questionnaire for which we obtained 245 answers, primarily from people inclined to use mobile technologies, with a "mobigeek" tendency; second, the exploitation of a customer database of a small firm selling mobile ringtones, to recruit 23 users who agreed to in-depth interviews on their uses. Our objective was not representativeness but rather a variety of profiles in which the variables were gender (14 men and 9 women), age (ranging from 15 to 40 years old), socioprofessional categories (4 SPC+, 10 intermediate, 9 SPC−), and intensity of use. Recruitment proved to be fairly difficult; users were clearly annoyed at being contacted via their mobile phones and very few were willing to agree to ethnographic protocols.

Three types of musical ringtone exist: monophonic, polyphonic, and hi-fi. Monophonic ringtones characterize the earliest mobile phones, whereas with hi-fi ringtones music can be rendered with sufficient quality for it to be listened to as a musical extract. Many different ways of accessing and installing mobile ringtones exist, from built-in features to various free-of-charge options (creation, downloading from the Web, or direct mobile-to-mobile exchange). Commercial offers are provided by way of wireless application protocol (WAP, used by eight respondents), SMS (used by two), and Audiotel (used by seven). In general, throughout the sample there was a preference for WAP and free options (calls to Audiotel numbers are included in parents' phone bills and are therefore considered by teenagers to be free), and a distinct orientation toward the new hi-fi format.

The Equivocal Nature of Musical Ringtones and the Reshaping of the Obligation to Answer

The possibility of personalizing ringtones allows the representations of their interactional functions and their situated uses to be refined. Users exploit that flexibility to 1) make their own ringtones recognizable, particularly in complex perceptive environments; 2) shape the sequential properties of telephone ringtones to design the way rings prompt correspondents to react; and 3) counterbalance compliance to the generic obligation to answer with a personal "treat."

Musical Ringtones in the Soundscape: To Be Answered or To Be Listened To?

Users mention that they are frequently in situations in which their attentional environment is so complex that a number of devices can solicit their attention at any moment. They consequently need to choose ringtones that can be distinguished from other attention-calling items in their soundscape and recognized as a ringing telephone, specifically as their own in perceptively saturated environments. One way to do so is to exploit the growing resources that mobile telephony offers to make ringtones more distinctive. This concern leads to a particular way to define the qualities of ringtones and to categorize them. Many users believe there is an inverse correlation between the sound quality of ringtones and their capacity to be recognized in noisy yet structured environments: "Okay, everyone has their own opinion, but I find that the more faithfully a piece of music is rendered, the more difficult it is to hear in your pocket" (man, 24). In this respect, ringtones in hi-fi format are believed to be less effective than polyphonic ones, and even less so than monophonic ones.

However, the growing use of hi-fi musical ringtones has raised a new kind of tension in which one has to decide whether one chooses to perceive, recognize, and treat the sound of one's mobile phone as a ringtone (for which the normatively expected response is to answer) or as a music tune (in which case a proper response is to listen to it). The more faithfully a ringtone approximates a piece of music, the more equivocal it becomes with respect to the appropriate responses it projects: "midi ringtones can be heard more easily and you realize more quickly that it's your phone if it's a midi ringtone than if it's a hi-fi one, where you feel like listening rather than answering" (user on a mobile technology forum). How should this ambiguity be managed?

In its traditional use, the ringing of a phone is treated as a summons. The initiation of phone conversations therefore displays the sequential organization that characterizes the conversational treatment of a summons. A summons is a sequentially ordered two-turn device characterized by the rule of conditional relevance (Schegloff 2002). The initial turn projects an answer as a relevant next action. If the initial turn is ignored it becomes legitimate for the participants to treat the corresponding silence as a particular type of action, the "absence of an answer" to a summons. The sequential organization therefore consists of *adjacent pairs*, summons-answer (S-A) or summons-no-answer (S-NA), pairs to which a rule of *nonterminality* also applies. They must be followed by a new action, often performed by the initiator of the summons sequence. This rule legitimizes the repetition of a summons when an earlier one was ignored, and allows for the accountable repetition of summons-no-answer sequences. Traditional telephone rings (in which a ring is followed by a silence and repeated until the person being called answers, the caller gives up, or, more recently, an answering machine automatically takes the call) incorporate these principles and, in a sense, reify the sequential order characterizing summonsing devices (Schegloff 2002). The same author also provides direct empirical evidence of how participants display normative expectancies about the number of rings or the time that it has taken the answerer to pick up the phone (Schegloff 1986).

With musical ringtones the sound of the telephone may also be experienced as music rather than as a straightforward invitation to answer, projecting its treatment as a summons. (In 2005 a ringtone called "Crazy frog" was on the hit parade—an event that received extensive media coverage.) Many users design their ringtones with such a musical experience in mind. What is important then is "to stick to the original music more faithfully" (expert user). High quality musical ringtones may then be chosen to induce the kind of listening experience associated with the experience of music: "With the MP3 certain rings remind me of lots of things, like hits that I heard in a nightclub during the holidays" (woman, 22). Sophisticated ringtones therefore seem to be able to be treated as music, in contrast with traditional rings that one user describes as the "shrill noises" of a traditional phone's ringing. Some users are concerned that the musical quality of ringtones may even overwhelm the possibility of their being treated as the first part of a summons, and that one might "feel like listening instead of answering. . . . In my opinion, if I had longer ringtones I'd get a bit stuck on them . . . so the person on the other end of the line would wait for quite a while [laughs]. So maybe it wouldn't be a good idea" (man, 35). Only the fact that the ringtone represents a form of address (where a caller is thought to stand behind the ringing) prevents the answerer, *in extremis*, from completely giving in to the pleasure of the musical experience.

The choice of musical ringtones reflects an acute sensitivity to this tension between a ringing sound and music, between its function as an invitation to answer and as an opportunity for a sensorial experience. Expert ringtone users embody this tension in the very design of their ringtones. The key notion here is that of *bouclage* (looping), that is, using short musical extracts that may be repeated without pause between each repetition. The principle that such a design embodies is precisely that of the sequential organization of the summons-in-interaction, with the additional idea that the more the repetition, the greater the force of the summons (that is the way the first turn projects an answer or makes a nonanswer more of a work to be accounted for). "It's true that the phenomenon of repetition of a small extract is a little like beep-beep rings; it spurs you to answer" (man, 35). On the other hand, the longer the excerpt of music, the more the ringtone leads toward inviting to a musical experience: "In general I choose fairly long extracts because it's less tedious than a ring that's repeated over and over. It's crazy but I don't feel like it's a phone..." (man, 21). Some even orient their design toward the deliberate maximization of the ambiguity of their musical ringtones. They opt for repeated musical extracts chosen so that the flow of music appears unbroken by their repetition: "Ideally, you shouldn't be able to hear when the ring loops" (man, 24). They create in the world of musical ringtones an equivalent to the ambiguous rabbit/antelope-like drawings used in Gestalt psychology, that is, a ringtone that can be perceived and treated as ring or music.

Turning the Summons into a "Treat": Cloaking the Obligation to Answer with New Meanings

Some users of customized ringtones develop very particular "politics" of ringtone design. They refuse turning the ringtone into a "treat" and shape them to display and strengthen their summoning potential. Some expert users thus play on the sequential organization characterizing the traditional telephone ring. Since lengthening the sound part of the ring with each repetition can be perceived as an intensification of their capacity to invite an answer (and to project their treatment as a summons), they compose rings of the beep—beepbeep—beepbeepbeep type (with each "ringing" longer and more shrill than the previous one). This makes call recipients more accountable for failing to answer. Some go even farther, proposing to intensify this already extreme sequential ringtone structure by replacing the "ringing" part with a baby's cry, which usually is treated as a quite compelling type of summons in most social situations: "It may be a baby's cry. It starts softly and with each repetition the cry is louder, in proportion to the time that it is ignored. I can't think of a surer way to be allowed to leave a meeting," (comment on an expert forum).

On the other hand, musical ringtones are used to shape and manage the summoning power of the ring in an opposite way. They are used as a resource to enable users to renegotiate their vulnerability to the irruption of telephone calls. For those users who are most sensitive to the intrusive nature of mobile calls, the musical ringtone provides a positive and personal compensation that strengthens the capacity of the ringtone to project its treatment as a summons by making one more willing to answer: "You'll answer in another frame of mind, you won't go shouting that you're over the moon but really, it's perceptible, it can make you want to answer more easily. . . . I answer but at least I heard my ringtone, I'm happy" (man, 24). Claire, a young woman of 25, feels that answering the mobile has become a burden because many people call her on her mobile to ask her what she is doing. She often screens incoming calls, thus avoiding the obligation to answer a telephone ring immediately. However, she feels guilty about doing so, which makes the normativity associated with the act of answering the phone perceptible. The musical ringtone opens another path, combining intruding and summoning with the fact of allowing oneself a little pleasure, a special treat: "the ring softens the constraint . . . with a constraint it's important to do yourself good." The ringtone maybe specifically tailored to provide a special kind of treat: "On my birthday, a few weeks ago, I put on the ringtone 'Happy Birthday' by Stevie Wonder for all my contacts, I thought that was a scream," (man, 26). As this last example shows, the pleasure the musical ringtone provides acts as a mediation between the world of the caller (who provides the opportunity of the treat by calling) and that of the answerer (who designs the ringtone).

The use of a musical ringtone therefore turns the phone's ring and the impersonal obligation to answer (a kind of contribution to the collective good) into a personal gift, an individual treat. A similar phenomenon has been observed in the area of ordinary consumption, where shoppers concerned with the domestic good occasionally allow themselves a treat (Miller 1998). As a treat, the musical ringtone, because it is a personal cultural musical experience addressed by the answerers to themselves, separates the collective and the individual. In that respect, musical cell phone ringtones may be taken as one of the recent artifacts in the ICT domain that contribute to a "society of individuals" (Elias 1987). The treat retrospectively constitutes the invitation to answer that is embedded in the first part of the summons as a generic obligation to a larger entity, distinct from the very individual and personal pleasure it is designed to provide the answerer with.

The success of musical ringtones then seems to point toward a change in the way the telephone ring may be treated a summons, with respect to how technology acts both as a consequence and a cause. A possible interpretation lies with the development of "connected presence," (Licoppe 2004; Licoppe and Smoreda 2006). In this communication mode, the participants maintain strong social bonds and ongoing projects by multiplying all forms of contact, especially telephone contact. *Phatic* calls ("I called just to say 'hi'"), which have no aim other than sustaining the bond and serving as a reminder of it for both the participants, are particularly characteristic of "connected presence" patterns. The "connected presence" model legitimizes impulsive forms of telephony in which the value of the call stems from the fact that it is made without any premeditation and attests to the caller's feelings at a particular time.

This shifts part of the burden of assessing the relevance of the calls more toward the answerer. The ringing of the phone may signal any type of call from anybody. Besides increasing the usefulness of caller identification devices (a topic to which we return), such an evolution tends to allow for a more abstract treatment of the ringing of the phone in which the answerer may feel compelled to answer out of a general obligation to be available and to participate in the generalized connectivity of a networked society. All the more so if many of one's incoming calls prove only vaguely relevant. This may explain why users may feel a weakening of the obligation to answer and design their ringtones according to different strategies for shaping the normative expectations that surround the interpretation of a ringtone as a summons and its subsequent treatment. In a networked society there is a politicization of connectivity and availability issues (Boltanski and Chiapello 1999): being available (in our case answering the phone) becomes a form of contribution to the collective good. To give oneself a treat to compensate for the effort sustained in keeping oneself available may then make sense.

Management of Musical Ringtones in the Public Sphere

Many uses of musical ringtones have the same public character as uses of the mobile phone in general. The equivocal nature of musical ringtones also comes into play in public situations with other co-present parties than the owner of the phone. As an invitation to answer, it is likely to capture the attention of the other participants, and to turn them into involuntary witnesses of a telephone call not addressed to them. As a form of music, it may be exploited as a way to display the user's personality and tastes, and as an interactional resource.

Use of Mobile Ringtones as a Device for Displaying Identity-Related Features

The way in which the use of the mobile ringtone as a display of personal characteristics is assessed varies according to user profiles. Technology-oriented expert users seek distinction. Distinction to them always means displaying their skills in using all the potentialities of the mobile (according to an "ethos of virtuosity"), even when their musical tastes are involved: "I far prefer the world of geeks who want something they own, who want to go and fiddle around inside to turn it into something that's theirs, to be able to explore it, really know how to use it. With the others it's more a matter of the image that it gives them" (man, 27).

Many youth (we will call them "expressive youth") are using their mobile phones as a way to assert and make public various identity claims. Musical ringtones are then a resource for distinguishing oneself by making one's tastes visible in the public sphere, usually in relation with some form of collective and recognizable identity claim, either with respect to an actual peer group (friends) or an imaginary one (everyone who likes a particular type of music): "Everyone knows that I'm a Nirvana fan. If I'm surrounded by people with a mobile and there's Nirvana music, clearly I recognize mine and also the others" (man, 35). This staging is sometimes colored by a claim of membership in, and solidarity with, a group considered to be a minority or oppressed. Choosing raï music for one's ringtone may correspond to an effort to publicly assert a Kabyle identity: "It happened that there was music that I liked but that my friends didn't like. For example, Nadiya. None of my friends like Nadiya but I love her style of music so even if they don't like it I put her on!" (woman, 21). When the display of individual tastes is involved it is often conceived as an effort to distinguish and assert oneself.

Some users, mostly women, which we will call "intimists," generally try to minimizes the public exposure of personal features for fear of attracting strangers' attention in public: "I really don't want to show my personal tastes to others" (woman, 22). This orientation usually pervades their use of musical ringtones in public spaces populated with strangers, such as trains: "I don't like drawing attention to myself because of the ringing and because of anything, in general" (girl, 15). This behavior may be attenuated in public places where they know other co-present members and are known to them.

The last group is composed of subjects quite involved in using and disseminating ringtones, and highly concerned, because of an acute lack of self-assurance, with the negative inferences that others could make on the basis of their public image. They combine an aversion for exhibiting themselves in the public sphere with the concern to display a more positive image. Several said they chose not to answer when their telephone rang in public because they thought that their modest social background would be too obvious to those overhearing the conversation. Their use of musical ringtones becomes a strategic "gloss" (Goffman 1971). They design them to project a positive image for themselves and elicit some form of sympathy: "People who see me will think 'hey, he's dressed like that' but when they hear my ringtone they'll say 'hey, he listens to that music that's not for his generation,' then they'll look at me with sympathy" (man, 24). Some are even ready to be deceitful and choose a musical ringtone that is totally unrelated to their tastes: "sure, if the person came to talk to me about Aznavour (his current ringtone features a sample of a song by Aznavour), I'd feel kind of stupid because I don't know anything about him" (man, 24).

Mobile ringtones put their users' different positions in the public sphere to the test. Whereas they seem to serve well the interests of those who are in a position of strength or feel quite self-assured with respect to public exposure, their use takes on a very different meaning for groups who feel insecure or even dominated in public. On one hand, women in the intimist group will try to avoid attention-catching use of ringtones, while men in the disadvantaged group, especially the youngest ones, design musical ringtones to construct a more positive public image.

Use of Musical Ringtones and Public Management of the Obligation to Answer

Mobile ringtones are a topic within a broader public debate on the intrusive nature of uses of mobile phones in situations of co-presence. The occurrence of a ringtone is liable to attract the ear and attention of co-present bystanders, and interrupt ongoing activities and interactions. They cannot avoid hearing the ringtone and are therefore also subjected indirectly to its interactional consequences. The situation in which a ringtone is heard and several people look for their mobile phones at the same time reflects this diffusion of the invitation to answer in the sound space, a well as normative expectations about the need to answer one's ringing phone and the proper time to do so, as ethnographic observations of the uses of mobile phones have shown (Murtagh 2002). As was the case with those who take the call, musical ringtones are also designed with an eye toward such public space situations, so as to provide a kind of treat to co-present, potentially disturbed, bystanders. They are chosen then so that their ringing may be recognized and treated as a sign of consideration to whoever might be around. They are used as one of these many offerings through which subjects express deference to one another, in line with Goffman's general model of politeness in the public sphere (Goffman 1963).

Some users thus deliberately choose their musical ringtone from a set of shared cultural resources (popular music, hits, famous TV jingles or movie soundtracks), which they know would be recognized by most people. Such a choice of musical ringtones is oriented toward the way they will be received by copresent parties. The underlying assumption is that these musical items are so common and well known that the soundtrack ringtones blend more and become less noticeable, less attention-catching, and easier to be dealt with: "Because you already listen to them on the radio you don't notice them so much [...] These ringtones don't trigger any particular reactions" (woman, 22). Some even adapt their ringtones to fit more easily in specific settings: "If I'm expecting a business call and I'm in the bus, I put on something that'll be well-received like a TV jingle or something funny because it goes down better" (woman, 27).

Because they evoke common sound registers (radio, TV) foreign to the telephone, and are based on generally acknowledged cultural referents, these choices are intended to render the ringtones less intrusive (listening to them prolongs a series of pleasant experiences of listening to the piece concerned) and more acceptable (they are a sign of consideration by the potential disturbee): "'Four to the Floor,' quite a lot of people have it, so when it rings they recognize it, and it's not too loud either..., so it's cool" (man, 36). The obligation to answer and the interactional disturbances to which those who are copresent expose themselves are more easily accepted: "If you're with a person and the phone rings, if it's music that is nice...it makes the fact of answering more acceptable" (man, 21). Some users of mobile ringtones explicitly try to trigger a positive reaction. They want a sign of appreciation of their ringtones to validate the orientation to other copresent their choice of ringtone. They try to elicit a laugh or some other emotional reaction: "The ringtone of the 'Mystérieuses citée d'or' rang out in the metro and people smiled when they heard it" (forum). They attach a particular significance to those situations in which those copresent actually display their appreciation for the efforts made to solve the interactional problems posed in public by the obligation to answer. The most noteworthy stories are those in which witnesses creatively and visibly appropriate the ringtone in their own interactional context: "In the metro, White Stripes started going off full blast. A group of girls carried on with it even after I'd answered. In the street as well, except it was Nirvana" (forum). This type of public reaction shows that the offering has been accepted, and the intrusiveness of the ring in their sound environment ignored. Those users most concerned about the reception of their ringtones circulate these anecdotes in their personal social circles and on specialized forums.

The findings of our research thus attest to mobile phone users' concern in their choice of ringtones about the reactions that they are likely to trigger in public. They assess the possible effects of the different musical ringtones and try to make their choice as discreet and acceptable as possible, especially by configuring it so that it can be appreciated as a small sign of consideration for others, as a treat for other co-present parties. This is not always the case in practice, but what is important here is this particular orientation in the choice and design of musical ringtones.

Through the possibility of choosing a noteworthy musical ringtone, mobile phone users make their own availability and openness to the occurrence of a call publicly accountable. They do it in a particular way that is selected among a set of alternatives, so that their design is socially and interactionally significant. For instance, opting for switching their phone onto vibrate mode would have been a way of displaying their concern with the potential threat of the ringing telephone, and making reparations in advance by tuning down the strength of the perturbation (thus also reducing the salience of the obligation to answer). By choosing a musical ringtone, they more blatantly display their openness to an incoming call, but they still display their concern for politeness by trying to domesticate or "civilize" the interactional consequences of the ring, and turning it into a pleasant experience and even a treat to bystanders.

Musical ringtones are therefore shaped as an act of "positive politeness" that displays a supposedly shared desire by others for the gratification and musical experience that they procure. The success of musical ringtones thus reveals a shift concerning the potentially disruptive and threatening nature of a ringing telephone. This threat to others' face, formerly treated by behaviors of "negative politeness" (conspicuously showing one's wish to be discreet, or excusing oneself for the intrusion), can now be treated in the lighter mode of "positive politeness" (Brown and Levinson 1987). In line with a "connected" social order founded on generic normative expectancies concerning mutual availability, mobile musical ringtones are used by the answerer so as to achieve two distinct and somewhat contradictory goals: to conspicuously ratify the importance of showing oneself open to the incoming call, and to reduce the threat posed by its materialization to the face of the answerer himself and that of those nearby. The individual has a duty to be open and to respond to such requests for engagements of the face, but these seem less preoccupying and may be treated with less caution.

Conclusion

Musical ringtones reshape the embodied experience of the ringing telephone, for their users and for others. With respect to the standardized mechanical ringing of the traditional phone they invite an ambiguous experience: either they project a summons (by inviting an answer) or a musical experience (by providing an elaborate soundtrack). We have shown how different types of users were exploiting that ambiguity in their design in all possible ways (exacerbating either the treatment of the ringtone as a summons or as the occasion of a musical experience, and even maximizing their ambiguity through seamless "looping." The moral issue that underlies these alternatives in the choice and design of ringtones is the possibility that the obligation to answer, which the ringtone qua summons entails, may be counterbalanced with a musical treat with which the answerers individually gratify themselves. The very possibility of perceiving the ringtone as a treat discriminates between the collective and the individual. The treat retrospectively constitutes shared expectation of the obligation to answer as a generic collective obligation, distinct from the very individuality it is designed by the answerers to provide themselves.

Mobile phones are expected to be used in public, and the choice of musical ringtones orients toward the fact they will be heard by co-present parties. The choice of musical ringtones is a form of self-expression, a projection of personal preferences in the public sphere. Ringtones will be used by self-assertive users to display their expertise or to claim membership in various social groups. However, for groups that feel insecure or dominated in public, their usage takes on very different meanings. Such is the case with women who play down their personae in the public sphere or with some male users from disadvantaged social categories who fear others' judgment and exploit musical ringtones to construct a more valorizing (and sometimes deceitful) image of themselves.

The ringing of musical ringtones in public spaces is also a source of interactional concern. Ringtones are assessed with respect to the effect they may have on bystanders. Users declare choosing well-known tunes or funny ones because they think it makes their ringing in public more discreet and acceptable. The pleasure they are supposed to provide is offered as a treat to other co-present parties. Musical ringtones are therefore shaped as an act of "positive politeness." This marks a significant departure from previous practices where the potential threat of a ringing phone to "innocent" bystanders was typically treated with acts of "negative politeness" (removing oneself from the interactional scene or explicitly excusing oneself). Since the latter are stronger forms of reparation, it looks as if the user is expected to be open to an incoming call and to have an obligation to treat the summons it projects, while its consequences both with respect to the answerer or other co-present parties seem less preoccupying and to require less caution to be properly handled.

The success of musical ringtones and the way they are used may be interpreted in the context of a transformation in the management of mediated sociability and "connected presence," where social bonds are maintained through continuous patterns of interaction (Licoppe 2004; Licoppe and Smoreda 2006). Impulsive calls in which the value of the call stems from the fact that it is made without any premeditation and attests to the caller's feelings at a particular time are especially characteristic of connected-presence-mediated sociability patterns. Maintaining the liveliness of a given bond through its accountable contact via contact management becomes an autonomous goal: users are consciously managing fields of connection and connectivity (Nardi 2005). The burden of assessing the relevance of the calls shifts more toward the answerer, for the ringing of the phone may signal any type of call from anybody. Such an evolution tends to allow for a more abstract treatment of the ring, in which the answerer may feel compelled to answer out of a general obligation to be available and to participate to the generalized connectivity of a networked society. All the more so if many of the incoming calls prove only vaguely relevant. This may explain why users may feel a weakening of the obligation to answer (which becomes more abstract and rational than in the standard model in which it was more prominently related to the specific nature of social ties). Since the latter is relevant to the collective good of a larger but more abstract entity, it may make sense (among many other strategies) to counterbalance it with something individual and personal, like a "treat." Musical ringtones therefore appear as a handy resource in the context of this shift in mediated sociability, which their very success contributes to entrench even more.

Finally, even if it has not been discussed here, it is important to note that the development of practices of assigning individual ringtones to correspondents or groups of correspondents accompanies the shift from the caller to the answerer of the process of evaluating the relevance of the call in relation to the answerer's current engagements. The sudden sound of a particular ringtone, its perception, and its recognition lighten the calculations and deliberations surrounding the decision to answer. The possibility of personalizing musical ringtones introduces a veritable management of availability by answerers. Availability becomes the object of a project, with aims and calculations, dependent in particular on the categorization and classification of social links.

References

Arminen, I. 2005. Sequential order and sequence structure: The case of incommensurable studies on mobile phone calls. *Discourse Studies* 7(6): 649–662.

Bertho-Lavenir, C. 1984. *Histoire des Télécommunications en France*. Paris: Eres.

Brown, P., and S. Levinson. 1987. *Politeness. Some Universals in Language Usage*. Cambridge: Cambridge University Press.

Boltanski, L., and E. Chiapello. 1999. *Le Nouvel Esprit du Capitalisme*. Paris: Gallimard.

Elias, N. 1987. *La Société des Individus*. Paris: Fayard.

Goffman, E. 1963. *Behavior in Public Places*. New York: The Free Press.

Goffman, E. 1971. *Relations in Public: Microstructure of the Public Order*. New York: Harper & Row.

Hutchby, I., and S. Barnett. 2005. Aspects of the sequential organization of mobile phone conversation. *Discourse Studies* 7(2): 147–171.

Katz, J. E., and M. Aakhus, eds. 2002. *Perpetual Contact: Mobile Communication, Private Talk, Public Performance*. Cambridge: Cambridge University Press.

Licoppe, C. 2004. Connected presence: The emergence of a new repertoire for managing social relationships in a changing communication technoscape. *Environment and Planning D: Society and Space* 22(1): 135–156.

Licoppe, C., and Z. Smoreda. 2006. Rhythms and ties: Towards a pragmatics of technologically-mediated sociability. In *Domesticating Information Technologies*, edited by R. Kraut, M. Brynin, and S. Kiesler. Oxford: Oxford University Press.

Ling, R. 2004. *The Mobile Connection: The Cell Phone's Impact on Society*. San Francisco: Morgan Kaufman Publishers.

Miller, D. 1998. *A Theory of Shopping*. Ithaca: Cornell University Press.

Murtagh, G. 2002. Seing the "rules": Preliminary observations of action, interaction, and mobile phone use. In *Wireless World: Social and Interactional Aspects of the Mobile Age*, edited by B. Brown, N. Green, and R. Harper. London: Springer.

Nardi, B. 2005. Beyond bandwidth: Dimensions of connection in interpersonal interaction. *The Journal of Computer-Supported Cooperative Work* 14: 91–130.

Plant, S. 2000. On the mobile: The effects of mobile telephones on social and individual life. www.motorola.com/mot/doc/0/234_MotDoc.pdf.

Relieu, M. 2002. Ouvrir la boîte noire: Identification et localisation dans les conversations mobiles. *Réseaux* 20(112–113): 19–47.

Schegloff, E. 1986. The routine as achievement. *Human Studies* 9: 111–151.

Schegloff, E. 2002. Beginnings in the telephone. In *Perpetual contact: Mobile communication, private talk, public performance*, edited by J. Katz & M. Aakhus. Cambridge: Cambridge University Press.

12 | Mobile Technology and the Body: Apparatgeist, Fashion, and Function

Scott Campbell

The use of technology has transformed the role of the body in the process of communication. Thanks to innovations such as the telegraph, the telephone, and e-mail, humans do not have to be physically together to interact. The social implications of interaction from a distance have been a topic of scholarly concern since ancient times. In the *Phaedrus*, Socrates criticized the written word because he felt it does not allow for the reciprocity needed for true love (Peters 1999). In contemporary times, e-mail has been scrutinized for its limited ability to establish a sense of presence due to a lack of nonverbal cues (Trevino, Lengel, and Daft 1987). These examples illustrate the role the body has played in much of the scholarship of new media—as something that is absent in the process of mediated communication. While some lament the depersonalization that may result from the absence of the body during communication, others celebrate the new possibilities that this absence affords. In either case, the role of the body in this line of research has largely been focused on the effects and processes associated with its absence.

The recent explosion of mobile phones and other wearable personal communication technologies (PCT) presents a challenge to the traditional view of the body for research on new communication technologies. No longer can the body simply be viewed as a component removed from the communication process. Research focusing on portable PCT must also recognize the body as an integral part of the technology, and vice versa. This is because the body now wears communication technology, and the technology is often like a second skin to its user. As a result, technology that is worn on the body can become an important part of one's sense and presentation of self. As the ensuing literature illustrates, one wearable technology, the mobile phone, is regarded for much more than its functional utility. For some, the aesthetics of a mobile phone is regarded as a reflection of their sense of style. In addition, it can become an important part of the physical self by extending the body. In fact, many users speak about and treat their mobile phones in ways that humanize and make them organic, like body parts.

Beyond personal display, the mobile phone is commonly used to accomplish tasks, build and maintain relationships, and provide a sense of security, especially in case of emergency. But what is the relationship between the rationales of display, on the one hand, and uses of the technology for communication, on the other? The purpose of this chapter is to address this question through an exploration of the fashion and function of the mobile phone. In my efforts to do so, I report on a study surveying how people think about their handsets from a fashion perspective and how they use them functionally. I present the findings not as proof, but rather as new evidence shedding light on the relationship between the fashion and function aspects of the technology.

An equally important goal of this chapter is to deepen the discussion of *Apparatgeist* theory, advanced by Katz and Aakhus (2002a) to make sense of the seemingly natural conceptualizations of PCT, such as the mobile phone, and resulting trends in their adoption and usage around the world. Drawing from Apparatgeist, I was able to predict that certain uses of the technology are linked in meaningful ways to perceptions of the technology as fashion, while others are not. But the findings from the study also provide something else—an opportunity to rethink certain assumptions that are deeply embedded in Apparatgeist. In this chapter, I share what I found about the fashion and function of the mobile phone as well as the insights I arrived at about the core assumption of Apparatgeist theory.

The Mobile Phone as Fashion

The mobile phone is not just a social technology, but a highly personal one as well. The close relationship between the mobile phone and the body contributes to the device's personal and symbolic significance. Some users perceive their handsets as extensions of their physical selves (Gant and Kiesler 2001; Hulme and Peters 2001). This mind-set is perhaps best illustrated by the Finns, who commonly refer to the mobile phone as *kännykkä*, which translates into English as "an extension of the hand" (Mäenpää 2000; Oksman and Rautiainen 2003a; Oksman and Rautiainen 2003b).

Ling (1996) explained that the social meaning of the mobile phone is linked to the fact that the medium is "almost by definition, individual and not attached to a physical location" (p. 10). Even when compared to other personal and portable technologies, the mobile phone is considered characteristically stylish, particularly among long-term owners (Katz, Aakhus, Kim, and Turner 2003). For many, style plays an important role in brand selection (Lobet-Maris 2003).

Perceptions of the mobile phone as fashion are especially high among young people (Alexander 2000; Green 2003; Lobet-Maris 2003; Skog 2002). For example, in a study of mobile communication in the United Kingdom, Green (2003) found that *all* teens interviewed had extensive knowledge of handset styles and designs, and that the youngest individuals were most interested in the fashion of the technology. As Kaiser (2003) explained, "young people articulate complex ideas visually through their appearance styles, using and adapting goods available in the marketplace. They also get ideas from technoculture" (p. 156). There is evidence that these attitudes are prevalent among young adults as well as adolescents (Katz and Sugiyama 2005).

In addition to personal flair, young people use the style of a mobile phone to represent group membership. Similar to the way clothing can mark the boundaries between groups (Douglas and Isherwood 1996) and demonstrate a joint spirit within groups (Levine 1971), the physical appearance of the mobile phone can demonstrate network membership (Taylor and Harper 2001). Indeed, what is fashionable and appropriate use of the technology is often negotiated through interaction within social networks (Campbell and Russo 2003).

Purposes for Mobile Phone Use

Another prominent area of research examines purposes for using the mobile telephone. Based on studies of mobile phone use in Norway, Ling and Yttri (1999, 2002; Ling 2004) categorized mobile phone use into three primary groupings: safety/security, instrumental use, and expressive use. The first category includes use for emergency situations, such as calling for an ambulance after an auto accident, or carrying a mobile phone for a general sense of security. Ling and Yttri found that older adults tend to emphasize safety/security when interviewed about their mobile phone use. Instrumental use of the technology may involve the coordination of basic logistics, the softening of schedules, or making arrangements "on the fly." Examples include redirecting trips already underway and calling to notify someone you will be late. Ling and Yttri dubbed this "micro-coordination" and reported that middle-aged adults, particularly two-career parents, tend to emphasize it when discussing their use of the technology. Their third category, also referred to as "hyper-coordination," includes safety/security and instrumental communication but also mobile phone use for self-expression. This form of use involves relational communication, such as chatting with friends and family, as well as following norms for what is stylish, in terms of display, and what is proper, in terms of use. Not surprisingly, young people have been found to embrace the mobile phone as a resource for expressive communication.

Note that Ling and Yttri included both display of handset style and relational communication in their category for expressive use of the technology, suggesting a significant relationship between the two. Conceivably, the fashion and function of the mobile phone could indeed be linked in this way because both reflect modes of self-expression. On the other hand, these modes of expression may be considered quite different in nature—one involving the symbolic display of an artifact to co-present others, and the other involving linguistic communication with friends and family members from a distance. To further our understanding of the extent to which these forms of expressive mobile phone use are related, I carried out an exploratory study on how mobile phone owners think about and use their handsets.

Theoretical Framing

Katz and Aakhus (2002a) advanced the theory of Apparatgeist to make sense of consistencies in the effects and uses of mobile phones and other PCT in very disparate cultures. Apparatgeist, which literally means "spirit of the machine," refers to a common human orientation toward PCT and coherent trends in adoption, use, and social transformations. Apparatgeist was conceived when Katz and Aakhus (2002a) observed parallel shifts in communication habits that came out of mobile phone adoption in Finland, Israel, Italy, Korea, the United States, France, the Netherlands, and Bulgaria. These trends appeared in the coordination of everyday activities, configuration of social networks, private use of public spaces, new forms of connections to the workplace, and numerous other areas of the social landscape. Apparatgeist refers to an underlying spirit that contributes to these consistencies. Katz and Aakhus attributed the spirit of Apparatgeist to a common *logic* that "informs the judgments people make about the utility or value of the technologies in their environment . . . and predictions scientists and technology producers might make about personal technologies" (p. 307). This is the logic of perpetual contact. According to the authors, perpetual contact is a "sociologic" derived from collective sense-making, and it "underwrites how we judge, invent, and use communication technologies" (p. 307).

On its surface, the logic of perpetual contact is shaped by a host of social factors, such as values and norms, as well as technological factors, such as size and design, which influence how people think about and use their personal technologies. Peeling back the external layer of these social and technological factors exposes the core assumption of perpetual contact and the spirit of Apparatgeist—the ideal of *pure communication*. Katz and Aakhus (2002a) explained,

The compelling image of perpetual contact is the image of pure communication, which, as Peters (1999) argues, is an idealization of communication committed to the prospect of sharing one's mind with another, like the talk of angels that occurs without the constraints of the body (p. 307).

Peters (1999) invoked the teachings of Socrates to illustrate the ideal of pure communication. Socrates was an advocate of face-to-face dialogue as the paragon for communication because it offers the best chance for souls to be intertwined with reciprocity. This proclivity for reciprocity through dialogue is evidenced in Socrates' criticism of writing. According to Socrates, the written word is a barrier to reciprocity and can be dangerous because it may fall into the hands of unintended recipients. Socrates preferred dialogue between the corporeally present because he viewed it as more selective, intimate, and unmediated (Peters 1999).

Pure communication can be regarded as the merging of self and other in an attempt to establish a perfect social connection. Along with our differences (i.e., otherness), time and distance are also obstacles to this perfect connection, and Peters (1999)

argued that overcoming these obstacles became a driving force behind the development of modern communication technologies such as telephony. "Watson, come here; I want you," is what Bell said to Watson during the very first telephone call. Despite the fact that this call was placed unknowingly, "this utterance is the symbol and type of all communication at a distance—an expression of desire for the presence of the absent other" (Peters 1999, p. 180). In fact, telecommunication was anticipated as early as 1641 by Bishop John Wilkins, who expressed ambition for privacy and speed in communication across long distances (Peters 1999). The theoretical lineage of the logic of perpetual contact and Apparatgeist can be traced to such ambitions.

Apparatgeist and its assumption of pure communication can be used as a framework for anticipating certain connections between the fashion and function of the mobile phone. The ideal of pure communication closely resonates with certain expressive uses of the mobile phone. That is, one can argue that individuals who use the mobile phone to exchange thoughts and feelings with others tend to idealize (at least latently) an unobstructed social connection, "like the talk of angels." Accordingly, it seems that individuals who use the mobile phone in this very expressive manner would tend to regard the technology not as a barrier between self and other, but rather as a bridging mechanism in the pursuit of pure communication, even as an extension of the self. As noted, the mobile phone is indeed regarded as an extension of the self by some users (Gant and Kiesler 2001; Hulme and Peters 2001; Mäenpää 2000; Oksman and Rautiainen 2003a; Oksman and Rautiainen 2003b). From this vantage point, one can see how the lines separating subject from object become blurred, to the extent that the mobile phone is considered part of the self, both as a relational bridge and as a reflection of one's style. In other words, one might expect that socially expressive use of the mobile phone is linked to perceptions of the technology as fashion, much more so than safety/security and instrumental use. To put this expectation to the test, I conducted an empirical study of mobile phone users.

An Empirical Study

Two hundred seventy-six mobile phone users (63 percent female, 37 percent male) volunteered for my study by completing a short survey about their perceptions and uses of the mobile phone. Participants for the study were graduate and undergraduate students of mine and my colleagues at a private university in Hawaii. On average, they were twenty-five years old, owned a mobile phone for four years, and used about eight hundred minutes per month. The survey assessed the extent to which the style of a handset is important as well as uses of the technology for safety/security, logistical coordination, and relational communication. Factor analysis and reliability tests were used to ensure the survey items clustered into independent, reliable factors, so that statistical procedures could be performed to examine the relationship between uses of the

Table 12.1

Analysis of Factor Data

Factors	Eigenvalue	Alpha	M	SD	Range
Relational use	4.87	.81	3.91	.84	1.5–5.00
Fashion	2.39	.74	3.17	.82	1.00–5.00
Safety/security	2.29	.75	3.56	.84	1.00–5.00
Instrumental use	1.14	.62	4.16	.61	2.00–5.00

Factor eigenvalues, percents of variance, scale reliabilities, and summary statistics

Table 12.2

Predictors of Fashion Index

Predictors	Beta	Correlation between predictor and fashion index	Correlation between predictor and fashion index controlling
Relational use	.24	.23*	.21*
Instrumental use	−.02	.09	−.02
Safety/security	.01	.05	.01

Betas, bivariate correlations, and partial correlations of use predictors with fashion index (each predictor controlling for other predictors)

*$p<.01$

mobile and perceptions of it as fashion. Summary statistics for the factors are reported in table 12.1. A multiple regression test showed that the combination of all three uses was significantly related to the fashion index, $F(2, 272) = 5.20$, $p = .002$. However, when examining the relationships to each factor individually, only the relationship between relational communication and fashion was statistically significant (see table 12.2). In other words, use of the mobile phone for relational communication was significantly linked to perceptions of the technology as fashion, while safety/security and instrumental use were not. In general, my expectation held up.

Implications

Despite its limitations in scope and sampling, this investigation helps illuminate the extent to which relational use of the mobile phone and perceptions of the technology as fashion are linked. On one hand, they may be viewed as distinct constructs because they loaded on two separate factors. On the other hand, statistical analysis indicates they are related in ways that safety/security and instrumental use are not. Put simply, it appears that grouping expressive use and fashion into one category (i.e., hyper-coordination) is warranted, but this conclusion must be qualified by a moderate effect size for the relationship between expressive use and fashion. Beyond shedding light on the relationship between the fashion and function of the mobile phone, the study, in conjunction with other mobile communication research, provides a platform for deepening the discussion of Apparatgeist and theory building.

The first and most obvious theoretical implication of the study is that it demonstrates the utility of Apparatgeist as an analytic tool for understanding and predicting how certain perceptions and uses of PCT are related. Katz and Aakhus (2002a) acknowledged that so far only "the rudiments to establish a formal theory" (p. 312) are in place and that they see emerging from "the rudiments a theory that explains, and makes testable predictions about, the phenomenon of personal communication technology" (p. 315). The hypothesis that expressive use of the mobile phone is significantly related to perceptions of fashion was supported, showing that Apparatgeist can be used as means for making predictions and understanding the phenomenon of personal communication technology. Other recent studies I have conducted also demonstrate the utility of this theoretical perspective. For example, I used Apparatgeist to frame a cross-cultural comparison of perceptions and uses of mobile telephony, revealing several international patterns in the adoption and use of the technology (see Campbell, 2007). In addition, I employed Apparatgeist to frame an investigation of mobile phone use in Alcoholics Anonymous (AA) recovery networks. Drawing from some of the technological and social factors identified in Katz and Aakhus's (2002a, p. 311) explication of Apparatgeist, my coauthor and I successfully predicted that expressive and instrumental mobile phone use plays a substantial role in the recovery process of AA members. Remarkably, the sample reported that 67 percent of their total mobile phone use was for recovery-related interactions (see Campbell and Kelley 2006). These investigations along with the study reported in this chapter show the value of Apparatgeist in framing research, predicting results, and explaining mobile communication practices.

Beyond this, results of the study conducted for this chapter also provide a starting point for deepening the discussion of Apparatgeist and its core assumption of pure communication. A closer look at the survey data reveals that nearly 20 percent of the participants ($n = 49$) in my fashion/function study scored fairly low on the expressive use measure, with mean scores between 1.50/5.00 and 3.00/5.00. Although 20 percent may not seem like a large portion of the sample, it is noteworthy considering the sample primarily consisted of young people. Not all mobile phone users in all social contexts are driven by a fundamental desire for a complete social connection. In other words, pure communication may not be an all-encompassing driving force behind the adoption, conceptualization, and use of the mobile phone and other PCT.

The notion of pure communication, as an ideal, seems to resonate with certain expressive uses of the technology and much less so with safety/security and instrumental uses. However, even in the expressive arena, one can see a duality in that the technology is simultaneously used to overcome barriers separating self from other *and* to accentuate these differences. This can be seen in the mobile communication practices of young people. Adolescents are known for expressive use of the mobile phone to demonstrate social network membership (Johnsen 2003; Kasesniemi and Rautiainen 2002; Ling and Yttri 1999, 2002; Taylor and Harper 2001). Teens frequently make short calls or send brief text messages that on the surface appear to be meaningless, but in reality carry symbolically meaningful messages of social fellowship (Johnsen 2003; Licoppe 2003). Furthermore, teen peer groups are known to develop their own special language and characters when text messaging—a practice that heightens the sense of belongingness to one's peer group (Taylor and Harper 2001). While these mobile communication practices strengthen in-group membership, they also help mark the boundaries that display one's status as an outsider. While social network members are privy to text messages with distinctive language, nonmembers are often relegated to less inclusive forms of mobile phone use, such as voice mail. In their study of teens, Taylor and Harper (2001) observed that "Text messaging was used . . . to consolidate a community of peers and to differentiate themselves and their peers from others, such as adults" (p. 1). As Green (2003) explained, "The 'insiders' and 'outsiders' thereby created are not so much 'excluded' as they are 'unconnected' to some of the ongoing conversations among peers (in the form of text messages) that provide resources for building and maintaining peer relationships" (p. 209). So, while mobile communication technology is used to make connections between self and other, it is at the same time also used as an instrument for demarcation of these boundaries. This may be viewed as evidence that pure communication plays a bounded role as an underlying influence on mobile communication practices.

Perhaps instead of pure communication, it is *possible communication* that underlies all forms of mobile phone use and reasons for adoption. Used in this way, the phrase *possible communication* refers to the ability to connect and not the connection itself. Possible communication suggests all forms of potential connection—expressive, instrumental, safety/security, or otherwise. The idea here is that sometimes people are indifferent to and even conscientiously avoid "a union of the souls" in their use of PCT to interact with others. For instance, using e-mail can be an effective way of keeping one's distance from, and yet still connecting with, an unsavory work colleague. This familiar scenario highlights circumstances where social presence (Short, Williams, and Christy 1976) is intentionally minimized through PCT use, a practice that is contrary to the ideal of pure communication, yet in line with that of possible communication.

What is being proposed here is an alternative fundamental influence on the adoption and use of the mobile phone. To better understand the driving force(s) (e.g., pure communication, possible communication) behind mobile phone adoption, I re-examined interview transcripts from my dissertation research on how mobile phones are conceptualized and used in social networks (see Campbell and Russo 2003). The transcripts were pulled from a subsample of twenty-seven participants who were asked to explain their reasons for acquiring a mobile phone. Thematic analysis of the transcripts resulted in three key categories, which loosely parallel Ling and Yttri's (1999, 2002) notions of safety/security, hyper-coordination, and micro-coordination. More than half the respondents said they adopted a mobile phone for reasons of safety/security. Most of these individuals explained they acquired a mobile phone in case of car trouble or because it provides a general sense of security. As one participant explained, "You need a [mobile] phone in case your car breaks down. So it was basically for emergencies at first." Just under half said they originally adopted a mobile phone to keep in touch with friends and family, especially those relationships that involve long distance conversations. One college student in the study remarked, "I got a mobile phone because I came here to school. I'm from out of state . . . when I got here I was using my parents' MCI long distance card. Well, I was racking up a $500 a month phone bill!" One third of the respondents provided general convenience as a primary reason for mobile phone adoption. For example, one participant said he adopted a mobile phone,

. . . so I could be accessible, being that I was in the military and going to school. I don't spend a lot of time at home, and my wife and I only have one car, so it's easier just to let her take the car to work while I run around and do my things with the cell phone. That way she can always get ahold of me.

These interview responses support the idea that the potential to communicate is an all-encompassing influence on mobile phone adoption, whether the potential communication is for security, coordinating daily activities, or maintaining personal relationships. Expressive use of the mobile phone was mentioned as an explicit reason for mobile phone adoption by less than half of the participants, yet all participants provided reasons related to the possibility of communicating with others, indicating that possible communication may be a (dare I say *the*?) universal underlying influence on how people think about and use mobile communication technology. That is, instead of pure communication, perhaps possible communication "is the image deeply embedded in the logic of perpetual contact that underwrites how we judge, invent and use communication technology" (Katz and Aakhus 2002a, p. 307).

Advancing possible communication as a core assumption should not impede upon the objectives for Apparatgeist as a theoretical orientation. In fact, it should strengthen the theory's ability to predict and explain trends in both expressive and not-so-expressive mobile communication practices. For example, the assumption of possible communication may better equip Apparatgeist as a framework for explaining recent transformations in how real estate agents carry out their daily work activities—transformations that are common throughout this profession and appear to be far removed from the ideal of pure communication. Perhaps it is possible communication that lies at the heart of these and other social consequences of mobile communication.

Admittedly, the ideas I present in this chapter are bolstered by exploratory and, in some cases, anecdotal evidence. Additional research is needed to test the viability of possible communication as a universal, underlying influence on perceptions and uses of PCT. Additional research will also help in the continuing development of Apparatgeist. Katz and Aakhus (2002a) acknowledged that this perspective is in the early stage of theory development. In order to blossom into a formal theory with both explanatory and predictive power, Apparatgeist needs more testing, application, and explication. This chapter makes one contribution toward this end, while shedding more light on the connections between the fashion and function of mobile communication technology.

References

Alexander, P. S. 2000. Teens and mobile phones growing-up together: Understanding the reciprocal influences on the development of identity. Paper presented at the Wireless World Workshop, University of Surrey.

Campbell, S. W. 2007. A cross-cultural comparison of perceptions and uses of mobile telephony. *New Media and Society* 9(2): 343–363.

Campbell, S. W., and M. J. Kelley. 2006. Mobile phone use in AA networks: An exploratory study. *Journal of Applied Communication Research* 34(2): 191–208.

Campbell, S. W., and T. C. Russo. 2003. The social construction of mobile telephony: An application of the social influence model to perceptions and uses of mobile phones in personal communication networks. *Communication Monographs* 70(4): 317–334.

Douglas, M., and B. Isherwood. 1996. *The World of Goods: Towards an Anthropology of Consumption.* London: Routledge.

Gant, D., and S. Kiesler. 2001. Blurring the boundaries: Cell phones, mobility, and the line between work and personal life. In *Wireless World: Social and Interactional Aspects of the Mobile Age*, edited by B. Brown, N. Green, and R. Harper. London: Springer.

Green, N. 2003. Outwardly mobile: Young people and mobile technologies. In *Machines that become us: The social context of communication technology*, edited by J. Katz. New Brunswick, N.J.: Transaction Publishers.

Hulme, M., and S. Peters. 2001. Me, my phone, and I: The role of the mobile phone. CHI 2001 Workshop: Mobile Communications: Understanding Users, Adoption, and Design. Seattle, Wash. http://www.cs.colorado.edu/~palen/chi_workshop.

Johnsen, T. E. 2003. The social context of the mobile phone use of Norwegian teens. In *Machines that Become Us: The Social Context of Communication Technology*, edited by J. Katz. New Brunswick, N.J.: Transaction Publishers.

Kaiser, S. B. 2003. Fashion, media, and cultural anxiety: Visual representations of childhood. In *Mediating the Human Body: Technology, Communication, and Fashion*, edited by L. Fortunati, J. Katz, and R. Riccini. Mahwah, N.J.: Lawrence Erlbaum.

Kasesniemi, E. L., and P. Rautiainen. 2002. Mobile culture of children and teenagers in Finland. In *Perpetual Contact: Mobile Communication, Private Talk, Public Performance*, edited by J. Katz and M. Aakhus. Cambridge: Cambridge University Press.

Katz, J. E., and M. A. Aakhus. 2002a. Conclusion: Making meaning of mobiles—a theory of Apparatgeist. In *Perpetual Contact: Mobile Communication, Private Talk, Public Performance*, edited by J. Katz and M. Aakhus. Cambridge: Cambridge University Press.

Katz, J. E., and M. A. Aakhus, eds. 2002b. *Perpetual Contact: Mobile Communication, Private Talk, Public Performance*. Cambridge: Cambridge University Press.

Katz, J. E., M. A. Aakhus, H. D. Kim, and M. Turner. 2003. Cross-cultural comparison of ICTs. In *Mediating the Human Body: Technology, Communication, and Fashion*, edited by L. Fortunati, J. Katz, and R. Riccini. Mahwah, N.J.: Lawrence Erlbaum.

Katz, J. E., and S. Sugiyama. 2005. Mobile phones as fashion statements: The co-creation of mobile communication's public meaning. In *Mobile Communications: Re-negotiation of the Social Sphere*, edited by R. Ling and P. Pedersen. London: Springer.

Levine, D. N., ed. 1971. *Georg Simmel on Individuality and Social Forms*. Chicago: University of Chicago Press.

Licoppe, C. 2003. Two modes of maintaining interpersonal relations through telephone: From the domestic to the mobile phone. In *Machines that Become Us: The Social Context of Communication Technology*, edited by J. Katz. New Brunswick, N.J.: Transaction Publishers.

Ling, R. 1996. One can talk about common manners!: The use of mobile telephones in inappropriate situations (Report 32/96). Kjeller, Norway: Telenor Research and Development.

Ling, R. 2004. *The Mobile Connection: The Cell Phone's Impact on Society*. San Francisco: Morgan Kaufman.

Ling, R., and B. Yttri. 1999. Nobody sits at home and waits for the telephone to ring: Micro and hyper-coordination through the use of the mobile phone (Report 30/99). Kjeller, Norway: Telenor Research and Development.

Ling, R., and B. Yttri. 2002. Hyper-coordination via mobile phones in Norway. In *Perpetual Contact: Mobile Communication, Private Talk, Public Performance*, edited by J. Katz and M. Aakhus. Cambridge: Cambridge University Press.

Lobet-Maris, C. 2003. Mobile phone tribes: Youth and social identity. In *Mediating the Human Body: Technology, Communication, and Fashion*, edited by L. Fortunati, J. Katz, and R. Riccini. Mahwah, N.J.: Lawrence Erlbaum.

Mäenpää, P. 2000. Digitaalisen arjen ituja: Kännykkä ja urbaani elämäntapa. In *2000-luvun elämä: Sosiologisia teorioita vuosituhannen vaihteesta*, edited by T. Hoikkala and J. P. Roos. Roos, Tampere: Gaudeamus.

Oksman, V., and P. Rautiainen. 2003a. Extension of the hand: Children's and teenagers' relationship with the mobile phone in Finland. In *Mediating the Human Body: Technology, Communication, and Fashion*, edited by L. Fortunati, J. Katz, and R. Riccini. Mahwah, N.J.: Lawrence Erlbaum.

Oksman, V. and P. Rautiainen. 2003b. "Perhaps it is a body part": How the mobile phone became an organic part of the everyday lives of Finnish children and teenagers. In *Machines that Become Us: The Social Context of Communication Technology*, edited by J. Katz. New Brunswick, N.J.: Transaction Publishers.

Peters, J. D. 1999. *Speaking into the Air: A History of the Idea of Communication*. Chicago: University of Chicago Press.

Rice, R. E., and J. E. Katz. 2003. Comparing Internet and mobile phone usage: Digital divides of usage, adoption, and dropouts. *Telecommunications Policy* 27(8/9): 597–623.

Short, J., E. Williams, and B. Christy. 1976. *The Social Psychology of Telecommunications*. London: John Wiley.

Skog, B. 2002. Mobiles and the Norwegian teen: identity, gender, and class. In *Perpetual Contact: Mobile Communication, Private Talk, Public Performance*, edited by J. Katz and M. Aakhus. Cambridge: Cambridge University Press.

Taylor, A. S., and R. Harper. 2001. Talking 'activity': Young people and mobile phones. CHI 2001 Workshop: Mobile communications: Understanding users, adoption, and design, Seattle, Wash. http://www.cs.colorado.edu/~palen/chi_workshop.

Trevino, L. K., R. H. Lengel, and R. L. Daft. 1987. Media symbolism, media richness, and media choice in organizations. *Communication Research* 14(5): 553–574.

13 | The Mediation of Ritual Interaction via the Mobile Telephone

Richard S. Ling

Introduction

Consider the impact of mobile communication on sociation, the social impulse. As an amateur or professional sociologist, one might ask what holds society together and how is the social order created and maintained? What is the interaction between the ideational world and our everyday concrete interactions with others and with our physical world? How does ritual—be it in the context of religion or, as is perhaps more often the case, in the context of mundane life—shape our experience of social solidarity, and, the point of the present chapter, how does mediated interaction play into social ritual?

Save mediated interaction, these are the issues that Emile Durkheim focused on in his analysis of ritual in religion and its implications for social solidarity (1995). Durkheim saw ritual including the assembly of a group, a mutual focus, a shared mood, and an entrainment that fosters solidarity. Ritual is also Goffman's subject of study when he considered everyday life (1959; 1963; 1967; 1971). Finally, Randall Collins brings up the same issues in his analysis of interaction ritual chains (1998, 2004). The focus in the work of Durkheim, Goffman, and Collins is, however, on the co-present face-to-face interaction. In the case of Durkheim, much of his examination of ritual derives from his examination of Australian Aboriginal religions. Thus, there is no opportunity for mediated interaction to enter the picture. Goffman made passing references to mediated interaction giving the reader with that turn of mind the sense that the door was left slightly ajar. However, with a single enticing exception and an innuendo in a footnote, he limits himself to the co-present. In spite of this, those working in the area of mediated communication have called on Goffman. Joshua Meyrowitz has applied Goffmanian concepts to broadcast communication (1985). Their application to ICT-based interaction is to be found in the work of Joachim Höflich (2003) and in my own analysis of mobile communication (Ling 2004b, 2008).

Collins is quite clear in his exclusion of mediated interaction in the development and maintenance of social solidarity via what he calls interaction ritual chains. Collins goes beyond the omission of Durkheim and Goffman. Indeed he is explicit in saying that physical co-presence is generally a requisite aspect of ritual interaction.

The issue of social solidarity in mediated interaction has, however, received a new urgency. In the period since Goffman's writing we have witnessed the explosion of mediated interaction. It is a period that has seen the rise of the Internet and the broad acceptance of mobile telephony. Using the context I know best, that of Norway, one sees the explosion of ICTs onto the scene. In 1997 only 7 percent of Norwegians said they were daily users of the Internet. As of 2006 this was up to 79 percent (Vaage 2006). Looking at mobile telephony, we see that more than 66 percent of the calls were to mobile telephones in 2004. An average mobile phone user sent more than two SMS messages per day and some social groups like teen girls sent up to ten per day. This is 30 percent more than in a similar period in the year before (PT 2004).

Beyond the raw statistics, the technologies also play on and play into our social interactions. We use them as we navigate through daily life and their use is a type of marker of our technical competence in the eyes of others. In this chapter I focus on the mobile telephone that is a multidimensional object to be sure. It is, in its physical form, an object that can be read as an indicator of an individual's status or position (Fortunati, Katz, and Riccini 2003). At the same time, it seemingly has a life of its own. Indeed, the mobile phone can seemingly ring, peep, squawk, or play a Beethoven piano concerto without provocation. Finally, the mobile telephone can be used as a portal for interacting with another person. Thus, it takes the individual's attention away from the co-present situation and directs it to other corners of the universe that can be physically or temporally separate.

The development of interactive systems of mediation has given new life to the question of how we achieve and maintain interpersonal interaction. Indeed, it is clear that the rise of mediation systems such as the Internet and mobile telephony has given us new cause to think about the mechanics of social interaction. It is quite possible that the recent interest in, for example, social capital, is motivated by a concern that, in the words of Yeats, "The center will not hold." When considering Internet use, there is an active discussion as to its impact on social interaction. The work of Nie (2001) seems to indicate that excessive use of the Internet is detrimental to social interaction, while the analysis done by Katz and Rice points in the opposite direction (2002). At the same time, work on the mobile telephone seems to indicate that moderate use correlates with an active social life (Ling et al. 2003) and a more closely knit peer group (Reid and Reid 2004; Rheingold 2002).

There is a boundary here that begs to be examined, namely the way in which ritual transcends the co-present and indeed the co-temporal. To what degree do Durkheim's effervescence or Collin's entrainment transcend the here and now? Is it possible that it can be done there and then instead of here and now?

I do not mean to say, however, that co-present rituals are on their way out, nor am I saying that it is either mediated or co-present. It is clear from the historical record and the focus of those who have examined this that an important aspect of many mediated rituals is corollary co-present interaction. Indeed, much of the discussion in this chapter examines the use of mobile communication in co-present situations. We are not on the edge of an era ruled by simple mediated interaction. We will still meet face-to-face, or in Fortunati's term "body-to-body" (2005), and engage in our interactions. In spite of this, mediation is taking a larger portion of the pie. We are doing more and more interacting with others via various types of electronic mediation. We are becoming more adept at its use and we are entrusting the mediated messages, conversations, and interactions with ever more nuanced forms of interpretation and meaning. They help us to maintain our social contacts and they modify, readjust, and displace social interaction. Thus, at this point co-present interaction is the locus of social ritual. However, it is increasingly being modified by mediated interaction. In addition, we are developing mediated rituals that further play on the understandings that have been built on the co-present sphere.

The Ritual Aspects of Mobile Telephone Use

In the preceding discussion I outline how Durkheim, Goffman, and Collins develop the issues surrounding ritual interaction. I focus on the application of these issues to the use of mobile telephony. The telephone and specifically the mobile telephone is a significant object in modern society. Its physical form is the focus of comment. Simply knowing the type of mobile phone that others own provides us with insight into their social position and status. We can recognize the teeny bopper with the Nokia, and the advanced business user with the Trio, Blackberry, or the Sony-Ericsson, et cetera. Indeed, the very form of the mobile telephone has become a type of minor cultural icon. The very object itself is not neutral. Its use in co-present situations is symbolically invested and can be seen in terms of its contribution to ritual interaction.

I explore how mediated contact via the mobile telephone can also be seen in the context of ritual interaction. The ways we greet one another over the phone, the ways we relate stories, and the ways we use the telephone to organize our daily life show that, in many respects, ritual interaction can be carried out via interactive media. Further, there are particular forms of interaction and parlance that seem to occur only via the mobile telephone (e.g., "Where are you") that can be seen as mediated ritual interaction. Thus, I am interested in expanding Collins sense of ritual interaction into the area of mediated communication.

Mediated Ritual Interaction in Co-present Situations

The exigencies of mediated interaction play into the way people interact in co-present situations. The mobile telephone has forced this issue upon us. Observations also point to the emphasis of mediated rituals at the expense of the co-present rituals. While the individuals being observed pay heed to the co-present, the material indicates that the mediated interaction has an equal if not superior place in the minds of the individuals. This is seen in the following observation.

Observation A woman who was casually dressed walked out of the area around the Nationalteater station. She started walking east on Karl Johan Avenue (at this time there is no traffic on the street since it is being renovated.) She walked eastward on the southern sidewalk somewhat slowly and tried to make a call two or three times on her mobile telephone as she walked. None of the calls went through and so she started to text; indeed, her focus was mostly on composing SMS text through the rest of the observation. She continued east, walking near the edge of the curb. A woman on crutches along with two other women walking abreast approached her going west. The woman using the crutches was also quite near the curb and actually somewhat hemmed in by her two friends who were walking to her left. As the woman who was texting and the woman on crutches approached, the "texting" woman gave a navigation glance and edged gently to her left and the line of women including the woman on crutches going west also edged slightly to their left. The texting woman and the woman on crutches passed each other with a small but uncomplicated margin. The texting woman next crossed the street at an angle—while continuing eastward and eventually came to the sidewalk on the north side of the street. All the while she continued to attend to the texting on her telephone that was in her right hand.

In this observation the woman balanced between an engrossment in her texting and at the same time a minimal but adequate awareness of her co-present situation. As she composed text on her mobile telephone she also maneuvered past the women on crutches and entourage. The two realms were parallel. While in some ways there was the danger that, in the words of Goffman, her "heart might not lie where the social occasion requires it to" (1963, p. 38), she was able to manage both situations. She shows an adroit ability to manage the texting and co-present interaction just as those with whom she interacts show a certain tolerance.

A central feature of this observation is the texting activity of the woman. She was clearly in the process of creating text in an interaction with an unseen interlocutor. Texting is an interaction with another person who is physically as well as temporally remote. The message(s) she created were in all likelihood a part of an ongoing interaction with a friend or family member. Other analysis has shown that the form of language, the linguistic conventions, the use of openings and closings, et cetera, all draw on various pre-existing traditions. The specific phrasings and the way the woman developed her text were in all likelihood the product of earlier interactions as well as being a part of the general genre of SMS (Hård af Segerstaad 2005; Ling 2005; Ling, Julsrud, and Yttri 2005).

The attention afforded by the woman to the mobile telephone is also something that is relatively new in the urban scene. It is rare for persons to walk along the street with their attention so acutely focused on an artifact. While we might read a map in an unfamiliar city, we do not often, for example, read books or become as completely absorbed in the manipulation of a physical object for such an extended period of time as we walk along the street. However, in the contemporary scene, observations such as the one reported here are far from unique. The observation of the woman texting here is not a rare sighting but rather a common part of the contemporary urban cityscape in Scandinavia. We see that the mobile telephone is not simply another object that we have on our person as we move through the urban sphere. Rather, it is a conduit through which we have physically and often temporally removed contact with others. Given this ability to mediate contact whenever and wherever we find ourselves, we have to pay our interlocutors the proper heed just as we have to pay the proper heed to those with whom we are co-present. In an expansion of Goffman's insight, we have a shared responsibility for maintaining two lines. We have, in effect, two front stages upon which to act.

While the interactions with the mediated partner are less evident to the observer, the interactions with other persons on the street are obvious. The subtle signaling and body language vis-à-vis the crutch lady and her entourage relied on a shared repertoire of signals. These are a part of our cultural ballast that have been developed, refined, and re-energized through common use. All the while, the subject engaged in the newer gesture of texting. As with the situations described, she was requesting and being afforded civil inattention as she walked and texted.

The interesting issue here is that the woman's nearly unwavering focus on the composition of text bespoke the entrainment in that activity. She was in the process of maintaining a ritual interaction chain. In both realms the texting woman and the people with whom she interacted were presumably observing the ceremonial aspects in their negotiation of everyday life (Goffman 1967, pp. 55–56). The degree to which she carried this off was determined by her sophistication in texting and the tolerance of the various persons with whom she interacted.

Focused Mediated Ritual

It is important to think of the broader role of mediated interaction, however. First, in the case of traditional rituals, mediated interaction can serve to set the scene beforehand and then allow the reliving of the event later. Indeed, Ito describes how Japanese youth en route to a date use mobile text messages to interact with their boyfriend or girlfriend (Ito 2005). In a similar way, after the date, they start a new texting session with each other that draws out the interaction. Second, the focus on heavily imbued rituals such as a wedding or a funeral is seemingly a step back from the direction pointed out by Goffman. They are not the small, everyday rituals such as telling a joke, relating a story, or exchanging a greeting, or Goffman's "brief rituals one individual performs for and to another, attesting to civility and good will on the performer's part and the recipient's possession of a small patrimony of sacredness" (1971, p. 61).

Up to this point I have looked at ritual in co-present situations albeit with a good dose of mediation in the picture. Now I look more directly at the potential for inter-action ritual in mediated interaction. Just as Goffman took the basic themes from Durkheim and reapplied them to the microsituations of everyday life, I want to be audacious enough to reapply Collins (along with his Durkheimian and Goffmanian baggage) to mediated interaction. As has been noted, Durkheim did his work before mediated interaction became an issue. In addition, his work with aboriginal tribes took co-presence for granted. Goffman consciously focused only on co-present encoun-ters. Indeed he explicitly excluded mediated encounters or "mediated engagements" from his discussion (Goffman 1963, p. 89 n12). Nonetheless, he recognizes that encounters can be mediated (Goffman 1963, p. 89 n12). Collins, however, is clear in his assertion that ritual interaction chains are possible only in co-present situations. Collins draws his reading of Goffman out of the need for physical co-presence (2004, p. 23). He posits that, as a general rule, it is not possible to have successful social rituals via mediated channels. He discusses the possibility of, for example, a wedding or a funeral that is conducted over the telephone and notes that the lack of feedback would diminish the experience and limit the effervescence of the situation. These, however, are quite fully developed rituals that have their roots and tradition in the co-present world. Indeed there is a slightly absurd nature to those examples of weddings that take place between couples over the phone—or in a case reported by Standage—over the telegraph (Standage 1998).

I am not convinced that physical co-presence is always necessary for a well-developed social interaction. Co-presence is a powerful generator of entrainment and can often be called upon as an ingredient for social interaction. However, I suggest that mediated interaction also has the potential to enhance the ritual dimensions of subsequent co-present interaction. Mediated interaction can draw on the symbols developed and rejuvenated in co-present ritual interaction and revitalize them. Fur-ther, ritual interaction can develop in exclusively mediated interactions. Micro-level social rituals can be coined via mediated interaction.

To be fair, Collins considers the degree to which mediation can come into ritual in-teraction. However, he asserts that in general, ritual interaction is best when done co-presently. I agree with this general assertion but I feel that there also needs to be a more rounded analysis of co-present versus mediated interaction. It is easy to think of examples, and indeed empirical evidence, showing people using mediated ritual forms to do exactly the work of ritual interaction.

We can also see ritual interaction chains in the way that teens negotiate the early portions of a romantic relationship. After meeting and establishing contact and exchanging mobile phone numbers the nascent couple engages in a more or less SMS-based courtship. In this period the form of the interaction is highly scripted in that the messages are carefully written and edited. This indirect form of interaction is calculated to allow the individuals to carefully work through their utterances and to cover over some of the pitfalls that might weigh heavily in the nascent period of the romance.

Rita (18): . . . if you meet a guy when you are out, for example, then it is a lot easier to send a message instead of talking like. Somebody you don't really know. It is more relaxed.

Anne (15): It is easier to tell if you like a person.

Interviewer: Via SMS?

Ida (18): Then your voice will not either shout or disappear. You need time to think [when constructing your turns].

In addition, the timing of the messages is carefully calibrated. If one answers too quickly, one is perhaps seen as being overeager. If one waits too long, one is not interested. In the era of telephone-based courting there was much thought put into how many times the phone could ring before answering it. Now, each turn in an SMS "conversation" is a new round of the same game. The strategic use of the mobile telephone speaks to the development and use of new ritual forms of interaction here. There is the assembly of the dyad with a mutual focus and a shared mood. There is an entrainment that "plays on" and "plays into" the evolving solidarity of the couple. In addition, there is the establishment of sacred objects in the form of the SMS texts that are in some cases saved, transcribed, embroidered, and placed into the mythology of the couple.

Another example of mediated ritual integration is seen in play and joking via mobile telephones. Much of mediated communication is associated with this type of interaction. Gag e-mails, emoticons, greetings, and ASCII-based representations (Danet 2001). In this case all the conditions of ritual interaction—baring co-presence—are fulfilled. There is the mutual focus of attention and collective engrossment that results in a type of solidarity. The joking can contribute to the revitalization of group identity and finally the joke can go flat. Indeed in many cases, the specific forms of humor and playfulness trade specifically on the fact that it is mediated.

There are other forms of mediated phatic interaction. There are, for example, various types of locutions that are particular to SMS. As reported elsewhere, Norwegian use of the word *Koz* as a closing in SMS messages is a case in point (Ling 2004c; Prøitz, forthcoming). The word draws on the culturally familiar word *kos* (hug). However, its use in SMS messages as a closing phrase and its peculiar spelling (with a *z* instead of the expected *s*) makes it unique.

Mobile communication technology allows us to exchange mediated greetings regardless of time and place, as seen in the following sequence:

Interviewer: Do you get SMS messages in the middle of the night?

Per (17): Yeah, one time a friend of mine had a birthday and so I called him at 5 a.m. or something like that.

In this example, Per's spontaneous call—like the shivaree from earlier periods—was probably not appreciated at the time, but it was a celebration of Per's friendship. The call was probably commented on the next time the two met and it may have become a part of the lore surrounding their friendship. Indeed, Per possibly can expect a similar call from his friend on his birthday. In this way the interaction is a type of interaction ritual—sans co-presence—in the sense outlined by Collins. There is mutual, if somewhat groggy, focus of attention and bleary engrossment that results in a type of solidarity, particularly when the episode is discussed later. The interaction also contributes to the revitalization of group identity. The point here is that this form of interaction was facilitated by mobile telephony. Should Per have physically knocked on his friend's door he would have disturbed the whole household. The individualization of communication afforded by the mobile telephone allowed for the form of celebration Per chose.

Another example of this type of mediated interaction is the exchange of SMS-based endearments by lovers who still live with their parents, as in the case of teens, who are only in the formative portion of their relationship or who are in a routinized relationship but are away from one another. This is seen in "good night" or "good morning" greetings. Examples range from the utilitarian "G'nite" (female, 15), or "Have a good night, hug" (man, 47), to the rakish "Do you want to spend the night? Hug" (female, 17) or "Good night sex bomb" (female, 35). For some young couples this an obligatory ritual that is disregarded only at great risk to the relationship.

This is a partial catalogue of the ways that mediated communication can be seen as ritual interaction. There are other analyses of these issues. For example, others have examined the historical development of using "hello" as a telephone greeting (Bakke 1996; Fischer 1992, pp. 70–71; Martin 1991, pp. 155–163; Marvin 1988), the specific identification and greeting sequences that are used in telephonic interaction (Saks, Schegloff, and Jefferson 1974; Schegloff and Saks 1973) and children's acquisition of these cultural artifacts (Ling and Helmersen 2000; Veach 1981). Deference is seen in the ways we address one another in telephone greetings (Bakke 1996) but also in the way we use various inflections in our greeting and parting sequences in order to give fuller meaning.

This material indicates that mobile communication can be used to elaborate, embroider, and even engender interaction rituals. I say this with humility, however. As noted, the types of powerful rituals that often serve as the basis of social solidarity are co-present. This said, we should not exclude the possibility that these types of social interactions can be extended to the mediated world.

Conclusion

Returning to the point of departure, this chapter is framed by the broader issue of how ICTs, and in particular mobile communication, play into the development and maintenance of social capital. The intention has been to look at the work of Durkheim, Goffman, and Collins to glean insight into how small-group ritual interaction results in symbolically imbued cultural features that in turn provide us with a type of social integration. Thus, the focus here has been on the impulse toward sociation.

At the microsocial level, focused ritual encounters are the basis upon which social solidarity is built. We work through these encounters, be they co-present or mediated, using a repertoire of ritual devices. These encounters can be spontaneous (meeting an old friend on the street), institutionalized (packing the kids off to school), expansive (a lecture to the local PTA), or discreet (a quick wink to an acquaintance you pass in the hallway). And, even though they are often co-present, they can also be mediated. In all these cases we rely on ritual interactions and formulations.

Our interactions often balance between discord and order. We can make clever jokes or come close to doing so but end up embarrassing ourselves. Social encounters are dynamic and indeed perilous. A focused social encounter can become an entrained event from which we draw social integration. But these events come in all sizes and shapes. These are the items upon which the specific social encounter and the more elaborate social solidarity rests. The bonding can be among people in broadly similar circumstances (teens in a school) or it can be bonding based on hierarchical differences (interaction between students and teachers in the school).

Our common willingness to use these symbolically imbued strategies, facades, manners, rituals, poses, and savoir faire, as well as our indulgences in others' use of these, is central. Indeed, the degree to which a social group is bounded together is seen in the elaboration of these strategies. The ritual interaction can be done "correctly" or it can be muddled. We can, in effect, engineer the use of certain social devices but at the same time "give off" other unguarded signs that alert others to alternative strategies and different sides to our feelings in a situation. While being perhaps uncomfortable for the persons involved in the situation, these slips provide the situation with its dynamic nature. It is when we slip or when others slip—that is, when we get a peek into Goffman's backstage area—that there is the need for maintenance work. It is also in these situations where there is a threat to the symbolic solidarity of the group that we see the breadth and depth of the social structure.

One type of slip is to have an indeterminate status vis-à-vis the situation. Social interaction is difficult when we are not sure as to the status of the others in the situation. That is, there is a pressure to either be clearly in or clearly outside a social interaction. To be half in or half out is the cause of anxiety since the other partners in the interaction are unclear as to the status of the individual (Goffman 1963, p. 102). The mobile telephone brings out this issue.

Use of the device puts us into a type of social limbo wherein others cannot tell as to our true status. More generally, we must pay attention to our lines of action, even when there are competing lines of activity. In effect, we must let others know what we are up to. We must disclose to them just how much time and energy we can afford them. Regardless of the other being a beggar on the street or a lover, our status as a social individual means that we need to give the other a sense of how open or closed we are.

Slips are also important in that they threaten to disturb our common sense of the situation. Slips require repair. The greater the slip, or in other words the deeper the engrossment, the greater the need for repair work. This said, disturbances can become normalized as can the strategies for dealing with them. Indeed we have seen this with the adoption of the mobile telephone. On the one hand, we are not as disturbed by the use of the mobile telephone in the public sphere as compared to the time of its introduction (Palen 2002). In addition, users have developed routines that help us to stage-manage their use.

In the case of the telephone, and in particular the mobile telephone, there is the possibility that there is a type of dual situation. Goffman discusses this possibility in *Relations in Public*, albeit from the perspective of the co-present situation as opposed to the telephonic one (1971, pp. 220–221). In this case, Goffman discusses the strategies used to maintain the co-present situation. This is difficult in telephonic situations that lay claim to the individual's attention. The question becomes which line is to be followed in these situations. Goffman describes how the telephonic line is reduced in importance through the use of various facial gestures indicating to the co-present individual that the telephone conversation is only a temporary digression. However, it is easy to observe exactly the opposite, namely the focus on the telephonic at the expense of the co-present. I have observed a woman who gave her friend a hug while actually concluding her texting. For that brief period she seemingly set a higher priority on finishing the textual interaction than on giving herself completely over to the co-present interaction. The mobile telephone has expanded the geographical range for telephonic interaction and made it into a more omnipresent social factor.

In sum, I posit that mediated ritual interaction with its attendant elements of mutual focus, collective engrossment, sense of solidarity, symbolic imbuement, and group revitalization can help to support and maintain social interaction. Mobile communication is a factor in this process that has only recently arrived on the scene. However, it will play into ritual interactions in new and unexpected ways.

References

Bakke, J. W. 1996. Technologies and interpretations: The case of the telephone. *Knowledge and Society* 10: 87–107.

Collins, R. 1998. *The Sociology of Philosophies: A Global Theory of Intellectual Change*. Cambridge: Belknap.

Collins, R. 2004. *Interaction Ritual Chains*. Princeton: Princeton University Press.

Danet, B. 2001. *Cyberpl@y: Communicating Online*. Oxford and New York: Berg.

Durkheim, E. 1995. *The Elementary Forms of Religious Life*. K. E. Fields, trans. Glencoe, Ill.: The Free Press.

Fischer, C. 1992. *America Calling: A Social History of the Telephone to 1940*. Berkeley, Calif.: University of California Press.

Fortunati, L. 2005. Is Body-to-Body Communication Still the Prototype? *The Information Society* 21(1): 53–61.

Fortunati, L., J. E. Katz, and R. Riccini. 2003. *Mediating the Human Body: Technology, Communication and Fashion*. London: Lawrence Erlbaum.

Goffman, E. 1959. *The Presentation of Self in Everyday Life*. New York: Doubleday Anchor Books.

Goffman, E. 1963. *Behavior in Public Places: Notes on the Social Organization of Gatherings*. New York: The Free Press.

Goffman, E. 1967. *Interaction Ritual: Essays on Face-to-Face Behavior*. New York: Pantheon.

Goffman, E. 1971. *Relations in Public: Microstudies of the Public Order*. New York: Harper.

Hård af Segerstaad, Y. 2005. Language use in Swedish mobile text messaging. In Front Stage/Back Stage: Mobile Communication and the Renegotiation of the Social Sphere conference, edited by R. Ling and P. Pederson. Grimstad, Norway.

Höflich, J. 2003. *Mensch, Computer und Kommunikation*. Frankfurt: Peter Lang.

Ito, M. 2005. Mobile phones, Japanese youth and the re-placement of social contact. In *Mobil Communication: Re-negotiation of the Social Sphere*, edited by R. Ling and P. Pedersen. London: Springer.

Katz, J. E., and R. E. Rice. 2002. *Social Consequences of Internet Use*. Cambridge: The MIT Press.

Ling, R. 2003. Mobile communication and social capital in Europe. In *Mobile Democracy: Essays on Society, Self and Politics*, edited by K. Nyíri. Vienna: Passagen Verlag.

Ling, R. 2004a. "Goffman er gud": Thoughts on Goffman's usefulness in the analysis of mobile telephony. In Wireless communication workshop: Inspirations from unusual sources, edited by H. Sawhney and C. Sandvig. Ann Arbor, Michigan: University of Michigan, Department of Communication.

Ling, R. 2004b. *The Mobile Connection: The Cell Phone's Impact on Society*. San Francisco: Morgan Kaufmann.

Ling, R. 2004c. Where is mobile communication causing social change? In *Mobile Communication and Social Change*, edited by S. D. Kim. Seoul, South Korea.

Ling, R., ed. 2004d. *Report of Literature and Data Review, Including Conceptual Framework and Implications for IST*. Brussels: EU.

Ling, R. 2005. The socio-linguistics of SMS: An analysis of SMS use by a random sample of Norwegians. In *Mobile Communications: Re-negotiation of the Social Sphere*, edited by R. Ling and P. Pedersen. London: Springer.

Ling, R. 2008. *New Tech, New Ties*. Cambridge, Mass.: The MIT Press.

Ling, R., and P. Helmersen. 2000. "It must be necessary, it has to cover a need": The adoption of mobile telephony among pre-adolescents and adolescents. Paper presented at the Conference on the Social Consequences of Mobile Telephony, Oslo, Norway.

Ling, R., T. Julsrud, and B. Yttri. 2005. Nascent communication genres within SMS and MMS. In *The Inside Text: Social Perspectives on SMS in the Mobile Age*, edited by R. Harper, A. Taylor, and L. Palen. London: Klewer.

Ling, R., B. Yttri, B. Anderson, and D. Diduca. 2002. *E-living Deliverable 6. Family, Gender and Youth: Wave One Analysis*. IST.

Ling, R., B. Yttri, B. Anderson, and D. Diduca. 2003. "Mobile communication and social capital in Europe. In *Mobile Democracy: Essays on Society, Self, and Politics*, edited by K. Nyri. Vienna: Passagen Verlag.

Martin, M. 1991. *Hello, Central?: Gender, Technology and Culture in the Formation of Telephone Systems*. Montreal: McGill-Queens.

Marvin, C. 1988. *When Old Technologies Were New: Thinking about Electric Communication in the Late Nineteenth Century*. New York: Oxford University Press.

Meyrowitz, J. 1985. *No Sense of Place: The Impact of Electronic Media on Social Behavior*. New York: Oxford University Press.

Nie, N. H. 2001. Sociability, interpersonal relations, and the Internet: Reconciling conflicting findings. *American Behavioral Scientist* 45: 420–435.

Palen, L. 2002. Mobile telephony in a connected life. *Communications of the ACM* 45: 78–82.

Prøitz, L. 2006. "Cute boys or game boys? The embodiment of femininity and masculinity in young Norwegians' text message love-projects. *Fibreculture Journal*. http://journal.fibreculture.org/issue6/issue6_proitz.html.

PT. 2004. *Det Norske Telemarked: 1. halvår 2004*. Post-og Teletilsyn, Oslo.

Reid, D., and F. Reid. 2004. Insights into the social and psychological effects of SMS text messaging. http://www.160characters.org/documents/SocialEffectsOfTextMessaging.pdf.

Rheingold, H. 2002. *Smart Mobs*. Cambridge, Mass.: Perseus.

Saks, H., E. A. Schegloff, and G. Jefferson. 1974. The simplest systematics for the organization of turn-taking for conversations. *Language* 50: 696–735.

Schegloff, E., and H. Saks. 1973. Opening up closings. *Semiotica* 8: 289–327.

Standage, T. 1998. *The Victorian Internet*. London: Weidenfeld and Nicolson.

Vaage, O. 2006. *Mediabruks Undersøkelse*. Oslo: Statistics Norway.

Veach, S. R. 1981. *Children's Telephone Conversations*. Ann Arbor: University Microfilms International.

14 Adjusting the Volume: Technology and Multitasking in Discourse Control

Naomi S. Baron

Consider two images associated with contemporary information and communication technologies (ICTs). The first: "Always on, always connected," the slogan of RIM (Research in Motion), makers of the BlackBerry. The second: A college student in conversation with her professor (in this case, me) when the girl's mobile phone rings. Rummaging in her backpack to retrieve the device, she glances at the number on the display screen. Dismissively, the student lets the call go to voice mail. "It's just my mom," she says shrugging.

An apt metaphor for analyzing technologies and techniques for manipulating participation in linguistic discourse is that of the volume control on a radio or television. Users can "turn up the volume" as with BlackBerries that are constantly checked. Alternatively, users might "turn down the volume," as did the student ignoring her mother's call. In talking about communication, the volume control image alludes less to physical noise level than to frequency of contact or restrictions on access to an interlocutor.

The notion of volume control is applicable to any sort of linguistic interchange, be it face-to-face speech, traditional writing, or language mediated through an ICT. In-place ICTs include landline telephones and stationary computers used for composing and receiving e-mail or instant messages. Among mobile technologies are personal digital assistants (PDAs) and cell phones, in both speech and texting capacities. All these devices provide ample opportunities for users to control discourse in a variety of ways— increasing or decreasing contact with interlocutors, or manipulating the form that interaction takes.

Along with the volume-control metaphor, another useful concept is that of affordances, originally developed by the psychologist James Gibson (1979) and later applied to technology issues (e.g., Gaver 1991). Affordances are the physical properties of objects that enable people perceiving or using those objects to function in particular ways (Sellen and Harper 2002, p. 17). For example, an affordance of paper (as opposed to computers) is that it needs no electrical power source to record the written word. An affordance of mobile telephone systems is that they extend the physical circumstances under which communication is initiated and received.

Language users magnify the affordances of ICT when they engage in multitasking behavior. Multitasking (that is, involvement in more than one activity at a time) takes many forms—from simultaneously talking on the phone and reading e-mail to participating in multiple IM conversations. Multitasking is widespread in contemporary society, especially when using ICTs. This chapter represents an attempt to understand multitasking with ICTs as a venue for managing interpersonal communication.

We begin by looking at discourse control in traditional face-to-face and written-communication settings, and then at the affordances ICTs add for adjusting the conversational volume. We then turn to multitasking in both its cognitive and social dimensions, specifically in the context of instant messaging. Drawing upon additional data from both the IM study and separate research on mobile phone usage, the chapter closes by considering how technology-driven linguistic control stands to affect social communication and, derivatively, the social fabric.

Traditional User Control Over Communication Volume: Adjusting Communication Volume through Access, Avoidance, and Manipulation

Although speech and writing are social activities, people are not inherently "always on, always connected." Rather, individuals have long found avenues for controlling their communicative exchange, including establishing zones of privacy, even in societies that offer little physical space for seclusion (Westin 1967). What have shifted over time are the amount of control and the mechanisms for effecting it, reflecting both technological developments and concomitant opportunities for multitasking.

Language users can maneuver their communicative behaviors in three ways. The first of these is *access* to potential interlocutors, enabling them to increase the chances of a linguistic exchange taking place. The second is *avoidance mechanisms*, through which potential interlocutors avert linguistic encounters. And the third is *manipulation*, whereby a message sender or recipient alters the expected discourse relationship (e.g., violating status hierarchy or assumptions of veracity). All these maneuvers can be used for adjusting the volume on spoken and written language.

Affordances of Face-to-Face Speech for Volume Control

Speakers and listeners have historically been at the mercy of the laws of physics and the social pecking order. As for access, the human voice only projects so far, even with cupped hands or megaphone. Regarding avoidance, those in positions of authority could typically control face-to-face access, while the rest of the populace was more exposed to unregulated encounters—on the street, at the market, or in church. In response, people have devised social avoidance mechanisms—crossing the street or looking in shop windows when attempting to circumvent conversation with someone heading their way, or offering a brief greeting before dashing off due to fabricated time constraints. Conversely, individuals sometimes manipulate social conditions to become privy to the conversations of others. Eavesdropping is an age-old practice—whether consciously arranged or arising serendipitously.

Affordances of Offline Writing for Volume Control

Writing that predates modern ICTs can also be characterized with respect to affordances for volume control. Access was often limited by physical or economic circumstances: ships carrying the mail sometimes sank, roads on which mail coaches traveled were filled with brigands, and postal rates were high (Baron 2002). At the same time, both letter writers and recipients exercised some control over access. Senders, for example, could pay for mailing options such as "return receipt" or express delivery to speed transmission or to increase the chances of getting the recipient's attention. Recipients, of course, could limit access by delaying a response or ignoring the missive outright.

Written letters or memoranda enable potential interlocutors to avoid face-to-face encounters. From classic "Dear John" letters breaking off romantic relationships to impersonal termination notices delivered in the workplace, written communication has provided a social shield for people not wishing to deliver unwelcome news in person.

Finally, writing provides opportunities for deception or gossip. Rather than accurately presenting themselves, correspondents sometimes misrepresent their physical appearance or academic credentials. On the receiving end, instead of maintaining the presumed confidentiality of a letter addressed to a specific person, recipients may share documents with others for whom the writing was not intended.

Volume Control with Information and Communication Technologies

What do ICTs contribute to traditional mechanisms for managing spoken or written discourse control? In this section, we look at usage parameters for several fixed and mobile technologies: landline phones, e-mail and instant messaging (both prototypically fixed), and mobile phones. Later in the chapter, we turn to situations in which multitasking behavior involving ICTs augments users' ability to adjust the volume on social discourse.

Landline Phones

For most of the technology's existence, telephones have had ringers that could not be turned off, placing their owners at the social mercy of callers (Baron 2000). A ringing phone remained a summons that nearly always took precedence over an on-going face-to-face conversation. Those within hailing distance of telephones were "always on" and potentially "always connected." This situation began to ease only in the

1970s with the widespread availability of telephone answering machines (Morton 2000).

Over the past thirty years, technological developments have provided callers and recipients with increased opportunities for controlling their conversations (Katz 1999). As for access, voice mail enables us to leave messages, regardless of the availability of our intended interlocutors. Using the Internet, we procure direct telephone numbers, enabling us to bypass traditional call screeners such as secretaries. Through call-waiting features, we can queue up for our desired interlocutor's attention.

Telephone technologies are also decreasing access. Intended interlocutors avoid contact by using caller ID, screening calls to go to voice mail, or blocking calls from specific numbers. Call initiators can avoid conversations by using express messaging, whereby a call goes directly to voice mail without the recipient's phone ever ringing. In the business and professional worlds, a burgeoning number of telephone systems preclude our speaking with a human being, shunting us instead to telephone trees, voice recognition systems, and recorded messages.

Modern landline telephony also enables users to manipulate conversational audience. With speaker phones, for example, callers may have no choice—or even knowledge—of how many others are privy to a conversation intended only for the addressee.

E-mail

E-mail builds upon the volume-control affordances of modern landlines: we leave asynchronous e-mail messages and use the Internet to locate electronic addresses of strangers we wish to contact. Anecdotal evidence suggests that many individuals are more likely to reply to e-mail from an unknown correspondent than to an unsolicited letter or phone call. Not only is the effort in answering online relatively small, but the social distance afforded by the medium makes responding a less-personal act than a face-to-face or voice-to-voice encounter.

Again as with contemporary telephony, e-mail affords recipients opportunities to avoid or manipulate communication from message senders. E-mail systems empower users to leave incoming messages unread (or unanswered) for as long as we please or local social conventions permit. As for manipulation, e-mail can be forwarded—more potential for gossip—to unintended recipients, reminiscent of callers unknowingly being placed on speakerphones.

Instant Messaging (IM)

Unlike e-mail, instant messaging is designed for synchronous communication. Moreover, IM systems contain features that e-mail platforms lack, such as personal profiles, social affinity groups (known as Buddy Lists), away messages (informing Buddies that although your IM program is still on, you are temporarily away from your computer), and the ability to engage in multiple conversations simultaneously.

Access to interlocutors using IM can be controlled in a number of ways. Even if you know a person's IM screen name (less publicly available than an e-mail address), that individual has the option of blocking your messages. A more subtle avoidance mechanism is to read the profiles or away messages that people on your Buddy List have posted rather than directly contacting them through an IM or a phone call (Baron et al. 2005). Similarly, users wanting to avoid face-to-face or even telephone voice contact with acquaintances may choose to IM them instead (Baym, Zhang, and Lin 2004).

Instant messaging is particularly well suited to manipulation of conversational volume control. As with phone calls and e-mail, IMs can be forwarded to other recipients, unbeknownst to the original sender. Other practices (especially popular with adolescents and young adults) are unique to IM. For example, some individuals post an away message when actually sitting at their computers, enabling them to screen which incoming IMs to respond to and which to ignore.

Mobile Phones

Mobile telephony enables users to be in "perpetual contact" with one another (Katz and Aakhus 2002). However, having your mobile phone turned on need not imply you welcome being generally available. In studying mobile phone usage by American college students, my colleagues and I found that although the attribute students like most about mobile phones is being able to contact others, the aspect they like least is their continuing accessibility to people (Baron and Ling 2007; Young, Deal, and DiMarco 2005).

As speaking devices, mobile phones afford a variety of discourse control features transcending those available on landlines (e.g., caller ID, call waiting, speakerphone). Distinct ringtones can be assigned to each person in an address book, making it unnecessary to view the phone display panel before deciding whether to take a call. Moreover, camouflage services are available that, for example, provide background noise from a traffic jam, enabling a user to say with authority, "Sorry, I'll be two hours late. I'm stuck in traffic," while actually sitting at a café. These services can also generate a ringtone in the middle of a conversation, appearing to signal an incoming call and providing a plausible excuse for terminating the current exchange.

Text messages on mobile phones share affordances with e-mail (both are asynchronous) and IM (both tend to be social media). Access and avoidance issues with texting are similar to those with e-mail and IM in that users can identify the message sender before deciding how and when to respond. As with IM (but unlike e-mail), mobile phone users nearly always know the message sender, increasing the likelihood of a timely reply.

Multitasking

ICTs afford users considerable control over the ways in which they interact with other people. However, control can be augmented by combining communicative exchanges with other tasks—that is, by multitasking. Because recipients of ICT-based messages cannot see us (Webcam technology excluded), they typically are unaware when we engage in additional activities. On the phone and on the Internet, nobody knows you're multitasking.

Reasons for Multitasking

Multitasking commonly occurs in ordinary tasks that make simultaneous demands of our cognitive or physical faculties. For example, in driving a car, we must look three ways (ahead, in the rearview mirror, and peripherally) while controlling the speed and direction of the vehicle and perhaps conversing or listening to the radio. Another real-world example is playing the piano or the organ, for which we need to read multiple lines of musical notation and control two hands, along with one or two feet.

A second reason for multitasking is perceived time demands. Time-driven multitasking (e.g., house cleaning plus child care; commuting plus reading) is prevalent in every-day life (Damos 1991; Floro and Miles 2003; Ironmonger 2003; Michelson 2005; Ruuskanen 2004; sciam.com 2004). Using data from time-interval diary studies in the UK, Susan Kenyon and Glenn Lyons (2007) report that through multitasking, people "add" nearly seven hours of activity to each day.

Thirdly, multitasking may be a response to an emotional state such as loneliness or boredom. Many people turn on a radio, music player, or TV upon returning home or entering a hotel room, even though their primary activity is neither listening nor viewing. As we will see, impatience is a motivation for some college students to multitask while using ICTs.

Cognitive versus Social Multitasking

Multitasking may affect us either cognitively or socially. The classical psychological literature has explored whether multitasking degrades performance and has also attempted to explain how the mind handles multiple tasks or task shift. More recent inquiries have considered multitasking situations that involve social communication as one of the "tasks" (e.g., driving a car while conversing on a mobile phone).

We use the term *cognitive multitasking* to refer to performance of two or more mental tasks, where all tasks are primarily cognitive in nature (e.g., doing a crossword puzzle while completing a questionnaire). In contrast, we speak of *social multitasking* when the activities are primarily social-interactive (e.g., alternating between a face-to-face conversation and typing an IM). We extend the term *cognitive multitasking* to situations in which the tasks include both cognitive and social activities, but where the research interest is in cognitive performance. Similarly, we extend the term *social multitasking* to combined social and cognitive undertakings, but in which concern focuses on the social consequences of mingling these activities. The language we use in social interaction is itself grounded in cognitive activity, but for our purposes, we focus on its social dimension.

Cognitive Issues in Multitasking

Multitasking behavior has long been of interest to psychologists (Manhart 2004; Stroop 1935). Most studies have suggested that engaging in simultaneous activities (particularly involving unfamiliar or unpracticed tasks) decreases performance level. For example, watching TV while simultaneously recalling sets of digits (Armstrong and Sopory 1997) or while doing homework (Pool, Koolstra, and Van der Voort 2003) makes for poorer recall (and homework results) than when focusing on a single task. Similarly, switching between tasks (such as alternating between solving mathematics problems and classifying geometric objects) has been shown to degrade performance (Rogers and Monsell 1995; Rubenstein, Meyer, and Evans 2001).

Tests done under laboratory conditions indicate that if the tasks at issue tap different modalities (e.g., visual versus auditory), the amount of performance degradation may be less than if both tasks rely upon the same modality (Brooks 1968). However, in the real world other factors may come into play such as the amount of experience a person has in processing particular multiple stimuli—for example, students who typically study with background music versus those who do not (Daoussis and McKelvie 1986)—and the nature and difficulty of the tasks at hand.

Cognitive Multitasking while Using ICTs

Recent experiments involving ICTs confirm that multitasking commonly degrades cognitive performance. Even when individuals attempt to attend strictly to a single task, they are often distracted by ICT demands such as phone calls or e-mail messages—with dramatic results. Psychologist Glenn Wilson administered a variety of tasks, including IQ tests, to ninety subjects in the UK. When these tasks were performed in the presence of communication distracters such as a ringing telephone, average performance on the IQ test fell ten points—essentially the equivalent of missing an entire night's sleep (Hewlett-Packard 2005).

Several investigations have explored the relationship between academic accomplishments and use of the Internet. Hembrooke and Gay (2003) report degraded memory for lecture content when students simultaneously listened to classroom lectures and accessed the Internet to do searches or communicate with peers online. Crook and Barrowcliff (2001) found not only that undergraduate students with Internet access in their dormitory rooms engaged in considerable multitasking but also that the ratio of school work to recreational computer-based activity was roughly one to four—hardly an efficient way to complete assignments.

Another cluster of experiments (e.g., Adamczyk and Bailey 2004; Cutrell, Czerwinski, and Horvitz 2001; Dabbish and Kraut 2004) has explored the cognitive effects of interrupting a person's work flow, for example, sending an IM to someone engaged in an online search task. Research suggests that the timing and form of such interruptions are critical in determining how disruptive the incoming message is.

Is multitasking with ICTs necessarily detrimental to cognitive performance? The answer may reflect the extent to which users perceive themselves to be doing multitasking. One of my students, Tim Clem, argues that it makes little sense to talk about multitasking on a computer. In his eyes, computers are naturally multitasking devices. (By analogy, recall how driving an automobile or playing the organ puts simultaneous demands on our cognitive and physical faculties.) Having grown up with the technology, Clem does not notice a degradation of performance by engaging in simultaneous computer-based activities (e.g., surfing the Web and writing a paper or carrying on an IM conversation). Drawing upon Roger Silverstein and Leslie Haddon's notion of domestication (1996), computers are domesticated technologies for much of Clem's generation—though whether computer multitasking actually degrades performance within this age cohort remains an empirical question.

Putting aside teenagers and young adults, what about the rest of us? In learning to drive a car, the ability to look three places at once develops with experience. It is, then, possible that learning to multitask on computers without cognitive degradation is a matter of training and experience. A growing literature documents how practicing complex skills, such as taxi drivers do in navigating the streets of London (Maguire et al. 2000) or novices do by learning to juggle balls (Draganski et al. 2004) leads to changes in adult brains. Plausibly, if we practice multitasking with ICTs, our brains will adapt, and performance degradation will diminish.

Social Issues in Multitasking

But what about social multitasking? Does it degrade social performance? Consider a person talking on the telephone while doing a Web search or engaging in an IM conversation. Does the quality of the IM conversation or the spoken exchange suffer? Unlike the case of cognitive multitasking, there is little research on the interpersonal effects of multitasking while communicating with others. What is clear, though, is that social multitasking involves volume control over the communication, for example, saying just "uh huh" to an interlocutor so you can focus on making an online purchase or deciding which of three IM messages to respond to first.

Multitasking and Volume Control while Using Instant Messaging

To objectively gauge the impact of social multitasking on human interaction, we need empirical information on contemporary multitasking behavior entailing ICTs such as e-mail, instant messaging, and mobile phones. A few studies (e.g., Baym, Zhang, and Lin 2004; Lenhart, Madden, and Hitlin 2005; Shiu and Lenhart 2004) have asked participants to note their multitasking behavior while using these technologies. However, self-reporting of behavior is notoriously problematic. To help address this methodological challenge as well as to gather data specifically involving instant messaging, my stu-

Table 14.1
Multitasking of U.S. College Students

Activities	Percent
Computer-based activities	
Web-based activities	70.3%
Computer-based media player	47.5%
Word processing	38.6%
Offline activities	
Face-to-face conversation	41.1%
Eating or drinking	36.7%
Watching TV	28.5%
Talking on the telephone	21.5%

N=158

dents and I undertook research on undergraduate multitasking while communicating via an ICT.

Multitasking Patterns of American College Students

Using online questionnaires, we explored the multitasking behavior of undergraduates who were engaged in IM conversations. The data were collected in fall 2004 and spring 2005 at American University, in Washington, DC. We knew that all subjects were participating in at least one IM conversation at the time they completed the questionnaire, since IM was the medium through which student experimenters distributed the URL for the questionnaire Web site to subjects, who were members of their Buddy Lists. Therefore, the other online and offline activities in which the subjects were engaged entailed cognitive or social multitasking. Since most subjects were in their dormitory rooms at the time they participated in the experiment, they had ample opportunities for involvement in multiple activities.

The results revealed a high level of multitasking. In our first study, summarized in table 14.1, out of 158 subjects (half male, half female), 98 percent were engaged in at least one other computer-based or offline behavior while IM-ing.

Subjects often participated in multiple examples of the same activity (e.g., having three Web applications open, being involved in more than one IM conversation). Students in this study averaged 2.67 simultaneous IM conversations, with a range from 1 to 12 conversations.

Logic dictates that users cannot literally engage in multiple IM conversations simultaneously. And indeed they do not. Subsequent focus groups revealed that many college students use IM both synchronously and asynchronously, that is, turning the volume up or down on particular conversations. Decisions depend upon such factors as how good the "gossip" is in a conversation, how serious the conversation is, and individual communication habits. A few students found it rude to hold simultaneous IM conversations, though they were by far in the minority.

Why Multitask while Doing IM?

We used both informal focus groups and a revised online questionnaire (this time with fifty-one subjects) to probe why students multitask while using a computer. Most respondents spoke of time pressures: multitasking enabled them to accomplish several activities simultaneously. Time pressures were also specifically invoked to justify multiple concurrent IM conversations. Interestingly, several students commented that IM is not, by nature, a stand-alone activity. When asked whether they ever held a single IM conversation during which they did not engage in any other online or offline activity, the overwhelming response was "no." Such behavior, said one participant, would be "too weird," because IM conversations are (she said) conducted as background activity to other endeavors. Several students from the second online study noted they multitasked on computers because the technology enabled them to do so. As one student put it, "There is no reason not to when everything is accessible at once."

Ten of the fifty-one students in the second online study indicated they multitasked while using computers because they were bored. Boredom sometimes resulted from having to wait for the person with whom they were IM-ing to respond. Other students spoke of "get[ing] bored with just one activity" or "having too short an attention span to only do one thing at a time."

The ability to control both their individual activities and their social networks figures as a significant motivation for multitasking while doing IM. A student from the second study said he multitasks on the computer "because i can." Focus-group members observed that with IM, participants are in control of how dynamic a given IM conversation is. With lengthy IM dialogues, interlocutors may go through spurts of communication interlaced with periods of inactivity. One student aptly described IM as "language under the radar," meaning it resides in the background of other online or offline endeavors. Users control whether to make a particular conversation active (i.e., synchronous) or let it lie dormant (i.e., asynchronous) without formally closing the interchange.

Why Social Multitasking and Volume Control Matter

We have argued that people find ways to "control the volume" in spoken and written discourse. ICTs augment the options for control because of their technological affordances, which include opportunities for multitasking. Multitasking often depresses cognitive performance, and may compromise social interaction by reducing interlocutors' level of interpersonal engagement. To expand our understanding of multitasking while using ICTs, we studied American college students' multitasking while doing instant messaging on computers.

Are there social consequences of contemporary ICT usage patterns? While technologies such as computers and mobile phones may be too new for us to answer this question definitively, two studies hint at trends that bear watching.

Acceptable and Unacceptable Social Multitasking: IM Multitasking Study
As part of the second administration of the IM multitasking study, we asked a series of free-response questions regarding multitasking behaviors the students felt were or were not suitable. A typical response to the question "For which computer-based activities is multitasking appropriate? Why?" was "IMing, listening to music, browsing the web. Those are all things that do not interfere with one another."

A content analysis revealed that 86 percent of the fifty students responding to this question specifically mentioned IM or e-mail—both forms of interpersonal communication—or indicated that any type of multitasking behavior is acceptable.

In an earlier portion of the questionnaire, 43 percent of the participants claimed they often engaged in only one IM conversation at a time, suggesting that even if the students were multitasking, perhaps they were not doing social multitasking. However, in their real-time responses to questions about specific online and offline multitasking activities, the same subject cohort averaged 2.5 simultaneous IM conversations, leading us to question whether the 43 percent statistic reflects perceptions of appropriate social etiquette rather than actual practice.

Another free-response question asked "For which non-computer activities is multitasking not appropriate? Why?" Of the forty-four students responding, 59.1 percent singled out face-to face or telephone conversations as inappropriate contexts for multitasking. This number stands in stark contrast to the 86 percent who felt that carrying on an IM or e-mail conversation while using the computer for other functions was appropriate.

Students offered various explanations for avoiding multitasking while speaking face-to-face or by phone. The most prevalent answer was that such behavior was simply wrong. One male student said, "It's rude not to give your full attention to someone face to face," while a female observed that "talking on the phone and talking to people on the computer [i.e., IM] isn't appropriate because the person on the other phone line usually feels left out or unattended to."

Similar feelings of personal abandonment were reported in a study conducted by Sprint (2004). More than 50 percent of the respondents said they felt unimportant when a friend or colleague interrupted a face-to-face conversation with them to answer a mobile phone. Hewlett-Packard (2005) reported that 89 percent of office workers felt that colleagues who responded to e-mails or text messages during a face-to-face meeting were being rude. However, 30 percent of the same respondents indicated that such behavior was both acceptable and an efficient use of time. Apparently, at least 19 percent of those surveyed (i.e., the difference between the 11 percent who were not bothered by the behavior in others and the 30 percent who justified the behavior—perhaps in themselves) had yet to resolve the conflicting demands of social etiquette and work pressure.

Some respondents in our IM study said multitasking was precluded only if the topic of a face-to-face or telephone conversation was important. Of the twenty-six students who were against multitasking while face-to-face or on the phone, six disapproved of such behavior only if the conversation was particularly serious or important. Other explanations for avoiding multitasking while face-to-face or on the phone were strictly pragmatic. For example, you might be found out. As one male remarked, "people [on the phone with you] get pissy about hearing a keyboard clicking." Another said, "You should devote attention to someone who can see what you are doing." This second response came from a student who believed that reading while talking on the phone, or cooking while talking on the phone, was an appropriate type of multitasking because "If they don't know, it won't hurt them."

Four students (all female) eschewed multitasking while talking face-to-face or on the phone because they were not good at it. Another complained that she was disturbed when other people with whom she was speaking were multitasking:

Talking on the phone—I [don't] want to listen to someone else's TV while I'm having a conversation with them. Nor do I want to hear their music. It is distracting.

To what degree do college undergraduates actually multitask with ICTs while engaging in face-to-face or telephone conversations? Baym, Zhang, and Lin (2004) found that 73.9 percent of their nearly 500 subjects reported multitasking on an ICT while in face-to-face conversation. Of the 158 students in our initial multitasking study, 41.1 percent were engaged in at least one computer activity while talking face-to-face, and 21.5 percent were simultaneously on the computer and on the phone. Clearly, many American college students control the volume on their face-to-face and telephone conversations by multitasking on computers.

Attitudes and Practices Regarding Volume Control: Mobile Phone Study

We turn now from social multitasking and volume control on computers to social issues surrounding mobile phone use. The results reported here are from a questionnaire administered to sixty-eight American University undergraduates (half male, half female) in fall 2005 (Baron and Ling 2007).

ICT researchers (e.g., Katz 2003; Ling 2004) have been analyzing the use of mobile phones in public spaces in many parts of the world. Our mobile phone study tried to get a statistical fix on practices and attitudes in the United States. The majority of females in the study responded "yes" to the question "Does it ever bother you when other people talk on their cell phones in public places?" Fewer males reported feeling bothered. When asked to identify their complaints, both males and females indicated they were most troubled by volume level. The second biggest grievance for females was use of phones in inappropriate locations (e.g., houses of worship, public restrooms), while second on the male list of objections was hearing about other people's private business.

We also inquired whether students felt other people spoke more loudly on cell phones than in face-to-face conversation, and whether they themselves did so. Nearly 70 percent of both males and females perceived other people to be louder on cell phones than in face-to-face discourse, implying an ICT-based imposition of control over the auditory space of innocent passersby. However, while 79.4 percent of males acknowledged speaking more loudly on mobiles than face-to-face, only 35.3 percent of females reported doing so.

Another control issue we explored was whether students ever pretended to be talking on their phones when actually they were not. Out of sixty-eight subjects, 35.3 percent replied affirmatively (with equal numbers of males and females). Of the twenty-four students who engaged in this behavior, 83.3 percent (i.e., twenty students) did so "to avoid talking with someone I see." This unexpected finding suggests ICTs (here, mobile phones) may be replacing the more low-tech move across the street to circumvent conversation.

From the Amish to iPod Nation

Interpersonal communication is progressively reshaped by technological contrivances that enable us to adjust the volume on spoken and written discourse. From the earliest landline phones to today's mobiles; from the first phonograph through earphones, the Sony Walkman, and now MP3 players; from the ENIAC to the modern wireless laptop, we have created both physical devices and social practices by which to orchestrate when and how we say (or write) what to whom.

Alongside the technological developments, some voices have questioned the effects technology might have upon the social fabric. A few years ago, Howard Rheingold began thinking about the tools that allow him to be "Always on, always connected" and asked "What kind of person am I becoming as a result of all this stuff?" (1999). His search for an answer led him to Lancaster, Pennsylvania, for a series of conversations with the Amish.

For more than a century, the Amish have struggled with the question of whether their members should be allowed to have telephones (Umble 1996). The issue of adopting new technologies is not as simple as outsiders may think. Today's Amish use disposable diapers, gas barbecue grills, and even some diesel-powered machinery. Each new contrivance must be evaluated by the Amish bishops, with one fundamental query in mind: "Does it bring us together, or draw us apart?" Diesel machinery is not allowed in working the fields, since use of the technology might jeopardize the social connection of families laboring cooperatively. Just so, having a telephone in the house is forbidden. In the words of one Amish man whom Rheingold interviewed,

What would that lead to? We don't want to be the kind of people who will interrupt a conversation at home to answer a telephone. It's not just how you use the technology that concerns us. We're also concerned about what kind of person you become when you use it.

Fast forward to the end of 2005. Writing in the *New York Times*, David Carr began a story about his new video iPod with more than a hint of guilt:

Last Tuesday night, I took my place in the bus queue for the commute home. Further up the line, I saw a neighbor—a smart, funny woman I would normally love to share the dismal ride with. I ducked instead, racing to the back of the bus because season one of the ABC mystery-adventure "Lost" was waiting on my iPod. (Carr 2005)

Like my student at the opening of this chapter who "turned down the volume" on her mother's cell phone call, Carr "turned down the volume" on his neighbor—at least this time, in favor of watching reruns on a 2.5-inch screen.

Admittedly, few of us would go as far as the Amish in banishing ICTs from our social space, and Carr has, I suspect, started chatting again on the bus. Kevin Kelly, former executive editor of *Wired* magazine, claims that many ICT-savvy people are making conscious choices—like the Amish—about which technologies to employ, which to eschew, and when. Referring to such individuals as the neo-Amish, Kelly suggests that a growing number of would-be power users are laying down their own individual and social ground rules, such as no personal e-mail at work, turn off the BlackBerry when you get home, or only give your mobile phone number to your spouse (Vargas 2006).

Realistically, though, it takes considerable self-discipline to be neo-Amish. Given the lure of advertising, the perception of always being pressed for time, and the feeling of empowerment that comes with conversational control, there is little incentive to put the metaphoric brakes on our use of ICTs.

As communication technologies become increasingly integrated into all aspects of life, two diametrical outcomes are plausible. The first is that as a society, we will determine to put the ICT genie back in the bottle whenever it seriously threatens communal propriety. Much as public spitting—and spittoons—are now banished from polite society, we can visualize some influential neo-Amish redirecting general ICT usage conventions. Admittedly, though, outside of giving up the gun in Tokugawa Japan (Perrin 1979), historical precedence for suppressing technology is scant. A more likely scenario is that we will find ourselves redefining acceptable patterns of interpersonal communication, with progressive amnesia about the way things used to be.

If practice makes us more adept at multitasking, then multitasking with ICTs may soon become no more challenging than playing the piano with both hands, and no more worthy of remark. But unlike the piano, language is at base a tool for social interaction. As of yet, many people still feel discomfited by multitasking (their own or that of their interlocutor) when speaking face-to-face or on the phone. Take away visual and voice contact (as with e-mail, IM, or text messaging), and our standards for demanding (and giving) exclusive attention promptly drop off.

Will IM or texting edge out physical (or voice-phone) conversations in the name of the increased efficiency and control that multitasking on ICTs affords? Given the success of mobile phones and Internet-based calling, telephony is not going away any time soon. However, ICTs do potentially degrade live face time. We need to ask ourselves what is unique about two people meeting and talking face-to-face, and how important it is to preserve uninterrupted live contact. Not easy questions, but their answers are part of what makes us human.

References

Adamczyk, P., and B. Bailey. 2004. If not now, when?: The effects of interruption at different moments within task execution. Proceedings of the SIGCHI Conference on Human Factors in Computing Systems (CHI '04). New York: ACM Press.

Armstrong, G. B., and P. Sopory. 1997. Effects of background television on phonological and visuo-spatial working memory. *Communication Research* 24: 459–480.

Baron, N. S. 2000. *Alphabet to Email: How Written English Evolved and Where It's Heading.* London: Routledge.

Baron, N. S. 2002. Who sets email style: Prescriptivism, coping strategies, and democratizing communication access. *The Information Society* 18: 403–413.

Baron, N. S., and R. Ling. 2007. Emerging patterns of American mobile phone use: Electronically mediated communication in transition. In *Mobile Media 2007*, edited by G. Goggin and L. Hjroth. Sydney: University of Sydney.

Baron, N. S., L. Squires, S. Tench, and M. Thompson. 2005. Tethered or mobile: Use of away messages in instant messaging by American college students. In *Mobile Communications: Re-negotiation of the Social Sphere*, edited by R. Ling and P. Pederson. London: Springer-Verlag.

Baym, N. K., Y. B. Zhang, and M.-C. Lin. 2004. Social interactions across media: Interpersonal communication on the Internet, face-to-face, and the telephone. *New Media & Society* 6: 299–318.

Brooks, L. R. 1968. Spatial and verbal components of the act of recall. *Canadian Journal of Psychology* 22: 349–367.

Carr, D. 2005. Taken to a new place, by a TV in the palm. *New York Times*, December 18, Section 4 (Week in Review), p. 3.

Crook, C., and D. Barrowcliff. 2001. Ubiquitous computing on campus: Patterns of engagement by university students. *International Journal of Human-Computer Interaction* 13: 245–256.

Cutrell, E., M. Czerwinski, and E. Horvitz. 2001. Notification, disruption, and memory: Effects of messaging interruptions on memory and performance. In *Human-Computer Interaction (INTERACT '01)*, edited by M. Hirose. Tokyo: IOS Press (for IFIP).

Dabbish, L., and R. Kraut. 2004. Controlling interruptions: Awareness displays and social motivation for coordination. Proceedings of the 2004 ACM Conference on Computer Supported Cooperative Work (CSCW '04). New York: ACM Press.

Damos, D. L., ed. 1991. *Multiple-task Performance*. London: Taylor & Francis.

Daoussis, L., and S. J. McKelvie. 1986. Musical preferences and effects of music on a reading comprehension test for extraverts and introverts. *Perceptual and Motor Skills* 62: 283–289.

Draganski, B., C. Gaser, V. Busch, G. Schuierer, U. Bogdahn, and A. May. 2004. Changes in grey matter induced by training. *Nature* 427 (January 22): 311–312.

Floro, M. S., and M. Miles. 2003. Time use, work and overlapping activities: Evidence from Australia. *Cambridge Journal of Economics* 27: 881–904.

Gaver, W. 1991. Technology affordances. Proceedings of the SIGCHI Conference on Human Factors in Computer Systems (CHI '91). New York: ACM Press.

Gibson, J. J. 1979. *The Ecological Approach to Visual Perception*. New York: Houghton Mifflin.

Hembrooke, H., and G. Gay. 2003. The laptop and the lecture: The effects of multitasking in learning environments. *Journal of Computing in Higher Education* 15: 46–64.

Hewlett-Packard. 2005. Abuse of technology can reduce UK workers' intelligence. *Small & Medium Business press release*.

Ironmonger, D. 2003. There are only 24 hours in a day! Solving the problematic of simultaneous time. Paper presented at the 25th IATUR Conference on Time Use Research, Brussels, Belgium.

Katz, J. E. 1999. *Connections: Social and Cultural Studies of the Telephone in American Life*. New Brunswick, N.J.: Transaction Publishers.

Katz, J. E. 2003. A nation of ghosts? Choreography of mobile communication in public spaces. In *Mobile Democracy: Essays on Society, Self, and Politics*, edited by K. Nyíri. Vienna: Passagen.

Katz, J. E., and M. Aakhus. 2002. *Perpetual Contact: Mobile Communication, Private Talk, Public Performance*. Cambridge: Cambridge University Press.

Kenyon, S., and G. Lyons. 2007. Introducing multitasking to the study of travel and ICT. *Transportation Research Part A* 41(2): 161–175.

Lenhart, A., M. Madden, and P. Hitlin. 2005. Teens and technology. *Pew Internet & American Life Project*. http://www.pewinternet.org/pdfs/PIP_Teens_Tech_July2005web.pdf.

Ling, R. 2004. *The Mobile Connection: The Cell Phone's Impact on Society*. San Francisco: Morgan Kaufmann.

Maguire, E. A., D. G. Gadian, I. S. Johnsrude, C. D. Good, J. Ashburner, R. S. J. Frackowiak, and C. D. Frith. 2000. Navigation-related structure change in the hippocampi of taxi drivers. *Proceedings of the National Academy of Sciences* 97(8): 4398–4403.

Manhart, K. 2004. The limits of multitasking. *Scientific American Mind* 1: 62–67.

Michelson, W. 2005. *Time Use: Expanding Explanation in the Social Sciences*. Boulder: Paradigm Publishers.

Morton, D. 2000. *Off the Record: The Technology and Culture of Sound Recording in America*. New Brunswick, N.J.: Rutgers University Press.

Perrin, N. 1979. *Giving Up the Gun: Japan's Reversion to the Sword, 1543–1879*. Boston: David R. Godine.

Pool, M. M., C. M. Koolstra, and T. H. A. Van der Voort. 2003. Distraction effects of background soap operas on homework performance. *Educational Psychology* 23: 361–380.

Rheingold, H. 1999. Look who's talking. *Wired* 7.01 (January).

Rogers, R., and S. Monsell. 1995. Costs of a predictable switch between simple cognitive tasks. *Journal of Experimental Psychology—General* 124: 207–231.

Rubenstein, J. S., D. E. Meyer, and J. E. Evans. 2001. Executive control of cognitive processes in task switching. *Journal of Experimental Psychology—Human Perception and Performance* 27: 763–797.

Ruuskanen, O. P. 2004. Essay 4. More than two hands: Is multitasking an answer to stress? In An economic analysis of time use in Finnish households. Doctoral dissertation, Helsinki School of Economics, 2004, *HeSE print 2004*, 188–229.

sciam.com. 2004. Scientific American Mind poll: 90% of American adults are multitaskers. http://pr.sciam.com/release.cfm?site=sciammind&date=2004-12-20.

Sellen, A., and R. Harper. 2002. *The Myth of the Paperless Office*. Cambridge, Mass.: The MIT Press.

Shiu, E., and A. Lenhart. 2004. How Americans use instant messaging. *Pew Internet & American Life Project*. http://www.pewinternet.org/PPF/r/133/report_display.asp.

Silverstein, R., and L. Haddon. 1996. Design and domestication of information and communication technologies. In *Communication by Design*, edited by R. Silverstein and R. Mansell. Oxford: Oxford University Press.

Sprint. 2004. Sprint survey finds nearly two-thirds of Americans are uncomfortable overhearing wireless conversations in public. http://www2.sprint.com/mr/news_dtl.do?id=2073.

Stroop, J. R. 1935. Studies of interference in serial verbal reactions. *Journal of Experimental Psychology* 18: 643–662.

Umble, D. Z. 1996. *Holding the Line: The Telephone in Old Order Mennonite and Amish Life*. Baltimore: Johns Hopkins.

Vargas, J. A. 2006. One answer to too much tech: Sorry, I'm not here. *Washington Post*, January 16, Section 2, p. 1.

Westin, A. F. 1967. *Privacy and Freedom*. New York: Atheneum.

Young, K., L. Deal, and G. DiMarco. 2005. R U Thr? Text messaging and cell phone use by American college students. Unpublished paper, American University, Washington, DC.

15 Maintaining Co-presence: Tourists and Mobile Communication in New Zealand

Peter B. White and Naomi Rosh White

The tourism and telephone industries are following similar trajectories. Technological innovation, mass production, marketing, and mass consumption have resulted in tourism and mobile and fixed telephone services becoming more readily available and accessible to a wider range of consumers than in the past. Tourism has become democratized by low-cost transportation to what were once remote parts of the globe. The Internet has made it possible for individual tourists to research and identify destinations and modes of travel and to make arrangements for travel of all kinds. Parallel patterns have emerged with communications. The spread of mobile telephone services and their interoperability with other mobile and fixed telephone networks means that tourists can contact and be reached by their friends, family, and colleagues in an increasingly diverse range of tourist destinations. Mobile phone–based texting services also provide a low-cost way for tourists to keep in touch while on the move.

These transformations in the communications and tourism industries have led to a growing commercial interest in the provision of communications services for tourists. Most major airports and international train terminals have vendors offering short-term mobile phone packages designed specifically for tourists, as well as prepaid cards for access to the fixed telephone network. The geographical location of participants in these voice and text communications is no longer important. The availability of these services means that tourists are able to stay connected while being physically separated, often by large distances.

Changes in mobility and tourists' use of communications services are the starting point for the present chapter, which examines tourists' communications using mobile and fixed telephones. In particular, notions of co-presence, location, and distance-independent communication are considered through tourists' accounts of their experience of telephone use during their travels. This exploration of how the possibility of "perpetual contact" (Katz and Aakhus 2002) using mobile telephones has been incorporated into the daily lives of tourists. This chapter can be seen as a contribution to an understanding of the "domestication" of mobile telephony (Haddon 2003).

Framing the Research

Postmodernity, with its transience and fragmentation, is metaphorically configured in frequent—and for many, taken for granted—mobility. The issues raised by the links between travel, the new communications technologies, and a diminished spatial-time divide have been explored by John Urry (2002). According to Urry, mobile electronic devices make it possible for people "to leave traces of their selves in informational space" (Urry 2002, p. 266). Using these informational traces, mobile telephone technologies "track" the movements of tourists, enabling them to send and receive telephone calls and text messages synchronously. One impact of the irrelevance of geographical location is the separation of "body" and "information," with people becoming nodes in multiple networks of communication and mobility (Urry 2003). This view of people as "communication nodes" suggests new ways of thinking about relationships in a globalized world of mediated interpersonal communication.

A shift in the meaning of social connection is another consequence of readily available communication independent of location. That is, the domain of social connections is both expanded and intensified (Gergen 2000). Communication technology enables individuals to remain part of their social networks while being physically absent and distant from them (Johnsen 2003; Makimoto and Manners 1997). These technologies "dramatically expand and intensify the domain of [tourists'] social connection" (Gergen 2000, p. 136) and expand the "symbolic world that may be little related to the immediate practical surroundings of either speaker" (Gergen 2002, p. 239) by providing immediate access to individuals wherever they are located. In other words, social life can now be networked (Urry 2003). The result is a capacity for what has been called a "nomadic intimacy" in a social environment characterized by extended spaces of everyday life and individual freedom to move around in these spaces (Fortunati 2002).

The notion of "nomadic intimacy" has particular relevance to tourists. They use mobile communications services and a range of other communication strategies to maintain a "symbolic proximity" with family, friends, and colleagues (Wurtzel and Turner 1977) and to promote a sense of presence while absent or co-presence. Central to the original notion of co-presence was that it was contingent on those involved in a given communication both being and feeling close enough to perceive each other in the course of their activities (Goffman 1963). That is, the notion of co-presence initially referred to *physical presence* in face-to-face contact and interactions. However, increasing use of mobile phones by tourists has meant that this sense of connection can be affirmed at a distance. Tourists can be physically absent while socially present (Short, Williams, and Christie 1976; Lury 1997; Urry 2002).

Whence Our Data?

People were included in the study if they used mobile and landline phones to contact their family, friends, or colleagues while traveling in New Zealand. The participants were approached in holiday parks. They were traveling in campervans or caravans, by car or public transport. Twenty-seven people were interviewed. Of those, fourteen were men and thirteen women. Nineteen were traveling with partners or spouses, three were traveling with spouses and children, four were traveling with friends, and one was traveling alone. They ranged in age from nineteen to sixty-seven years, and origi- nated from countries other than New Zealand, including Australia, Germany, Holland, the United States, Canada, and the United Kingdom. The duration of participants' trav- els varied from several months to several years, with only two participants traveling for three weeks. Therefore, the majority of participants were traveling for substantial periods.

Participants were interviewed using a semistructured interview schedule. The inter- views were transcribed and then thematically coded with respect to regularly articu- lated points of view. Where points of view were distinctive, they were noted during the coding process as contrasting instances. While the relatively small sample size lim- its generalizability, the issues raised by the respondents provide insights into the signif- icance of phone use for tourists.

What We Learned

The interviews showed that access to communication services was a significant issue when planning travel. However, once tourists' journeys and communication with dis- tant friends and families had commenced, tourists made clear distinctions between the uses, benefits, and drawbacks of phones and texting. Both phones and texting had sim- ilar impacts on tourists' sense of an ongoing integration into relationships from which they were temporarily physically distant. However, the two modes of communication differed with respect to what they were seen to offer. That is, phone communication was characterized by its "emotionality," while texting in particular was seen to offer distinctive opportunities for spontaneous and sometimes playful contact.

The Telephone
The social and emotional function of communication was highlighted in tourists' accounts of their use of the telephone. When compared with e-mail and mobile phone texting, phone conversations were seen by tourists as the communication mode that allowed emotion to be communicated. Conversations enabled tourists to obtain emo- tional support from those at home and to monitor the emotions of the people with

whom they were speaking. They could satisfy themselves that they were accurately informed about how people at home were feeling. The emotional function of the telephone had another positively valued emotional dimension arising from its role in enhancing a sense of security. As shown in previous research, mobile phones particularly were seen to ensure contactability and therefore security in times of emergency (Rakow and Navarro 1993). These positively viewed functions were undercut by the unsettling aspects of the emotional content of communication via telephone. That is, these interactions sometimes unleashed in tourists unexpected emotions of homesickness or feelings related to ambivalence about relationships with parents, children, or work colleagues.

The interviewees weighed the perceived costs and utility of keeping touch in various ways. They were very clear about why they chose different ways of keeping in touch. Mobile phones were seen as expensive for voice calls, and mobile-based voice calls were often reserved for emergencies. Tourists with mobile phones almost always left them switched off unless they were making a call or sending or collecting text messages in order to avoid international roaming charges. In contrast, text messaging was seen as a cheap and effective way of keeping in touch, although North American tourists reported that few people back home knew how to use this function. The various communication modes were understood to affect the sense of physical distance between communicators, and to have quite different implications for the experience and control of emotions and the feeling of proximity or distance.

Cost was another factor affecting attitudes to the telephone, and contributed to a clear differentiation between mobile and landline phones. As noted, regular use of mobile phones for voice conversation was restricted because of cost. When extended conversations on mobile phones occurred they were initiated by those "at home," particularly if these callers had a low-cost calling plan, if they were not paying for the call, and if the tourist was not liable for international roaming charges. The perceived cost of telephone calls had an impact on the way that calls were viewed and the tourists' feelings after the calls had ended.

When you speak on the telephone you are always in a hurry because it is costing either you or your friend a lot of money. You can't speak for long so everything is rushed. So when I speak with my mother she says "I forget everything I want to ask you." . . . The telephone is too exciting and you can't remember everything you would like to say. (Male, 39, traveling with wife and children)

Largely because of the perceived cost of voice calls, mobile phones were primarily used for texting, a communication mode that enabled tourists to feel they were readily contactable, and that was seen to have a lesser emotional content and impact than voice communication. Texting was seen as inexpensive and was highly valued. So to maintain voice contact, extended voice conversations initiated by tourists were conducted from public telephones using prepaid (and relatively inexpensive) phone cards.

The tourists reported that the clarity of fixed-line telephone conversations was a source of wonder and pleasure. One tourist noted:

We phoned home from Tonga last Christmas, the line was so clear and with the time difference we were just finishing our Xmas day and they were just starting it. I felt very close. (Female, 42, traveling with partner)

Hearing the voice of family and friends was an intimate experience that altered perceptions of distance. It was possible to "laugh with someone" who was thousands of kilometers away. One tourist reported that he had made a telephone call the previous day because his wife wanted to hear their daughter's voice. Another reported:

Here we are on the other side of the world. We hear their voices very clearly. It feels that they are very close to you. (Female, 25, traveling with husband)

International calls of high quality led may participants to temporarily forget that they were separated by great distances and the cost factor. As conversations progressed, these international calls came to have the status of local calls, with some people acting as if their fellow conversationalist was "just around the corner."

...because the phone communications are so clear they talk about really mundane things and I have to cut them off all the time. I keep saying "it's getting a little expensive to hear about how you tied your shoes today." I know they forget because they aren't thinking. (Female, 34, traveling with partner)

The manifest content of communication was generally about the minutiae of daily life, such as the weather or pets. While the topics of these conversations might be seen as inconsequential, the conversations allowed the participants to deal with the fact of being physically separated.

While for some the sense of separation decreased, telephone conversations could also draw attention to the fact of physical distance. One tourist suggested that these kinds of communications

...made me feel the distance when they tell me about the negative-forty-degree weather. (Female, 21, traveling alone)

When tourists were homesick, phone calls sometimes reminded them how far they were from family and friends. Sometimes the feelings of closeness and distance were experienced concurrently. Being in touch with family and friends could mean that

On the one hand, [with the phone call] you felt close to them, but if anything it made you more aware of actually how far away you were. (Male, 40, traveling with partner)

The emotional aspect of phone communication was a key aspect of telephone use. The major strength of telephone conversations was that they were seen as a means by which emotion was authentically monitored and conveyed. One tourist contrasted the emotional significance of telephone conversations with those of e-mail.

[Speaking on the phone] brings me back to Europe and I don't have that feeling when I send an e-mail. Because of the direct contact, hearing the voice. It's much more personal. E-mail I think is impersonal.... (Male, 56, traveling with wife)

Another tourist observed:

With the phone you've got emotion. When...you hear their voice you know whether they are happy or not. With e-mail they can say that they are happy, but maybe they aren't happy and they don't want to worry us. When you speak to people you can tell if they are really happy. You can hear how they are feeling. (Female, 25, traveling with husband)

One woman commented:

With a telephone it is more emotional and you can get an immediate response. Excitement, Adrenaline. It's the voice and the stories. For a few minutes you feel that you are together with your friend even though you know that you are a long way apart. (Female, 39, traveling with husband and children)

While valued by some, the emotional content of telephone calls was perceived to be a problem by other interviewees. For example one reported that after a phone call

...[for a] few minutes you are in shock. My stomach is churning and I am thinking about what I heard, what they told me and I am trying to work out what I feel about the things they told me. (Female, 39, traveling with husband and children)

On the one hand, telephone conversations could be a source of emotional support and contact. These conversations, because of their immediacy, helped to reduce the sense of separation and distance between tourists, their friends, and families. On the other hand, interviewees were aware of the more difficult emotional impacts of telephone contact. The very immediacy of telephone conversations could also precipitate strong feelings that were less easily handled.

Text Messaging

The low cost and easy accessibility of text messages meant that text messages could be sent spontaneously and frequently. When time differences were not an issue, if both parties were awake at the same time, the texting interchanges often became multimessage conversations. The possibility of an instant response and the ability to create a real-time, low-cost, text-based "conversation" was often contrasted with e-mail, where there was often a significant time lag between communications. As one interviewee stated:

With text messaging it's like having a short conversation. But with e-mail there's a big delay. (Female, 21, traveling alone)

Text messaging allowed for spontaneity, the ability to "play" with distant friends and to affirm relationships. Text messages alerted family and friends to random details of travels undertaken and confirmed the tourists' ongoing place in the family and friendship networks. The conversational possibilities of texting encouraged playfulness. One couple reported that when they were driving from one site to another they would text their daughter to report on how many sheep they were seeing. Others spoke about how they would spend the day driving around "texting about nothing" (female, 34, traveling with partner), or reported using texting for, in their words, "nonsensical" conversations about the weather.

As with many telephone calls, the content of these communications was not the key issue. The immediacy and fact of sending a text message was confirmation that the correspondent was "thinking about you" at that instant. This is the kind of activity described by Dag Album as "contentless meaningful chat" characterized by a high degree of expressive-symbolic content (cited in Johnsen 2003, p. 165). Text messaging is also used to maintain intense relationships and simulate co-presence. One young man reported that he and his girlfriend, who was in Europe, would text each other many times each day.

She sends me about five [text messages] each day and I send her maybe two back. When she goes to bed she sends me text message: "Goodnight, I hope you have a nice day"—because it's morning for me. It was Valentine's Day yesterday so we used text messages. (Male, 22, traveling alone)

He thought that these text communications made it possible for his girlfriend to be reassured that he was thinking about her.

The immediacy of texting encouraged one tourist to send a text message to a predesignated individual each day.

I send a sort of postcard to one person so that they can then relay it around to interested parties. I'm sending a little report every day of what I have been doing and I'm also sending text to individual people along the lines of "How are you doing?" so they are getting a report and a personal message as well. (Female, 58, traveling with partner)

Another tourist used text messaging as a way of continuing the pattern of mutual support between him and a special friend. It was the equivalent of their daily pretravel telephone calls.

I have friend back home and we are both on a diet together. So I send him text messages like "Drink water," "Don't eat too much." I send him a few words so that I can support him. (Male, 39, traveling with wife and children)

The tourists' accounts of how texting was used suggested that the symbolic function of these communications was a form of "gifting"; that is, communications where the act of getting in touch matters as much as, if not more than, the content of the communication (COST269 Mobility Workgroup 2002, p. 26).

The almost perpetual availability made possible by telephone communication raised new issues that had to be dealt with by the tourists. They had to deal with the ways in which communications could be seen as an indicator of the health of their relationships with people back home. They also had to deal with ways to manage their availability so that they could distinguish the experience of their time spent traveling from their everyday lives back home.

Managing Availability

The possibility of telephone contact with family, friends, and colleagues back home posed a real problem for many tourists. How could they experience being "away" if they were expected to make regular contact with others at home, or if others at home could make contact with them? These communication possibilities created a problem that had to be managed and resolved. Tourists with strong views about protecting their solitude and "distance" but who wanted to remain in voice contact developed specific management strategies. They created schemes for "managing their availability" (Ling, Julsrud, and Krogh 1997). Some set specific hours when their mobile phone would be turned on. Some turned on their phones to check for text messages.

Some tourists, particularly those traveling for shorter periods, avoided using mobile phones at all and described them as "tyrants." Some used phone cards in public telephones—a strategy that enabled them to control the cost, timing of calls, and frequency of contact. Others didn't use phones at all. These tourists were pleased to be away from their phones and out of contact. Their constant accessibility by mobile telephone back home was contrasted with a life of limited and controlled contact while traveling.

In my old job I was responsible for a lot of people and I had a mobile phone and a pager and I hated it. I couldn't stand it when people said "Your mobile phone was off" and I would say "What did you do before you had a mobile phone? Was it a life or death situation?" It drove me crazy. (Female, 34, traveling with partner)

For another tourist,

At home in Germany you always look at your mobile phone and say "Oh, he called, I have to call back," or "He mailed, I have to mail back." If you don't have your mobile phone with you, you say "Where is it? Shit, I forgot it." Now I'm away I enjoy this feeling that I can do what I like. (Female, 28, traveling with friend)

For this tourist, being away was a chance to cut off from work, friends, and family. She was

...not interested in what is going on back home, except for an emergency. We don't want to hear about the weather in Germany. (Female, 28, traveling with friend)

She had planned to make two calls back home. In the first she would let her family know that she had arrived, and in the second she would tell them that she was boarding her homeward flight. She enjoyed the feeling that she was in control of the situation.

Other tourists expressed contrasting views, emphasizing their desire to be contactable at all times:

I want to be contactable. I think that's the best way to stay in reality. I wouldn't want to be away from home [for a time] and not know anything about anyone. I would like to stay informed, to know what is going on. (Male, 32, traveling alone)

Another said:

It is reassuring [to be accessible]. You always want to know if there is a problem in the family. (Female, 64, traveling with husband)

These interviewees expressed a strong need to be connected and kept abreast of events back home even though they recognized that both parties might withhold information as a way of protecting each other. While the mobile phone was not the dominant method for voice contact, many tourists who were traveling with mobile phones expressed anxiety about any forced break in mobile phone coverage:

In the beginning we were anxious about the telephone. Has anyone called? Is there mobile coverage here?...We were talking about it and worrying about it. (Male, 39, traveling with wife and children)

For those who relied on mobile phones for voice contact, being out of range was a real concern:

When we have been out of mobile coverage and I have not spoken to my mum for two weeks I don't have a good feeling. (Male, 31, traveling with wife)

The tourists interviewed were aware of the advantages of communicating by telephone. Telephone conversations could be used to eradicate feelings of distance and to confirm feelings of connectedness. Their remarks demonstrate how traveling with phones can be understood as an attempt to extend the parameters of our life worlds without sacrificing our home place, and how they enable the desire to remain at home, with its security of the familiar, to coexist with the desire to reach out into the world (Suvantola 2002).

Being Kept in Mind

The possibility of relatively easy contact engendered the view that the communications initiated by either the tourists or those back home were an indication of the significance of their relationship for each other. Given this understanding, most of the interviewees reported the need to maintain regular levels of contact with people "back home." They spoke about how important it was for them to feel that, despite their physical absence, they were still socially and emotionally integrated into relationships and events at home:

You feel that you belong to something. I think that feeling that you belong to a family and a group of friends is something that you need, even when you are traveling. Because all the time that you are traveling you know that you have a place to go back to. If you cut all your roots you might feel lost. We need to feel connected to something, to belong to something. (Female, 39, traveling with husband and children)

Integration was signified through the receipt of communications from family and friends whose purpose and consequence was to "include" tourists in events back home. These messages contained information about events and were sent either synchronously (voice phone contact, texting and sending pictures by phone) or asynchronously using e-mail messages and postings on Internet sites. For younger tourists, being remembered occurred in relation to parties, gossip about relationships, and the events associated with new marriages and babies. For older tourists, the focus of communications was children and friends. Many of the tourists who were in the workforce were equally interested in keeping abreast of the politics of the workplace. These communications enabled tourists to feel that they were participating in events "back home."

The interviewees articulated a need to share in the minutiae structuring the daily life of those back home. They believed that these communications would prevent a hiatus that might drive a wedge into their sense of continuing engagement:

We were here [in New Zealand] for the holidays and in my family Christmas is a really special time....My sister got one of those cell phones with a camera and she sent me the whole Christmas....There they were opening presents. There they were eating. It was really, really nice. If it hadn't happened I would have felt much more disconnected. (Female, 34, traveling with partner)

The ongoing receipt of information about the fine details of domestic events was understood by tourists to offer confirmation that despite physical separation and absence, those at home had not cut ties with them. That is, they could be reassured that they continued to have a presence in and salience to the lives of those with whom they were communicating. The importance of not being forgotten was articulated time and again by interviewees:

When my friends called [unexpectedly] it really made me feel appreciated and that they missed me. So when someone is making an effort to call long-distance and pays so many cents a minute it shows that you're worth it. (Female, 21, traveling alone)

Most interviewees were clear that they wished to be seen as a part of ongoing life back home. They also wanted their family, friends, and colleagues to feel that they were a part of their lives while they were away. Each wanted to be "remembered" and to be co-present in the lives of significant others. As Urry puts it, physical presence "cedes to the socialities involved in occasional co-presence, imagined co-presence and virtual co-presence" (2002, pp. 256–257). Keeping in touch showed both tourists and their correspondents that they were being kept in mind. The act of keeping in touch was as, if not more important than, what was said (Licoppe 2003). The mere fact of contact, together with the content of messages from home, let tourists know they were still remembered and missed. These interactions served a ritual function similar to that of the small talk engaged in during routine telephone calls between family members or friends (Drew and Chilton 2000).

Conclusions

The research findings suggest that tourists use the voice and texting services of telephones to create a sense of co-presence. The goal of being kept in mind was uppermost in the minds of tourists. They were keen to be demonstrably engaged in the social groups and relationships with people from whom they were temporarily separated by long distances. This desire to remain connected coexisted with a desire to protect the physical separation by controlling the frequency and nature of contact. Different uses of telephone voice and texting capabilities, and variations in the perceived meaning of these two communication modes emerged from the tourists' accounts. Telephone calls were seen to have an emotional content and impact that was both disarming and rewarding. The voice was regarded as an unambiguous and direct vehicle for the expression and monitoring of feelings. Consequently, telephone conversations were sometimes emotionally unsettling, producing unexpected feelings of homesickness or ambivalence about relationships with parents, children, or colleagues. The cost of mobile telephone calls and their enforced brevity meant that conversations were sometimes hurried and this was experienced as unsettling as well. However, these negative aspects of telephone communication were balanced by the support and enhanced sense of security tourists felt they received during their conversations with those at home. This is consistent with previous research that has shown that mobile phones in particular are seen to ensure contactability and, related to this, actual and perceived security in emergencies (Rakow and Navarro 1993). Texting was regarded as a valued option for a spontaneous, immediate affirmation of connection unencumbered by the possible interpersonal complexities of telephone communication.

The findings of this study provide a perspective on the meanings of mobile phone use that differs from previously conducted research dealing with perceptions of telephone versus e-mail communication. Previous research has found that mobile phones are seen to provide an effective way to make interpersonal contact whereas computers and the Internet are understood to damage the communication necessary for good social relations (Leonardi 2003). Rather than confirming a duality of "damaging" versus "functional" corresponding to particular communication modalities, the present study suggests a more complex range of meanings associated with telephone use. That is, telephones engender a range of feelings of connection, both positive and negative. The findings of the present study also challenge the notion that tourists can be seen as entering a state of liminality, which frees them from the structures that encumber their everyday lives back home (Baranowski and Furlough 2001; Harrison 2003; White and White 2004). For most of our respondents, continued communication with their networks of contacts were integral to the travel experience. When tourists have ready access to and use of mobile and other communication services, the liminal experience is transformed into a continuing engagement with established relationships. The *experience* of proximity is thus shown to be experienced in complex, intersecting ways to be independent of actual *physical* proximity, and the telephone is integral to this process.

References

Baranowski, S., and E. Furlough. 2001. Introduction. In *Being Elsewhere: Tourism Culture and Identity in Modern Europe and North America*, edited by S. Baranowski and E. Furlough. Ann Arbor: The University of Michigan Press.

COST269 Mobility Workgroup. 2002. *From Mobile to Mobility: The Consumption of ICTs and Mobility in Everyday Life*.

Drew, P., and K. Chilton. 2000. Calling just to keep in touch: regular and habitualised telephone calls as an environment for small talk. In *Small Talk*, edited by J. Coupland. Harlow: Longman.

Fortunati, L. 2002. The mobile phone: Towards new categories and social relations. *Information, Communication and Society* 5(2): 513–528.

Gergen, K. 2000. Technology, self and the moral project. In *Identity and Social Change*, edited by J. E. Davis. New Brunswick, N.J.: Transaction Publishers.

Gergen, K. 2002. The challenge of absence presence. In *Perpetual Contact: Mobile Communications, Private Talk, Public Performance*, edited by J. Katz. Cambridge: Cambridge University Press.

Goffman, E. 1963. *Behavior in Public Places: Notes on the Social Organization of Gatherings*. New York: Free Press of Glencoe.

Haddon, L. 2003. Domestication and mobile telephony. In *Machines that Become Us: The Social Context of Personal Communication Technology*, edited by J. E. Katz. New Brunswick, N.J.: Transaction Publishers.

Harrison, J. 2003. *Being a Tourist: Finding Meaning in Pleasure Travel*. Vancouver: University of British Columbia Press.

Johnsen, T. E. 2003. The social context of mobile use of Norwegian teens. In *Machines that Become Us: The Social Context of Personal Communication Technology*, edited by J. E. Katz. New Brunswick, N.J.: Transaction Publishers.

Katz, J. E., and M. A. Aakhus. 2002. Introduction. In *Perpetual Contact: Mobile Communication, Private Talk Public Performance*, edited by J. E. Katz and M. A. Aakhus. Cambridge: Cambridge University Press.

Leonardi, P. M. 2003. Problematizing "new media": Culturally based perceptions of cellphones, computers and the Internet among United States Latinos. *Critical Studies in Mass Communications* 20(2): 160–179.

Licoppe, C. 2003. Two modes of maintaining interpersonal relationships through telephone: From the domestic to the mobile phone. In *Machines that Become Us: The Social Context of Personal Communication Technology*, edited by J. E. Katz. New Brunswick, N.J.: Transaction.

Ling, R., T. Julsrud, and E. Krogh. 1997. The Gore-Tex principle: The *hytte* and mobile telephones in Norway. In *Communications on the Move: The Experience of Mobile Telephony in the 1990s* (Report of COST 248: The Future European Telecommunications User Mobile Workgroup), edited by L. Haddon. Farsta, Sweden: Telia AB.

Lury, C. 1997. The objects of travel. In *Touring Cultures: Transformations of Travel and Theory*, edited by C. Rojek and J. Urry. London: Routledge.

Makimoto, T., and D. Manners. 1997. *Digital Nomad*. Chichester: Wiley.

Rakow, L., and V. Navarro. 1993. Remote mothering and the parallel shift: Women meet the cellular phone. *Critical Studies in Mass Communications* 10(2): 144–157.

Short, J., E. Williams, and B. Christie. 1976. *The Social Psychology of Telecommunications*. New York: Wiley.

Suvantola, J. 2002. *Tourist's Experience of Place*. Aldershot: Ashgate.

Urry, J. 2002. Mobility and proximity. *Sociology* 36(2): 255–274.

Urry, J. 2003. Social networks, travel and talk. *British Journal of Sociology* 54(2): 155–175.

White, N. R., and P. B. White. 2004. Travel as transition: Identity and place. *Annals of Tourism Research* 31(1): 200–218.

White, P. B., and N. R. White. 2005. Keeping connected: Travelling with the telephone. *Convergence* 11(2): 103–118.

Wurtzel, A. H., and C. Turner. 1977. "Latent functions of the telephone: What missing the extension means." In *The social impact of the telephone*, edited by Ithiel de Sola Pool. Cambridge, MA: MIT Press.

16 The Social Effects of *Keitai* and Personal Computer E-mail in Japan

Kakuko Miyata, Jeffrey Boase, and Barry Wellman

Japanese often use *keitai* (Internet-enabled mobile phones) to communicate with their close friends and family. The small size and portability of the *keitai* makes it possible to send messages at almost any time and in any place—even Tokyo subway lines have been wired to enable connection underground. Moreover, the ability to type discreet messages makes it socially conducive to send messages quietly in public places where voiced conversation would be socially unacceptable. As in other countries, the heaviest users of this technology are young people, who often text message each other as a way to nurture relationships that might otherwise be hampered by parents and other authority figures. The kinds of text messages sent by *keitai* vary considerably, from the utility-oriented *keitai* e-mail that is used to coordinate in-person meetings, to the seemingly superfluous "I'm so bored" e-mail that promotes a sense of "ultra-connectedness" between lovers and confidants (Ito 2001).

While these studies have gone a long way toward understanding the *keitai* phenomenon, a number of unanswered questions remain about the long-term social implications of this technology. First, although it has been well documented that *keitai* use increases contact with close friends and family, it is unknown if it is used to develop new supportive relationships. It is possible that *keitai* communication only supports existing supportive relationships, and does not help develop new relationships. Second, it is not known if *keitai* e-mail has the potential to replace PC (personal computer) e-mail over time. While it is possible that both types of e-mail will be used together to contact supportive ties, the convenience and portability of the *keitai* might make PC e-mail redundant.

This chapter addresses these questions using longitudinal data collected in Japan's Yamanashi prefecture. Collected over three years, this is one of the first studies to collect information over time about social networks and e-mail use over time. It offers a rare opportunity to map out the adoption of *keitai* and how its use is associated with changes in relationships. In this chapter, we address two research questions:

1. Is *keitai* used to form new relationships that provide important kinds of social support?

2. Are *keitai* and PC e-mail used together as a single communication system to contact and form supportive relationships, or do they play different roles in developing and maintaining social networks?

We next discuss their implications for young people who use *keitai* as their only means of accessing e-mail.

Locating E-mail Use in Japan

Japan is a useful location for studying the social implications of mobile *and* PC e-mail. In the following section, we draw on two highly regarded reports sponsored by the Japanese government (Ministry of Public Management 2004, 2005) to provide statistics about PC and *keitai* Internet use in Japan.

Advanced mobile usage is popular in Japan: 86 percent of mobile phone users have a browser phone, 74 percent have a camera phone, and 30 percent use a combination of cutting-edge technologies called "IMT-2000." Although mobile service started with mere voice service, the diversification and advancement of mobile service is becoming more significant with each passing year. In 1999 the first browser phone service was launched, and in 2000 the camera phone appeared on the market. In 2001 commercial IMT-2000 service started and currently there are more than twenty-five million subscribers. Today, *keitai* are multifunction terminals that can be used to exchange pictures, record videos, play online games, watch TV, and carry out financial transactions.

Japan is also among the world's leaders in PC-based high-speed Internet usage. It has the second highest penetration of high-speed connections in the world, trailing only behind the United States. It also has the least expensive connections in the world. The number of Internet users at the end of 2004 was estimated to be 79.48 million with a penetration rate of 62 percent. The percentage of broadband households (using FTTH, DSL, cable Internet, or wireless access) to the total households accessing the Internet via home PCs was 62 percent at the end of 2004. Since the penetration rate exceeded 60 percent at the end of 2003, the rate of adoption is beginning to plateau, as it has in the United States.

In Japan, the Internet is most commonly accessed by both PCs and *keitai*. Fifty-two percent of Internet users access the Internet by both PC and *keitai*, 26 percent use only PCs, and 19 percent use only *keitai*. Hence, in this chapter we focus on Internet access by PC and *keitai*, and reveal the consequences of using these different types of access.

Rather than looking at all Internet activities conducted through *keitai* and PCs, our analysis focuses on the use of e-mail. Although people can use various services by accessing the Internet from PCs and *keitai*, the most common use of the Internet is e-mail. Fifty-seven percent of Internet users access their e-mail from their home PCs. Fifty-seven percent use home PCs to search for information on goods and services, 48 percent obtain news and other information, and 36 percent purchase goods and services. As with PCs, e-mail is the most common activity when using *keitai*. Seventy-four percent of *keitai* users send and receive e-mail using their *keitai*. Fifty percent use *keitai* to download and listen to online music, 32 percent use it to download images, and 26 percent use it to obtain news and other information.

Can *Keitai* Use Help Develop New Supportive Relationships?

Does the additional quantity of contact facilitated by *keitai* help people develop new supportive relationships? Although cross-sectional studies have shown that *keitai* use is associated with having supportive ties, it is not presently known if this technology is useful for developing new supportive ties. It is quite possible that *keitai* users just use the *keitai* to communicate more often with their existing supportive relationships. Thus, their adoption and use of the technology only reinforces their existing relationships, without actually helping to develop new supportive relationships. Without charting the longer term uses of *keitai*, it is impossible to know if this increased communication is helpful in developing new ties that provide support.

This chapter examines three dimensions of supportive ties: emotional, financial, and instrumental. Emotional support helps people cope with daily stresses, as well as more serious problems such as illness. A number of psychological studies show that having emotionally fulfilling relationships with friends and family generally improves people's access to other kinds of social support, their sense of meaning in life, their self-esteem, and their commitment to social norms and to their communities (e.g., Cohen and Wills 1985; Thoits 1983). Financial support can also be important for coping with stressful situations such as the loss of a job. Moreover, it may enable people to improve their quality of life by helping buy property or start a small business. Finally, instrumental support is the provision of help needed to do certain activities. This kind of help includes the common activities such as taking care of a child for a short period of time while a parent visits the grocery store. However, it might include the less common tasks of cooking for someone who is ill. In general, these three kinds of support lead to greater levels of overall psychological and physical well-being.

Media Multiplexity or a Division of Media?

Lacking longitudinal context also limits the knowledge that can be gathered about the potential multiplexity that occurs between *keitai* e-mail and PC e-mail over time. On the one hand, both *keitai* and PC e-mail may be used to foster relationships with supportive ties. Research conducted in North America shows that people contact social ties using a wide array of communication media (Haythornthwaite and Wellman 1998).

This "media multiplexity" means that people take advantage of the multiple affordances provided by specific media. Contrary to fears that e-mail would reduce other forms of contact, high levels of e-mail contact are associated with high levels of land-line and mobile phone use (Boase et al. 2006). The longitudinal analysis from a large national panel of Americans suggests that the links between communication media are asymmetric: visits drive more e-mail communication and phone calls drive more visits, but e-mail drives neither phone calls nor visits (Shklovski, Kraut, and Rainie 2004). On the other hand, research has also shown that the links between different communication media are complexly interwoven: visits drive e-mail, and phone calls drive visits, but e-mail contact is often associated with both phone calls and visits (Boase et al. 2006; Carrasco, Hogan, Wellman, and Miller in press).

On the other hand, people in Japan may reserve *keitai* e-mail exclusively to foster relationships with supportive ties. Unlike in North America, Japanese people send heavy amounts of e-mail by *keitai* phones. Among this population, e-mail sent by *keitai* phones might make PC e-mail redundant for contact with supportive ties. Although PC e-mail is similar to *keitai* e-mail in many ways, *keitai* e-mail has several features, or social affordances, that make it especially useful for maintaining supportive ties. There are a number of ways that *keitai* allows people to maintain frequent contact, which most often occurs with supportive ties. The discreet nature of communication allows contact to occur in many venues where face-to-face or vocal phone communication would not be socially acceptable. Moreover, the mobility of these phones makes it possible to carry out this communication in places that are visited during everyday activities, the most prominent example being during the commute between home and work. The use of emoticons enhances the possibility for emotionally expressive communication, where the message is based more on emotion than on instrumental content. While much of this frequent contact may be expressive, *keitai* phones are also used to coordinate social meeting times and places, allowing them to serve instrumental functions in these close relationships as well. These affordances make *keitai* e-mail a convenient way to keep in constant connection with supportive ties.

Although the advantages of sending e-mail by *keitai* may make PC e-mail redundant for contacting supportive ties, it is unlikely that *keitai* e-mail has the potential to replace the social function of PC e-mail completely. In contrast to *keitai*, e-mail sent from PCs may be best utilized to maintain contact with diverse social circles. These diverse ties may not be as strong as supportive ties that provide emotional, financial, or instrumental social support. However, their diversity makes them valuable sources of information. The more different types of people you know, the more social milieus you are likely to be connected with (Feld 1981). Accessing information from diverse social milieus can be useful when making important life decisions such as deciding how to invest money or where to apply for a job (Granovetter 1973, 1983). Moreover, the new ideas that these ties provide might open people to new ways of understanding the social world, helping them make informed political decisions (Côté and Erickson, forthcoming; Ikeda and Richey 2005).

Both *keitai* and PC e-mail share a number of features that are useful for staying in touch with diverse ties. However, PC e-mail provides one social affordance that makes it far more conducive for contacting diverse ties than *keitai* e-mail: the ability to quickly type long and involved messages. Because *keitai* messages are often limited in their length and can only be thumbed in at a slow pace, they require that the receiver have a great deal of prior knowledge about the sender. By contrast, PC messages can be much more detailed, explaining the purpose of the message and making explicit the large amount of information required to make a meaningful interpretation of a message when little is known about the sender. PC e-mail also has a number of additional features that are common to those of *keitai* and make it useful for keeping in contact with diverse ties. For example, people can contact many diverse people, with a great amount of control over the time that they spend making these contacts, and without concern that they will interrupt these people. Moreover, the lack of visual cues decreases the time and effort needed to change demeanor and appearance when making contact with those from a different socioeconomic status. Hence, there may be a division of e-mail, where *keitai* becomes the dominant way of maintaining and nurturing supportive ties while PC e-mail is best used to maintain contact with diverse social ties.

Data and Measures

To examine these research questions, we analyzed longitudinal survey data. We conducted a panel survey in Japan. Our first random sample survey of 1,320 adults was conducted in November 2002 in Yamanashi prefecture (which ranges from urban to rural and is similar to Japan overall in its distributions of Internet use, civic engagement, and political activity). These potential respondents were chosen from a voter's list of people aged twenty to sixty-five years old. Surveys were in paper form and delivered in-person. They were collected in-person three weeks after being dropped off. Three-quarters of the selected individuals completed the survey, giving us a total sample size of 1,002 respondents at Time 1. Of these respondents, 646 were interviewed in March 2004 at Time 2. In March 2005, 432 of the original 1,002 respondents completed a third survey at Time 3. Our analysis focuses on the 432 respondents that completed surveys at Time 1 and Time 3.

Between Time 1 and Time 3, there was a decrease in the number of younger respondents, especially those in their 20s. There were fewer single people in Time 3, but more with a partner and a child living together or with a partner and a child living apart. Single adults dropped from about 24 percent in Time 1 to 14 percent in Time 3.

Table 16.1

Amount of Keitai and PC E-mail Sent "Yesterday" at Time 1 and Time 3

	Time 1 (N=1002)		Time 3 (N=432)	
	Keitai e-mail	PC e-mail	Keitai e-mail	PC e-mail
Do not have	56	56	45	51
Do not use	2	13	3	14
Did not send e-mail yesterday	16	22	17	27
1–5 e-mails	21	8	32	7
6–10 e-mails	4	1	2	0
11–25 e-mails	0	0	1	0
26–50 e-mails	0	0	0	0
More than 51 e-mails	0	0	0	0
	100%	100%	100%	100%

Quantifying E-mail Use

Unlike cross-sectional surveys, the Yamanashi study allows us to measure fluctuations in the quantity of e-mail sent between two points in time. With this information, we are able to understand if changes in the quantity of *keitai* and PC e-mail sent are associated with changes in the formation of supportive and diverse ties. We begin our analysis by describing these changes for the people in our sample between Time 1 and Time 3.

To quantify e-mail use, respondents in both Times 1 and 3 were asked about the number of e-mails they sent yesterday by mobile phone and PC. A six-point scale was used to record their responses: 1) I do not send e-mails by mobile phone/PC; 2) I sent 1 to 5 e-mails; 3) 6 to 10 e-mails; 4) 11 to 25 e-mails; 5) 26 to 50 e-mails; and 6) More than 51 e-mails. See table 16.1 for a summary of the percent of respondents sending *keitai* and PC e-mail at Time 1 and Time 3.

Given the small percentage of respondents sending *keitai* and PC e-mail at Times 1 and Times 3, we decided to collapse the last five categories of these variables when using these variables in our multivariate analysis. The final categories for *keitai* and PC e-mail use are as follows: 0) do not have; 1) have, but did not send any e-mail yesterday; 2) have, and sent e-mail yesterday.

To show the degree of change in *keitai* and PC use between Time 1 and Time 3, we code *keitai* and PC e-mail use as remaining stable if the level of use did not change, increasing if there was a positive change, and decreasing if there was a negative change. We find that approximately 68 percent of the respondents have stable levels of *keitai* e-mail, while 22 percent have increasing levels and 10 percent have decreasing levels. We also find that approximately 72 percent of the respondents have stable levels of PC e-mail, while 15 percent have increasing levels and 13 percent have decreasing levels.

Measuring Supportive and Diverse Ties

Supportive Ties   We focus on three dimensions of support to measure these ties: emotional, financial, and instrumental. Respondents were asked to report whether network members would give them words of encouragement (emotional support), treat them to lunch if they did not have enough money (financial support), or aid them in tasks such as moving (instrumental support). In Time 1, approximately 82 percent of the respondents said that they could receive all three kinds of social support, while 18 percent could receive only two kinds of support or fewer. In Time 3, approximately 81 percent of the respondents received all three kinds of social support from their ties, while 19% received only two kinds of support or fewer. In the analysis that follows, we make the reception of social support into a dichotomous variable. Respondents with a score of 1 have all three kinds of support, while respondents with a score of 0 do not receive all three kinds of support.

Network Diversity   Network diversity has many different facets, such as occupational diversity, gender diversity, ethnic diversity, and so on. We focus on occupational diversity because people who work in different occupations often come from different social backgrounds (Lin 2001). Respondents were asked to indicate if they have any relatives, friends, or acquaintances in any of fifteen categories of diverse occupations. A count of the number of different occupation categories was made for each respondent, yielding a score from 0 to 15, with a higher score indicating greater diversity of contact.

In addition to these two types of social networks, we control for network size. In order to control respondents' total proper network size. Social network size was measured by asking respondents how many personal New Year's greeting cards they had sent that year. This yielded estimates of the total size of networks, including supportive and diverse ties.

The Impact of E-mail on the Formation of Supportive Ties

Does increasing the quantity of *keitai* e-mail help people develop supportive ties? We used logistic regression to examine this question with our longitudinal data. Model 1 in table 16.2 shows the result of our analysis.

The dependent variable is the provision of all three kinds of support from social ties at Time 3, and the main independent variable is the change in frequency of *keitai* e-mails from Time 1 to Time 3. We control for demographic factors of gender, age,

Table 16.2

Logistic Regression—People Who Received All Three Kinds of Supportive Ties at Time 3

	Model 1			Model 2		
	B	Wald	Exp(B)	B	Wald	Exp(B)
Gender (0=female, 1=male)	0.143	0.144	1.154	0.166	0.192	1.180
Age (reference=20–29)						
30–39	−0.679	0.654	0.507	−0.763	0.838	0.466
40–49	−1.070	1.540	0.343	−1.043	1.462	0.352
50–59	−0.767	0.842	0.464	−0.818	0.935	0.441
60–65	−0.632	0.482	0.532	−0.685	0.551	0.504
Education (reference=middle school)						
High school	−0.813	1.180	0.444	−0.790	1.107	0.454
College	−0.578	0.516	0.561	−0.571	0.503	0.565
Undergraduate degree or more	−1.668	4.311*	0.189	−1.708	4.457*	0.181
Employment status (0=no job, 1=have a job)	−0.079	0.040	0.924	0.000	0.000	1.000
Partner (0=no, 1=yes)	−0.940	1.888	0.391	−0.871	1.637	0.419
Kids living at home (0=no, 1=yes)	0.523	1.574	1.687	0.512	1.519	1.668
Number of groups participated in (T3)	0.279	0.802	1.321	0.227	0.546	1.255
Diversity of social networks (T3)	0.126	3.214	1.135	0.109	2.451	1.115
Size of social network (T3)	0.257	2.385	1.293	0.250	2.358	1.284
Presence of 3 kinds of support (T1) (reference=less than 3 kinds of support)	1.412	14.635**	4.104	1.420	14.866**	4.138
Change in the use of keitai e-mail (T3-T1) (reference=stable)						
More use	−0.306	0.653	0.736			
Less use	0.951	1.935	2.588			
Change in the use of PC e-mail (T3-T1) (reference=stable)						
More use				0.078	0.029	1.081
Less use				0.053	0.013	1.054
Constant	0.800	0.550	2.225	0.868	0.648	2.383
Cox & Snell R²	0.111			0.102		
Nagelkerke R²	0.186			0.171		

*p<0.05
**p<0.01
N=432

education, occupation, and marital status. Moreover, we control for voluntary activity and other network characteristics at Time 3, as they may be associated with the presence of supportive ties at Time 3.

The results show that changes in the frequency of sending *keitai* e-mail has no effect on increasing the variety of support available from ties at Time 3. This indicates that *keitai* messages are not used to develop relationships that yield new kinds of support.

We find that while *keitai* e-mail is used to contact existing supportive ties, increasing levels of using *keitai* e-mail is not associated with developing additional supportive ties over time. Our previous cross-sectional analysis of Time 1 shows a significant and positive association between the amounts of *keitai* e-mail sent at one point in time and the reception of social support (Miyata et al. 2005). This indicates that *keitai* is just one communication medium that supports existing supportive relationships, although our results presented here show that it is not used to develop new supportive relationships.

Media Multiplexity or Division of Media

Although *keitai* e-mail is used to contact existing supportive relationships, we have shown that *keitai* use is not helpful in forming new supportive relationships. However, we do not know if PC e-mail can be used to develop new supportive relationships. It may be that the ability to enter long and involved messages makes PC e-mail more useful for developing new supportive relationships. On the other hand, it may be that these two media are used to contact different components of the social network.

To better understand the social uses of these media, we again use logistic regression with our longitudinal data. We start by using changes in PC e-mail use between Time 1 and Time 3 as our main independent variable, and again use the provision of all three kinds of social support at Time 3 as our dependent variable. As in the previous analysis, we control for demographics, participation in voluntary groups, and other social network attributes. The results of this analysis are presented in table 16.2, model 2.

As with *keitai* e-mail, these results show that there also is no significant relationship between increasing the level of PC e-mail use and having new supportive relationships at Time 3. This suggests that PC e-mail is not used to develop new supportive relationships.

Does this mean that PC e-mail is redundant with *keitai*? We suspect that while *keitai* e-mail may be important for contacting existing supportive ties, PC e-mail may be important for contacting and forming diverse ties. Our previous cross-sectional analysis already shows there is a significant and positive relationship between PC e-mail and diversity of ties at Time 1 (Miyata et al. 2005). This is evidence that PC e-mail is used to contact diverse ties. But is PC e-mail also used to develop new diverse ties?

To measure the effect of PC e-mail on increasing the diversity of ties, we use change in level of PC e-mail from Time 1 to Time 3 as our main independent variable, and network diversity at Time 3 as our dependent variable. We control for demographic and network attributes. Levels of voluntary group participation are also controlled for because Erickson and Miyata (2004) found that participation in informal groups enhances the diversity of social ties. See model 1 and model 2 in table 16.3 for results.

As shown in model 1, increasing levels of PC e-mail between Time 1 and Time 3 is associated with greater network diversity at Time 3. By contrast, model 2 shows that increasing amounts of e-mail sent by *keitai* is not associated with increases in network diversity. This shows that increasing the amount of PC e-mail can lead to an increase in diversity of networks, while increasing the amount of *keitai* e-mail does not cause such a change.

These results show that *keitai* and PC e-mail each serve different social functions. Unlike media use in North America where most people use all available media to contact their social ties, there is a division of media in Japan. *Keitai* e-mail is used to contact existing *supportive* ties, while PC e-mail is used to contact existing *diverse* ties and to form new diverse ties. Both of these media are important. Quality ties improve mental well-being; diverse ties provide information that can be helpful when making important decisions.

Implications for the Future

These findings cause us to wonder what will happen to the large percent of young people that use *keitai* as their only means of accessing e-mail. Even though *keitai* is more effective for nurturing existing supportive ties than the PC, we show that it is not as effective as the PC when it comes to forming diverse ties. This means that, compared to with those that have both *keitai* and PC e-mail, the large percentage of young people that use only *keitai* are disadvantaged when it comes to forming diverse relationships.

Will the *keitai*-only users of this young generation remain dependent on the *keitai* as they grow older? It is too early to know for sure, but there is some reason to believe that this scenario may become reality. Although our study spanned only a short period of time, 28 percent of our respondents in their twenties relied on *keitai* as their only means of sending e-mail throughout the three years of our study. By contrast, 42 percent of people in their twenties continued to use both PC and *keitai* e-mail throughout this three-year period. Finally, 14 percent went from only *keitai* e-mail to having both PC and *keitai* e-mail after three years. Although it is encouraging to see 14 percent of these young people gain both media, we are still concerned about the 28 percent that continued to rely only on *keitai* e-mail. Even though the relatively short duration of

Table 16.3

Regression Analysis—Diversity of Ties at Time 3

	Model 1		Model 2	
	B	Beta	B	Beta
Gender (0=female, 1=male)	−0.097	−0.017	−0.018	−0.003
Age (reference=20–29)				
30–39	0.088	0.012	0.148	0.020
40–49	0.293	0.045	0.421	0.064
50–59	0.179	0.030	0.209	0.035
60–65	−0.369	−0.047	−0.407	−0.052
Education (reference=middle school)				
High school	0.122	0.022	0.169	0.030
College	0.044	0.007	0.076	0.012
Undergraduate degree or more	0.101	0.015	0.134	0.020
Employment status (0=no job, 1=have a job)	0.335	0.054	0.370	0.060
Partner (0=no, 1=yes)	1.060	0.142	1.174	0.158*
Kids living at home (0=no, 1=yes)	−0.905	−0.159*	−0.974	−0.171**
Number of groups participated in (T3)	0.420	0.090	0.427	0.092
Diversity of social networks (T1)	0.518	0.550**	0.520	0.552**
Size of social network (T3)	0.098	0.041	0.071	0.030
Presence of 3 kinds of support (T1) *(reference=less than 3 kinds of support)*	0.787	0.105*	0.843	0.113*
Change in the use of PC e-mail (T3-T1) *(reference=stable)*				
More use	0.757	0.098*		
Less use	−0.393	−0.050		
Change in the use of keitai e-mail (T3-T1) *(reference=stable)*				
More use			0.142	0.021
Less use			−0.315	−0.034
Constant	−0.491		−0.593	
R²	0.433		0.421	
Adjusted R²	0.403		0.391	

* *p*<.05
** *p*<.01
N=432

time and small sample of young people make it difficult to make any certain projections about the future, these preliminary results give some indication that this digital divide may continue as the younger generation matures.

A continual dependence on *keitai* means that supportive ties will remain at the center of social life for a substantial percentage of the population. We have several concerns about this scenario. First, a lack of diverse ties might put this generation at an economic disadvantage because diverse ties often supply new information that can be useful for important life decisions. Moreover, because diverse ties connect people of different backgrounds, they provide exposure to new ideas and ways of understanding the world. For this reason, we also fear that a *keitai*-dependent generation would lead insular lives, remaining ignorant of how people from different social strata live and interpret the world.

Conclusion

Although *keitai* e-mail is a useful tool for connecting to existing supportive relationships, we find that it is not used to form new supportive relationships. By contrast, PC e-mail can be instrumental in forming new relationships with people from diverse social strata. These findings indicate that there is a division of media in Japan, with *keitai* and PC e-mail serving different social functions. *Keitai* e-mail is used to contact supportive relationships, while PC e-mail helps people expand and diversify their networks.

Because both supportive and diverse ties help people improve their lives in different ways, we argue that people using both media are most advantaged. Although they do not use *keitai* e-mail to form new relationships, they still have more contact with ties that provide emotional, instrumental, and financial support than those that do not use *keitai* e-mail at all. Moreover, people that use PC e-mail are able to cultivate relationships that provide valuable ideas and information. For these reasons, we expect that people using both media will be better able to improve their lives.

These results have important implications for the substantial percent of young people that rely on the *keitai* as their only means of accessing e-mail. If our preliminary results are indicative of a larger social trend, this generation may be overly focused on their supportive relationships at the expense of diverse relationships. Because diverse relationships often expose people to new sets of knowledge, information, and ways of understanding the world, this group of young people may hold attitudes that are myopic and based on ignorance rather than on informed deliberation. Further research is needed to see if these speculations hold true when they are examined using a larger sample of young people. However, given that our observations are indicative of existing inequalities, we suspect that our evidence foreshadows an unfortunate reality.

References

Boase, J., J. Horrigan, B. Wellman, and L. Rainie. 2006. The strength of Internet ties. *Pew Internet & American Life Project.*

Carrasco, J. A., B. Hogan, B. Wellman, and E. J. Miller. In press. Agency in social activity and ICT interactions: The role of social networks in time and space. *Journal of Economic and Social Geography.*

Cohen, S., and T. A. Wills. 1985. Stress, social support, and the buffering hypothesis. *Psychological Bulletin* 98(2): 310–357.

Côté, R., and B. Erickson. Forthcoming. Untangling the roots of tolerance: How networks, voluntary associations, and personal attributes shape attitudes toward ethnic minorities and immigrants. *American Behavioral Scientist.*

Erickson, B., and K. Miyata. 2004. Macro and micro gender structures: gender stratification and social networks in Canada and Japan. Presented at American Sociological Association 99th Annual Meeting, San Francisco, Calif.

Feld, S. 1981. The focused organization of social ties. *American Journal of Sociology* 86(5): 1015–1035.

Granovetter, M. 1973. The strength of weak ties. *American Journal of Sociology* 78(6): 1360–1380.

Granovetter, M. 1983. The strength of weak ties: A network theory revisited. In *Social Structure and Network Analysis,* edited by P. Marsden and N. Lin. Beverly Hills, Calif.: Sage Publishers, Inc.

Haythornthwaite, C., and B. Wellman. 1998. Work, friendship, and media use for information exchange in a networked organization. *Journal of the American Society for Information Science* 49(12): 1101–1114.

Ikeda, K., and S. Richey. 2005. Japanese network capital: The impact of social networks on Japanese political participation. *Political Behavior* 27(3): 239–260.

Ito, M. 2001. Mobile phones, Japanese youth, and the re-placement of social contact. Presented at the annual meeting for the Society for the Social Studies of Science, Cambridge, Mass.

Ministry of Public Management, Home Affairs, Posts and Telecommunications. 2004. White paper: Information and communications in Japan. Tokyo: National Printing Bureau.

Ministry of Public Management, Home Affairs, Posts and Telecommunications. 2005. White paper: Information and communications in Japan. Tokyo: National Printing Bureau.

Lin, N. 2001. *Social Capital: A Theory of Social Structure and Action.* Cambridge: Cambridge University Press.

Miyata, K., J. Boase, B. Wellman, and K. Ikeda. 2005. The mobile-izing Japanese: connecting to the Internet by PC and webphone in Yamanashi. In *The Personal, Portable, Pedestrian: Mobile Phones in Japanese Life,* edited by M. Ito, M. Matsuda, and D. Okabe. Cambridge, Mass.: The MIT Press.

Thoits, P. 1983. Multiple identities and psychological well-being: A reformulation and test of the social isolation hypothesis. *American Sociological Review* 48(2): 174–187.

Shklovski, I., R. Kraut, and L. Rainie. 2004. The Internet and social relationships: Contrasting cross-sectional and longitudinal analyses. *Journal of Computer-Mediated Communication* 10(1).

Politics and Social Change

17   Mobile Media and Political Collective Action

Howard Rheingold

The use of mobile media to incite and organize collective action is only in its infancy. The following unsystematic survey of published reports and personal communication is intended to suggest how broadly today's earliest forms of mobile phone–assisted collective action are enabling people to effect significant political changes. As both the enabling technologies and the literacies that grow around their use in the political sphere evolve further, the first manifestations noted here might portend more radical phenomena to come. I call this survey unsystematic because of the way I have collected various news reports of political smart-mobbing tactics. As far as I know, no exhaustive inventory of such events has been compiled. Such an inventory would be useful. I offer little political context—commentators such as Rafael have noted that the Philippine People Power II, with its much-remarked use of SMS organizing, did not result in revolutionary political change; what is notable is the way in which the Philippine protests were organized. And of course an inventory of news reports is at best a good pointer to an emergent social phenomenon that requires more systematic data collection and analysis to yield supportable assertions.

Communication media, literacies, and political governance have coevolved for millennia. Much has been written about the role of print and literacy in the emergence of the democratic public sphere. A rich literature has grown around the role of the printing press in the Protestant Reformation and the emergence of constitutional democracies. Communication technologies and literacies possess a power that has, on many occasions, proven mightier than physical weaponry—the potential to amplify, leverage, transform, and shift political power by enabling people to persuade and inform the thoughts and beliefs of others.

The same technologies and literacies can also organize, plan, and coordinate direct political actions—elections, demonstrations, insurrections. It may also be the case that they can be used to stifle, misdirect, and demoralize those who would otherwise be involved in these activities. The power to persuade and communicate, joined with the power to organize and coordinate, multiplied by the three billion mobile telephones in the world today poses a disruptive political potential that could equal or surpass that of the printing press, landline telephone, television, or the Internet. But the possibility of neutralizing this potential is also evident from the countermeasures that political and law enforcement authorities have deployed in response to smart-mob tactics.

Just as people have used alphabets and computers in both socially beneficial and socially destructive ways, mobile devices are being used to keep elections honest, to self-organize peaceful political demonstrations, and to provide disaster relief services—and the same technologies and practices are also used to commit crimes, coordinate terrorist attacks, and summon people to riots. This survey concentrates on forms of collective action in the political sphere that have been instigated and abetted via mobile telephone—elections, demonstrations, and riots. While the uses of mobile communication to foster governmental political objectives or to blunt or counter the actions of opposition or nongovernmental groups are also an important topic, they largely fall outside the scope of the phenomena I analyze in this chapter, which focuses on collective action by citizens.

Electoral Smart Mobbing

The rapid diffusion of mobile telephones since the 1990s, the sudden emergence of SMS as a ubiquitous form of messaging, and the increasing interconnection between mobile phones and the Internet have made it possible for people to coordinate and organize political collective action with people they were not able to organize before, in places they weren't able to organize before, and at a speed they weren't able to muster before. This coordinating function has affected daily life in the form of what Ling calls "hyper-coordination" and what I call "smart mobs" (Ling and Yttri 2001; Rheingold 2002).

The role of mobile media in organizing political collective action has manifested worldwide in coordinating street demonstrations (which, in the Philippines and Spain some have asserted contributed directly to the downfall of regimes), monitoring elections, and augmenting get-out-the vote campaigns in Ghana and Korea. The use of mobile telephony and SMS, both by themselves and in coordination with Internet tools such as Listservs, blogs, meetup.com, and online fund-raising, is still young, but has had significant impacts in (at least) Ghana, Hungary, Italy, Kenya, Korea, Kuwait, the Philippines, Sierra Leone, Spain, and the United States.

In Africa, cell phones have been used in two notable recent instances to combat election fraud, and as political organizing tools. During a panel discussion I attended at the 2004 O'Reilly Emerging Technologies conference, Ethan Zuckerman, founder of Geekcorps, claimed (perhaps sarcastically) that the last Ghana election "went considerably more smoothly than the last US national election due to the use of cellphones and radio to report voting fraud" (Zuckerman 2004). This was because people at polling places used their mobiles to report fraud accusations to local radio stations, which would then air the accusation. Zuckerman said the police were then forced by the pressure of public opinion to investigate since they no longer had the excuse that they had not received reports (Zuckerman 2004).

Fear of fraud was a motivation for the use of mobile phones in other African elections. One commentator, Bill Kagai, said just after the 2002 Kenyan elections that mobile phones contributed not only to high voter turnout but also to the legitimacy of results. Mobile phones gave enhanced transparency of process, campaign effectiveness, and reduction of fraud (Kagai 2002).

Political groups developed cell phone number databases allowing people to contact each other and those at the polling stations to call for support when needed. Campaigns made use of short messaging services and election results were disseminated as soon as they were counted, even in the most remote areas. (Kagai 2002)

Ebba Kalondo (2005) describes how mobile phones also played a role in Kenya's 2004 elections. He notes that SMS was used by Kenya's electoral commission as well as local media to distribute news about polling. He also notes that voters used mobile phones to monitor the voting in more remote areas. Local radio stations even fielded callers who alerted the listening audience to "the level of traffic at polling stations" (Kalondo 2005).

In Hungary, SMS became a political propaganda tool during recent elections. Miklós Sükösd, associate professor at the Department of Political Science at Central European University, and Endre Dányi, sociologist, editor in chief of eDemocracy Newsletter, and member of the eDemocracy Association in Hungary, documented the wide-scale use of SMS and e-mail in the 2002 Hungarian election campaign:

In a country of 10 million, where ca. 53% of the population has mobile phones and 15% are Internet users, millions of political mobile text messages . . . and e-mails were exchanged by party supporters. (Daily SMS traffic has increased 20–30%, i.e., by ca. 1 million messages between the two rounds of the elections.) (Dányi and Sükösd 2003)

In Italy, the 2004 election was complicated by the fact that the incumbent prime minister, Silvio Berlusconi, was also the country's largest media owner. One reader of the Smart Mobs blog, David Ture, wrote me about a horizontal SMS campaign for Berlusconi's party: "The 'SMS' (Sostieni Molto Silvio) campaign asks people to send one (or more) of eleven prepared SMS messages containing promotional sentences to five friends." Another Italian reader of the Smart Mobs blog, Bernardo Parella, commented that the night before the election,

many users of major carriers (such as tim and Vodafone) received an SMS inviting them to go to vote. . . . no slogans or other stuff, but it was an indirect support campaign for Berlusconi, since the messages were officially signed by the prime minister office, making them appear as a sort of "public service."

Yet another Smart Mobs blog reader living in Rome, Sepp Hasselberger, reported that shortly after the election results came in, and Berlusconi's side had suffered a reversal, a "grass root" mocking SMS appeared: "Hello, I am Silvio Berlusconi, next time no way that I will tell you when you should go and vote."

The Korean presidential election of 2002 was a watershed for smart-mob tactics in an electoral campaign; many have claimed that the use of Internet and SMS technology enabled an underdog candidate's followers to tip the election in his favor. Jean K. Min, an editor for the Korean citizen-journalism site OhMyNews, wrote me an e-mail in response to a query about the election. Min explained that the traditional Korean newspapers, three of which dominate more than 70 percent of the Korean news market, exhibit a strong conservative bent. As a progressive candidate, Roh Moo-Hyun was "ferociously" opposed by the establishment media. Younger Koreans, however, favored OhMyNews—an online newspaper with professional editors and an army of thousands of volunteer citizen reporters. Roh, already trailing in the polls, suffered a shocking political blow when his campaign partner withdrew support on the eve of election day. Min wrote, "The sudden news got circulated quickly throughout the thousands of internet cafes, bulletin boards and other political webzines such as seoprise.com, where hundreds of thousands of anxious netizens were staying overnight watching the developing situation." Another storm of SMS and e-mails from Roh's supporters followed news that Roh was trailing his opponent in the early exit polls by 1 to 2 percent. While Min admitted it was impossible to determine how much of an effect the SMS and e-mail campaigns had, an exit poll later that day showed Roh leading by 2 to 3.

Teddy Casino, one of the organizers of the SMS-organized EDSA-2 demonstrations that helped bring down the Estrada regime, was quoted by Shakuntala Shantiran in *Focus Asia* about the continuing importance of text messaging in Philippines politics: "[For n]ext year in the 2004 elections, I've already seen software that integrates text messaging into a database in computers, so you could instantly have a quick count of the number of votes which are cast in a certain precinct. The maker of the software says his product is selling like hotcakes" (Shantiran 2003).

Reports concerning Sierra Leone elections parallel reports from Ghana and Kenya regarding the use of phones as a weapon against electoral corruption. During the election that was held in 2002, a U.S.-based organization called Search for Common Ground distributed messages to journalists' cell phones. This allowed the journalists to report hourly the results of local polls. These were then announced publicly on radio stations. According to Search for Common Ground, "regular updates calmed fears of corruption and vote rigging...permitting the vote to take place without outbreaks of violence" (Black 2003).

When the Spanish government attempted to blame the terrorist bombings at Madrid's Atocha train station on Basque separatists, thousands of citizens began circulating SMS messages that questioned the government's version of events and summoned people to mass demonstrations across Spain. Ultimately, the opposition party won the elections (Adelman 2004). According to Spanish journalist Eva Dominguez,

If there is a medium that has contributed most to make news run like hell during the last days of this strange and difficult Spanish campaign, it is text messaging (SMS) through mobile phones. It has been used to spread news among citizens as well as political parties. But the most impressive use happened the night before the elections. The spread of text messaging congregated some thousands of people in front of the political party running the country, Partido Popular, in just a couple of hours. Any protest with political meaning is forbidden in Spain on that day. Any organization behind such public demonstrations could be punished. But, what if it is just the result of a spontaneous crush of SMS messages? The use of big media, which have been greatly tendentious in some cases, has not been as powerful as text messaging to spread the news and ask for the truth in the most intensive days of democracy in Spain. (Dominguez 2004)

Scottish screenwriter Paul Laverty published his eyewitness account of a spontaneous protest that convened outside PP party headquarters. He received a message calling for a *cacerolada*, a protest by banging of pots and pans. Laverty writes:

By the time I arrived at the PP headquarters at 7 p.m. there were hundreds streaming from the metro and the road was already closed off. The police moved in and demanded identification papers but backed off as hundreds more arrived. Now there were around five thousand, chanting, "We want the truth before we vote. Our Dead—Your War." There was a continuous chant of "Liars! Liars! Liars!" followed by "Don't play with the Dead." More mobiles flashing—there were demonstrations outside the PP offices in all the big cities. Spirits rose. (Laverty 2004)

Laverty reported that when he got home his neighborhood was "a cacophony of unbelievable noise," with entire families assembled on their balconies, and the neighborhood square filled with young and old Africans, Moroccans, Latin Americans, and Spaniards banging on kitchen implements.

The most effective U.S. electoral smart mobs weren't organized at the grass roots but from the very top of the political hierarchy. Collective action can be mobilized and directed from above: The Republican Party's chief strategist, Karl Rove, coordinated the 2002 Republican Congressional victories via his BlackBerry communicator. He used his BlackBerry to send messages even during meetings with President Bush, according to *Time* magazine (Carney and Dickinson 2002). *Time* reports that one of Rove's colleagues said, "Sometimes we're in a meeting talking to each other and Black-Berrying each other at the same time."

Democratic Party campaigners seem to be BlackBerrying, too, according to a *Washington Post* article shortly before the U.S. 2004 elections. Dan Manatt, director of a Democratic political action committee aimed at candidates under forty, notes that the Internet and mobile phones are not instruments of persuasion aimed at converting voters to a cause, but are better used to "preach to the choir" and help coordinate electoral campaigns, especially on Election Day. In particular, Manatt's group plans to equip their volunteers with mobile messaging devices such as BlackBerries for get-out-the-vote campaigns in Congressional elections, allowing the volunteers to coordinate political activity at polling stations where otherwise they could not openly support a candidate (Krebs 2002). John Collias, the field director of a Congressional candidate in Kentucky, noted, "Everyone is bouncing around so much on Election Day that it's not uncommon for the phone lines to be jammed. We think the BlackBerry is the ultimate firewall against that problem, and it has the potential to be very effective for us" (Krebs 2002).

The *Post* reporter, Brian Krebs, hypothesized that the BlackBerry preference might have started on Capitol Hill. After 9/11, all members of Congress were equipped with BlackBerries so they could communicate in an emergency. The Howard Dean campaign used the UPOC service as a kind of texting Listserv—campaigners could subscribe and broadcast SMS messages to other subscribers of specific campaign-related groups (Teachout 2003).

Street Demonstrations, Riots, and Swarms

If elections are the formal, socially contracted, legally controlled exercises of political power, street demonstrations are the informal, ad hoc, uncontrolled outbursts that can tilt elections, as they did in Spain, or unseat an elected leader, as they did in the Philippines. The street demonstrations that brought mobile communications and swarming tactics to the world's attention were the protests against the 1999 meeting of the World Trade Organization in Seattle, Washington, famous as "The Battle of Seattle" (Armond 2000). Organizers used mobile phones and Web sites to coordinate swarming—clusters of demonstrators who emerged from the general crowd to shut down traffic at specific locations at agreed times, then melt back into the crowd. Seattle police, unable to respond effectively to the new tactic, responded inappropriately—attacking innocent citizens while failing to achieve their objective of clearing out demonstrators (Reynolds 1999). The Seattle chief of police resigned after the incident. It appears obvious that the New York police closely studied the incident in preparing for their far more successful containment of demonstrators during the Republican National Convention in the summer of 2004.

Texting is famously popular in the Philippines. EDSA-2, the popular "people power" demonstrations that brought down the Joseph Estrada regime in 2001, was instantly and broadly recognized to have been self-organized via SMS. The president was under impeachment. When the impeachment trial was suddenly aborted by senators regarded to be Estrada supporters, hundreds of thousands of demonstrators began to assemble at Epifanio de los Santos Avenue (known as "Edsa")—the same place demonstrators had assembled in 1986 to protest the Marcos regime in the famous "People

Power" revolt. The role of demonstrations at that location in the fall of the Marcos regime lent power to the assemblies. Between January 16 and 20, 2001, more than a million people gathered at Edsa to demand Estrada's resignation:

Aside from TV and radio, another communication medium was given credit for spurring the coup: the cell phone. Nearly all accounts of People Power II available to us come from middle class writers or by way of middle class controlled media with strong nationalist sentiments. And nearly all point to the crucial importance of the cell phone in the rapid mobilization of people. "The phone is our weapon now," one unemployed construction worker is quoted in a newspaper article. "The power of our cell phones and computers were among the things that lit the fuse which set off the second uprising, or People Power Revolution II, according to a college student in Manila. And a newspaper columnist relayed this advice to "would-be foot-soldiers in any future revolution: As long as you[r cell phone] is not low on battery, you are in the groove, in a fighting mood." A technological thing was thus idealized as an agent of change, invested with the power to bring forth new forms of sociality. (Rafael 2003)

In China, where political demonstrations are risky for participants, the *New York Times* reported that twelve thousand workers went on strike in Shenzen at the factory of a supplier of Wal-Mart. The article mentions that while "few of [the young, migrant workers] are unionized, communication and coordination among them is growing, often through the sending of coded messages to each other by cellphone" (French 2004).

In early 2003, when the Chinese government was trying to keep the lid on news of the epidemic that was breaking out in rural areas, news of "a fatal flu in Guangdong" reached 120 million people within a few days via SMS messages that spread like an epidemic of their own (Hoenig 2003). The Chinese government reacted by admitting that there was, indeed, an outbreak—and by making it illegal to spread SARS rumors via SMS.

In April, 2005, anti-Japanese demonstrations broke out in several Chinese cities. The *New York Times* reported that mobile phones and Web sites, once again, played a central role:

For several weeks as the protests grew larger and more unruly, China banned almost all coverage in the state media. It hardly mattered. An underground conversation was raging via e-mail, text message and instant online messaging that inflamed public opinion and served as an organizing tool for protesters.

The underground noise grew so loud that last Friday the Chinese government moved to silence it by banning the use of text messages or e-mail to organize protests. It was part of a broader curb on the anti-Japanese movement but it also seemed the Communist Party had self-interest in mind.

"They are afraid the Chinese people will think, O.K., today we protest Japan; tomorrow, Japan," said an Asian diplomat who has watched the protests closely. "But the day after tomorrow, how about we protest against the government?" (Yardley 2005).

The Chinese government has good reason to fear the power of phone-mobilized political action—with more than 350 million phones in private hands. The same *New York Times* article that covered the anti-Japanese demonstrations also noted that the authoritarian Chinese government employs as many as fifty thousand people to censor the Internet, and the Shanghai broadcast SMS messages to citizens during the protests, advising them to obey the law. Web sites, online community discussions, and private e-mail is monitored for keywords, and individual SMS messages are monitored.

Xiao Qiang, exiled Chinese human rights activist, now at UC Berkeley's School of Journalism, sees the ad hoc organization of the anti-Japanese protests as a kind of watershed. Although anti-Japanese Web sites had been officially tolerated, and in the early days of the demonstrations the police were supportive of the protestors, Qiang noted the grassroots nature of the political organizing—outside official Communist Party channels—in an editorial for the *Asian Wall Street Journal*:

It's no coincidence that the largest of China's recent anti-Japanese protests occurred in cities such as Beijing, Shanghai, Guangzhou and Shenzhen, where the use of the Internet, cell phones and online chatting is among the highest in China. That's because a notable feature of the recent protests was that they were almost exclusively organized through such modern communication technologies....Right after the first public demonstration in Beijing on April 9, eyewitness accounts, photos and video clips from the protests spread rapidly through Chinese cyberspace despite a complete blackout of coverage in the official media. At the same time, demands for a boycott of Japanese products, online petitions, and calls for street demonstrations in many cities throughout China were widely distributed by the Internet and cell phones. Many of these messages were extraordinarily detailed, giving logistical information such as the route of the protest march and even what slogans to chant. (Qiang 2005).

Qiang noted that the messages were sent out in chain letter form via e-mail and text messages, and were posted on BBSs. The April 16 protest was organized primarily this way, "defying calls from Shanghai authorities for students to stay within campus." At the same time, Qiang acknowledged, in response to a query from me, that he suspected "the extraordinarily detailed" messages were planted by Chinese government agents.

Ironically, considering the ideological origins of the ruling Communist Party, more and more incidents of urban unrest appear to be outbreaks of class-based conflict. In July 2005, according to the *Washington Post*, a wealthy businessman's expensive automobile collided with a bicycle rider in Chizhou, China. The businessman's bodyguards beat the bicyclist bloody. Onlookers used their mobile telephones to summon others, resulting eventually in a full-blown riot of an estimated ten thousand people (Cody 2005). When police took the businessman down to their station, motorcycle drivers and vegetable merchants who had witnessed the accident followed. The *Washington Post* article reports, "Members of the crowd pulled out their cell phones to call friends and relatives, swelling their numbers further. By 3:30, witnesses recalled, several thousand people were gathered around the station" (Cody 2005).

The *New York Times* reported in 2005 that Chinese paramilitary police killed as many as twenty people in rural China who had been protesting a power company's plans to build a coal-fueled generator, which they feared would be a source of dangerous pollution. The *Times* cited one of the reasons why such incidents are becoming more common as

cellphones have made it easier for people in rural China to organize, communicating news to one another by text messages, and increasingly allowing them to stay in touch with members of nongovernmental organizations in big cities who have been eager to advise them or to provide legal help. (French 2005)

In the United States, organizers of protests against the Republican National Convention in New York City telegraphed their intention to use SMS to organize swarming, and one politically sympathetic enterprise created TXTmob, which enables Web-organized groups to send and receive text messages via mobile phone. However, protestors ran into a technical glitch. Although many of the dissidents suspected that T-Mobile had shut them down for political reasons, it was later discovered that the operator's spam filter had always been set to block sites that broadcast hundreds of text messages to subscribers (Rojas 2004).

John Henry, of the Institute for Applied Autonomy, which created the TXTmob application, together with Tad Hirsch, of MIT's Media Lab, pointed out a significant difference between the RNC protest and previous demonstrations around the world in which text messaging had played a part in spontaneous self-organization:

Unlike the protests in Manila and Madrid, the mass mobilizations against the DNC and RNC were neither spontaneous nor unexpected. Activists and law enforcement officials alike had planned and trained for months in anticipation of open conflict in the streets of Boston and New York. Accordingly, text messaging was part of a broader communications strategy developed by organizers, and was put to a wider variety of uses than simply getting bodies into the street. (Henry and Hirsch 2005)

Henry and Hirsch point out that law enforcement authorities appear to have learned key tactical lessons from previous instances of mobile phone–enabled street-swarming:

After the J18 and Seattle protests, law enforcement has adopted a more aggressive approach to crowd control during large-scale demonstrations. Independent observers have come to call the current strategy "The Miami Model," named for its use during protests against the 2003 Free Trade Areas of the Americas (FTAA) summit. The Miami Model has been described as "the criminalization of dissent," and is characterized by restricting public access to large parts of the city, pre-emptive arrests of activist "leaders," widespread use of nonlethal weapons including tear gas, pepper spray, and rubber bullets, and the use of mass arrests or "sweeps" that often includes the detention of law-abiding citizens who are later released without charge. (Henry and Hirsch 2005)

TXTmob was created as a tool to support the counter-counterreaction of demonstration organizers, who recognized that even more radical decentralization of protest was

234 | Howard Rheingold

called for—requiring more effective mobile, ad hoc communications infrastructure. Individual demonstrators, affinity groups, and organizers could create their own mobile message distribution lists, populate them with their own members, and determine whether membership would be open or closed. The Institute for Applied Autonomy was able to collect statistics on the use of the tool: 5,459 people registered with TXTmob during the conventions, exchanging 1,757 messages among 322 "mobs." Henry and Hirsch analyzed message content, mob descriptions, and timestamp data. Remote users who were not on the scene were able to monitor activities as they happened. The authors noted the uniqueness of this situation in the annals of street protest:

Remote users consistently expressed solidarity with protesters, and used the phrase "it felt like being there" to describe their experience. This raises several questions:

1. Does this form of participation constitute a collective identity (a sense of "we-ness" among participants)?

2. If so, is it reciprocated by activists in the street?

3. How does remote identification contribute to the ongoing work of movement-building by activist groups?

4. Is there something unique about mobile devices that enhances such identification (in a way that, say, email or websites don't). (Henry and Hirsch 2005)

Blogger Jeff Vail (2005) writes about the essentials of swarming tactics in his blog. He argues that the anarchists at the Republican National Convention lost their advantage by planning their actions ahead of time around convention events:

In particular, the NYC police were able to deny the protesters their principle strength of elusiveness/mobility. Borrowing directly from the playbook of Alexander the Great, police rolled out mobile plastic-mesh fences to quickly create artificial terrain obstacles, trapping large groups of more violent protesters before they could blend away into the city masses. (Vail 2005)

That anarchists failed to successfully execute swarming tactics at the Republican National Convention should probably not be regarded as proof that the tactic is obsolete. RAND Corporation analysts Arquilla and Ronfeldt write incisively and ominously about the intersection of smart mobs and organized violence, either state-sponsored (military) or "non state actors" (peaceful activists or armed terrorists):

Swarming is seemingly amorphous, but it is a deliberately structured, coordinated, strategic way to strike from all directions, by means of a sustainable pulsing of force and/or fire, close-in as well as from stand-off positions. It will work best—perhaps it will only work—if it is designed mainly around the deployment of myriad, small, dispersed, networked maneuver units (what we call "pods" organized in "clusters"). (Arquilla and Ronfeldt 2005)

Arquilla and Ronfeldt point out that while swarming has existed for a long time, it has now emerged as a power in its own right, one that can be directed and coordinated on the fly. The authors argue, "That is largely because swarming depends on a devolution of power to small units and a capacity to interconnect those units that has only recently become feasible, due to the information revolution" (2005).

Another violent form of political collective action is the riot. *Time* magazine reported on the riots in Nigeria triggered by the Miss World pageant (Taylor 2003). Because riots appear to require a critical mass of people who are willing to go beyond social and legal bounds, it is possible that the Nigerian riots might not have happened if text messages had not been used to summon people to the scene (Granovetter 1978).

In November 2005, disaffected Muslim immigrant youth in Paris suburbs, reacting to the deaths of two adolescents who were electrocuted when hiding from police in an electrical substation, began riots that spread through France, abetted by both Internet and text message communications. Patrick Hamon, the national police spokesman noticed that "bands of youths are, little by little, getting more organized" and sending attack messages by mobile phone (Smith 2005). The *Belfast Telegraph* reported in September 2005 on a growing phenomenon in North Belfast—recreational rioting by teenagers and children, some as young as five, recruited in the playground by text messaging. The police discovered one particular text message being distributed on school playgrounds: "R U up for a riot 2 nite?" (McCambridge 2005).

Action Alerts and Boycotts

Nonviolent direct action ranges from lobbying to boycotts. Amnesty International uses SMS to broadcast action alerts and mobilize political pressure, according to blogger Emily Turrettini:

Participants in their campaigns, opt-in by signing up online and giving their mobile number. They will then receive an "action" SMS every two weeks, which is then invoiced directly to their phone bill at a premium rate of $0,28 (25 eurocents). The latest one sent out for instance, was concerning the plight of a 16 year old boy who has been abducted in Guatemala. The recipient replies to this "action" SMS with a simple "yes," which will serve as a digital signature, his name then added to a petition which Amnesty International will send off to the Guatemala government to pressure them into releasing the boy.

To date, 7102 people have signed up. The campaign not only pays for itself but is also a clever and personalized way of keeping its members informed of their efforts and allows them to be active participants in a cause. (Turrettini 2003)

Anneke Bosman reported on a Web site about "New Tactics in Human Rights" that Amnesty International's SMS action alerts have also been successful in the case of a political prisoner in Africa. She also reports on a high success rate at inducing people to make calls and write letters, and mentions an interesting angle—the young people who are the most avid users of SMS are an audience that political activists want to reach:

With these messages, protests can be gathered faster than ever, enabling Amnesty International to take action against torture and other abuses more quickly. About 39 percent of the cell-phone campaigns conducted by Amnesty in 2002 were successful. Prisoners of conscience were released, people who had "disappeared" were found and death sentences were not carried out. Cell-phone campaigning also has a special appeal for youth, and we found this campaign attracted new younger members into Amnesty in a way that other outreach and activities had not. (Bosman 2004)

Young Moroccans in the Netherlands are turned away from clubs, because of—they claim—racial profiling. A few of them started an initiative to address this issue, the Web site Geweigerd.nl—*geweigerd* is Dutch for "refused." The site has a forum that can be reached by i-mode. People who are refused at a certain club are encouraged to immediately send a message with their mobile phone to Geweigerd.nl. They hope that the collective action of reporting small incidents helps the public get a grasp on the problem, and will eventually lead to a solution.

One boycott in Nigeria, which was not extensively reported outside Africa, was the subject of a scholarly paper by Ebenezer Obadare:

On September 19, 2003, following weeks of concerted mobilisation, mobile phone subscribers in Nigeria took the unprecedented step of switching off their handsets en masse. The consumers took this symbolic measure in protest against perceived exploitation by the existing GSM phone companies-Zimbabwean-owned Econet Wireless Nigeria Limited, the South-African-owned MTN Limited, and the Nigerian state-owned NITEL. (Obadare 2004)

The Nigeria boycott paper further explains that the use of mobile phones as political tools ought to be seen in the context of the customer and citizen's mistrust of the transnational corporations and state. That the power of the protest came from turning the mobile phones off is support for the paper's conclusion:

Thus, for Nigeria, while mobile telephony has no doubt come to be seen as a veritable instrument of political struggle, its potential effectiveness is bound to be determined by the way in which it is used. And while it is definitely a welcome addition to civil society's arsenal, it may not necessarily fulfill the fondest telecommunicative fantasies about securing total victory in the contest for social and economic justice. (Obadare 2004)

Conclusion: Something Significant Is Happening

The rapid adoption of sophisticated multimedia communication media by a significant portion of the world's population already is giving rise to spontaneous social experiments of varied forms. In the political sphere, the power of persuasion, organization, and coordination have been democratized worldwide by the availability of mobile telephones and text messaging. The examples cited here are neither exhaustive nor analytical. There are no guarantees that future smart mobs will be peaceful, or that democratization of the power to organize collective action will lead to stronger democracies.

The reasons political smart mobs could lead to stronger democracies include the empowerment of citizens to self-organize popular demonstrations in protest of events such as the Estrada impeachment, the capability of inexpensive and publicly visible monitoring of elections for fraud, the increased ability for volunteers to coordinate get-out-the-vote activities, and the power to disseminate information that is suppressed by authoritarian regimes and controlled mass media. The reasons political smart mobs may weaken democracies include the speeding up of political decision making that would benefit from more slowly paced deliberation, the manipulation of populations by planted provocations and misdirection, the potential for violent outbursts at spontaneous gatherings, and the potential for rapidly disseminating misinformation and disinformation. Finally, it is possible that these technologies will strengthen the hand of centralized authorities, the views of Henry and Hirsch (2005) and Arquilla and Ronfeldt (2005) notwithstanding, if such authorities succeed in automating surveillance, jamming, and countermeasures such as "the Miami model," and/ or introducing political "noise" in oppositional communications.

Perhaps the most important question about the future of augmenting collective action through the use of the Internet and mobile communications is the degree to which trustworthy and accurate information can be distinguished and screened from misleading, false, missourced information. The technosocial capability of increasing the trustworthiness of information through many-to-many media could magnify the positive potential of populations using these technologies to achieve their ends in democratic, cooperative, or at least nonviolent ways. To the extent that accuracy of information cannot be determined, the positive potential of these powerful technologies may be blunted if not turned against itself.

References

Adelman, J. 2004. U say u want a revolution: Mobile phone text messaging is evolving into a political tool. *Time Asia*. http://www.time.com/time/asia/magazine/article/0,13673,501040712 -660984,00.html.

Armond, P. 2000. Black flag over Seattle. *Albion Monitor*. http://www.monitor.net/monitor/ seattlewto/index.html.

Arquilla, J., and D. Ronfeldt. 2005. Swarming and the future of conflict. *RAND*. http://www.rand .org/pubs/documented_briefings/2005/RAND_DB311.pdf.

Black, J. 2003. Technology with social skills. *BusinessWeek Online*. http://www.businessweek.com/ technology/content/aug2003/tc20030819_4587_tc126.htm.

Bosman, A. 2004. Sending out an SMS. *New Tactics in Human Rights*. The Center for Victims of Torture. Minneapolis, Minn., 2004. http://www.newtactics.org/main.php/SendingOutanSMS.

Carney, J., and J. F. Dickinson. 2002. W. and the "Boy Genius." *Time*. http://www.time.com/time/ nation/article/0,8599,388904,00.html.

Cody, E. 2005. A Chinese city's rage at the rich and powerful. *Washington Post.* http://www.washingtonpost.com/wp-dyn/content/article/2005/07/31/AR2005073101163.html.

Dányi, E., and M. Sükösd. 2003. M-Politics in the making: SMS and e-mail in the 2002 Hungarian election campaign. In *Mobile Communication: Essays on Cognition and Community,* edited by Kristóf Nyíri. Vienna: Passagen Verlag.

Dominguez, E. 2004. SMS, the star media of the Spanish elections. *Poynter Online E-Media Tidbits.* http://www.poynter.org/column.asp?id=31&aid=62558.

French, H. W. 2004. Workers demand union at Wal-Mart supplier in China. The *New York Times.* http://www.nytimes.com/2004/12/16/international/osia/16china.html.

French, H. W. 2005. Protestors say police in China killed up to 20. The *New York Times.* http://www.nytimes.com/2005/12/10/international/asia/10china.html.

Granovetter, M. 1978. Threshold models of collective behavior. *American Journal of Sociology* 83(6): 1420–1443.

Henry, J. and T. Hirsch. 2005. *TXTmob: Text messaging for protest swarms.* http://web.media.mit.edu/~tad/htm/txtmob.html.

Hoenig, H. 2003. Beijing goes high-tech to block Sars messages. The *New Zealand Herald.* http://www.nzherald.co.nz/index.cfm?ObjectID=3507534.

Kagai, B. 2002. Mobile phone plays role in free Kenya elections. *Communication for Development News.* http://www.comminit.com/C4DNews2003/sld-7120.html.

Kalondo, E. 2005. Kenyans hold peaceful referendum after bitter campaign. *Monsters and Critics News.* http://news.monstersandcritics.com/africa/article_1063516.php/Kenyans_hold_peaceful_referendum_after_bitter_campaigns.

Krebs, B. 2002. Technology shapes get-out-the-vote efforts. The *Washington Post.* http://www.washingtonpost.com/ac2/wp-dyn?pagename=article&node=&contentId=A2467-2002Oct9&notFound=true.

Laverty, P. 2004. Spanish tools against terror: Mobile phones, pots, and pans. *Peacework.* http://www.afsc.org/pwork/0404/040404.htm.

Ling, R., and B. Yttri. 2001. Hyper-coordination via mobile phones in Norway, nobody sits at home and waits for the telephone to ring: Micro and hyper-coordination through the use of mobile telephones. In *Perpetual contact,* edited by J. Katz and M. Aakhus. Cambridge: Cambridge University Press.

McCambridge, J. 2005. Weekend debate: Shame of the child rioters. *The Belfast Telegraph.* http://www.belfasttelegraph.co.uk/eceRedirect?articleId-980311/pubId-105953.

Obadare, E. 2004. The great GSM (cell phone) boycott: Civil society, big business and the state in Nigeria. *Isandla Institute.* http://www.isandla.org.za/dark_roast/DR18%20Obadare.pdf.

Qiang, X. 2005. China's first web-organized protests. *The Asian Wall Street Journal*. http://www.wsj.com/wsjgate?source-jopinaowsj&URJ-/article/0,,SB111455387951817586,00.html.

Rafael, V. 2003. The cell phone and the crowd: Messianic politics in the contemporary Philippines. *Public Culture*. http://communication.ucsd.edu/people/f_rafael_cellphonerev_files.htm.

Reynolds, P. 1999. Eyewitness: The battle of Seattle. *BBC News*. http://news.bbc.co.uk/1/hi/world/americas/547581.stm.

Rheingold, H. 2002. *Smart Mobs: The Next Social Revolution*. Cambridge, Mass.: Perseus.

Rheingold, H. 2003. First-hand report on Korean election. *Smart Mobs*. http://www.smartmobs.com/archive/2003/06/02/firsthand_repo.html.

Rheingold, H. 2004. Berlusconi's SMS electioneering. *Smart Mobs*. http://www.smartmobs.com/archive/2004/06/13/berlusconis_sm.html.

Rheingold, H. 2004. Phones, radio, elections in Ghana. *Smart Mobs*. http://www.smartmobs.com/archive/2004/02/09/phones_radio_.html.

Rojas, P. 2004. More on T-Mobile blocking TXTmob messages during last week's RNC. *Engadget*. http://www.engadget.com/entry/7820473160426794/.

Shantiran, S. 2003. Txt Craze. *Star TV Focus Asia*. http://focusasia.startv.com.

Smith, C. S. 2005. As rioting spreads, France maps tactics. The *New York Times*. http://www.iht.com/articles/2005/11/06/news/france.php.

Taylor, C. 2003. Day of the Smart Mobs. *Time*. http://www.time.com/time/archive/preview/0,10987,1004369,00.html.

Teachout, Z. 2003. Dean wireless tops NASA wireless. *Blog For America*. http://www.blogforamerica.com/archives/000435.html.

Texually.org. 2003. *SMS Messages Urge Consumers to Boycott US Products*. http://www.textually.org/textually/archives/2003/03/000134.htm.

Ture, David. 2004. Personal Communication.

Turrettini, Emily. 2003. In the Netherlands, Amnesty International uses SMS as an action tool. *Textually.org: All About Texting, SMS, and MMS*. http://www.textually.org/textually/archives/2003/05/00051.htm.

Vail, J. 2005. Swarming, open-source warfare and the black block. *Theory of Power*. http://www.jeffvail.net/2005/01/swarming-open-source-warfare-and-black.html.

Yardley, J. 2005. A hundred cellphones bloom, and Chinese take to the streets. The *New York Times*. http://www.nytimes.com/2005/04/25/international/asia/25china.html.

Zuckerman, E. 2004. My blog is in Cambridge, but my heart's in Accra. *Ethan's Weblog*. http://blogs.law.harvard.edu/ethan/2004/12/07#a626.

18   Mobile Multimedia: Uses and Social Consequences

Ilpo Koskinen

In chapter 22 of this volume, Kenneth Gergen analyzes how mobile telephony is transforming the political process and democracy. For him, the structure of political communication in Western societies has gone through a series of highly significant changes during the past fifty years. Between the government and the individual voter, there has always been a layer of face-to-face relationships in which people deliberate social issues and political issues. The fourth layer, a more recent addition, is mediated communication that was originally monological: the public was informed, but had only limited possibilities in participating in opinion formation in media. However, with mobile phones, the nature of mediated communication changes. For example, people are able to organize political protests, as the ousting of Joseph Estrada, the former president of the Philippines shows (see Rheingold 2003a). The history of how he left his position was largely a story of urban crowds organizing massive demonstrations with mobile phones and text messages. With messages such as "Go 2EDSA, Wear blck," people organized a series of demonstrations at Epifanio de los Santos Avenue (EDSA), a major Manila thoroughfare (Rheingold 2003a, pp. 157–160). We can call this picture optimistic. Mobile communication is an important constituent of what Gergen calls "the proactive *Mittelbau*," opinion-formation and action that is rooted in the independent realities of civil society rather than in the opinions of political elites or mass media.

However, the second picture Gergen paints is more somber. In this vision, the civil society is being slowly replaced by small communication clusters, which increasingly take the role previously played by face-to-face conversation in public venues. Political communication shifts from civil society to these "monadic clusters," as Gergen calls them. Instead of participating in society, people move through the day largely disengaged from those around them, turning instead to their friends when in trouble or in need of advice or encouragement. In these clusters, people focus on immediate life and microrelationships at the cost of civic concerns. If they focus on issues relevant to democracy, they construct their opinions with their friends and acquaintances rather than in political parties or by participating in community decision making. People are distanced from politics, disrupting dialogue necessary for a healthy democracy.

The first sociological studies on mobile multimedia tend to point toward Gergen's second picture. For example, Scifo (2005) firmly situates camera phones and mobile multimedia in ordinary communication among friends and acquaintances, and Koskinen (2007b) has characterized mobile multimedia as machinery that produces banality: people capture and share ordinary things, but account and explain them into something exciting to justify sending them. In a darker tone, Rivière (2005) builds on French psychoanalysis, and connects ordinary uses of multimedia to the pleasure-seeking primary processes of human psyche. For her, multimedia phones increase intimate and sensational, spectacular communication: "they increasingly use intimate play context, which have no rational purpose but rather aim at sensations, and in which the search for immediately shared pleasure is more and more visible" (Rivière 2005, p. 212). However, mobile multimedia may also contribute something of more lasting value to society. For example, it may increase the feeling of belonging, make people more aware of their visual and aural environment, ease occupational problem solving, and enable citizen journalism. Why should it not contribute to the proactive Mittelbau also?

This chapter explores mostly English-language literature on mobile multimedia and by doing so tries to fit this phenomenon to the pictures Gergen gives us. My aim is to open debate on what mobile multimedia is doing in society by opening its *Apparatgeist*, as Katz and Aakhus (2002, p. 11) have called frameworks aimed at understanding the uses and consequences of mobile technologies. I wish to include the views of not just experts like researchers and journalists, but also of lay people to bring these frameworks into view to understand mobile telephony.

With the exception of Japan, mobile multimedia has not achieved anything like the success of text messaging (SMS). Still, the question of its consequences is already significant. Multimedia phones—or colloquially, "cam phones"—are already virtually ubiquitous. In 2004, more than two hundred million camera phones were sold, and the figure for 2005 may be over three hundred million. When this chapter was written, there were no reliable figures for phones with a TV capability, but these phones are becoming increasingly more common with so-called third generation network technology (or 3G). If we believe the industry, multimedia will be one of the main forces that drive the adoption of these faster networks. In an attempt to answer the questions Gergen poses, this chapter explores a small but growing body of social science research on mobile multimedia rather than building on original data. There is an extensive body of literature on multimedia in engineering, but I refer to this literature only if it is based on extensive, well-documented user studies, with at least a few weeks' field period. Usability studies and design studies with mock-ups of user interfaces have not been included. For reasons of space, I skip legal concerns related to mobile photography, and also marketing studies exploring the acceptability of services (see www.textually .org/picturephoning).

In this chapter, "mobile multimedia" refers to a whole set of new devices and technologies that have entered our pockets during the past few years. There are several competing multimedia technologies on the market, but from the standpoint of the Apparatgeist theory, I divide mobile multimedia into *mobile multimedia messages*, *moblogs*, and *mobile mass media*. The first term covers personal technologies that people can use to capture and share photographs, text, audio files, and, sometimes, short video clips (the best known technology is MMS, multimedia messaging system). Moblogs refer to messages shared through Web sites. The last category covers mobile television and mobile movies aimed to distribute professionally created and edited content to mobile phones that function as terminals for accessing this content.

Personal and Social Uses of Mobile Multimedia

By far, the best studied area of mobile multimedia is its personal and social uses. In her study of how young Milanese use MMS, Scifo (2005) learned that it is primarily a small group activity. Based on her interviews, she notes that multimedia messaging almost exclusively takes place in a network of strong relationships, in which it primarily is used for sentimental purposes. With multimedia messages, people give others access to places, individual and social situations, and emotions. For example, they share images of familiar objects and people, private life (as in objects, relatives, and haunted places), and social networks. If Scifo is right, multimedia roots small groups tighter to Gergen's immediate life, whether physical or social, and makes it an object of joint concern.

This interpretation gets support from several other sources. For instance, in terms of contents, images captured with multimedia phones focus on familiar objects such as family members, friends, self, pets, and travels (Okabe and Ito 2004; Kindberg et al. 2004). By and large, mobile multimedia seems to continue the tradition of ordinary snapshot photography, but makes it even more ad hoc in terms of what people choose to shoot (Chalfen 1987; Koskinen, Kurvinen, and Lehtonen 2002, pp. 21–26). For example, as Kindberg and colleagues (2004) argue, people tend to capture issues relevant for small groups rather than society at large. They classified 82 percent of 349 messages they gathered from British and American phone users as affective rather than functional. Emotional content and entertainment dominates audio and video, too, although people may use these features for functional tasks (such as sending driving instructions) as well (see Kasesniemi et al. 2003; Koskinen 2005). Also, it appears that as relationships get more intimate, messaging tends to get even more mundane. While friends and acquaintances tend to capture and share things, events and observations that are at least minimally interesting for the recipient, couples share pictures and sounds about almost anything they happen to see or hear just to maintain "visual co-presence" (Ito 2005). However, although most multimedia pictures focus on immediate life, this does not imply that immediate life comes in simple packages. What people see as important may result from years of symbolic and imaginary work: while "Paris" may be a sign on the map for one person, for another it may be an elaborate, cherished experience created over several years (Battarbee and Koskinen 2004). Also, messages may be designed using complex constructs. For example, people often take advantage of artifacts they find from media and culture, including documents, snapshots, post-cards, greetings, and chain messages, which are sometimes downloaded from the Web (Ling and Julsrud 2005).

A good deal of multimedia content seems to be for private purposes only. In their phones, people carry a mobile archive of memories. This archive is "always within easy reach, something to look at again and again, when feeling nostalgic, or just to pass an interstitial moment in one's daily routine" (Scifo 2005, pp. 365–366). In a more social vein, camera phones have been characterized as "capture and show" rather than "capture and send" devices (Kindberg et al. 2004, p. 12). In this function, camera phones work better than digital cameras simply because many people carry mobile phones with them practically all the time.

However, some contents are sent from one phone to another: they become messages. Do they follow different premises than those messages that are kept private or shared from the screen only?

One body of research has explored this question. Koskinen and his colleagues have studied how people use multimedia phones to interact with their fellows. In *Mobile Image* (Koskinen, Kurvinen, and Lehtonen 2002), they gave an advanced mobile phone and a camera to four groups of five people for approximately two months each. People could beam photographs from the camera to the phone via an infrared link, and send them as an attachment in mobile e-mails. In this study, images became methods used in ordinary interaction. For example, in Message 1, Johan requests a photograph from his friends. He had been in a floorball tournament a day before (floorball is a form of indoor hockey played mostly in Scandinavia), which his team had won. He knew that his friends had been taking pictures all along the tournament. In Message 2, Erik sends the requested picture and continues to ask for a lunch companion. The example shows how multimedia messages can "chain" in sequences that function much like turns in conversation (see figure 18.1).

In *Mobile Multimedia*, a sequel to *Mobile Image*, Koskinen and colleagues have analyzed interaction in more detail. For example, Kurvinen (2002) shows how people share not just serious emotions—such as love—with multimedia, but also more fleeting ones, such as the apathy following a party. In another paper, he has focused on how people tease one another with pictures (Kurvinen 2003). In this analysis, pictures and other multimedia elements are something people use in interaction for practical

Messages 1–2. Prize (2000/11/08 11:46:54 and 2000/11/08 12:17:08)		
	Multimedia content:	*Text:*
Request	–	Hi! Does anyone have good pictures from Yesterday? For example, at least 100 pictures were taken with Toni's camera. Has anyone a picture of me with the trophy? Johan
Response		It's easy to smile when you win. Lunch companion is available near to the School of Business. Erik

Figure 18.1
Messaging with multimedia. Source: Author.

purposes, not something that drives use from behind the backs of people. It goes without saying that in these studies of multimedia-in-action, messaging overwhelmingly takes place between couples, friends, and acquaintances (Koskinen 2007a). Following Simmel (1949), it can be described as sociability, interaction with no purpose outside itself (see Koskinen, Kurvinen, and Lehtonen 2002). Thus, the main conclusion from these studies corroborates things learned from interview and diary studies: mobile multimedia binds already closely knit networks tighter together (Koskinen 2007a).

I take these observations to mean that mobile multimedia provides people a sociable channel through which they can entertain each other in Gergen's monadic clusters. People do send information and solve problems through messages, but these exchanges by and large take place in the same tightly knit networks as more common emotional and entertaining exchanges. In occupational uses, of course, mobile multimedia may maintain occupational rather than only ordinary practices. For example, carpenters studied by Ling and Julsrud (2005) photographed details of their work not just for later reference but also for clarifying their problems and solutions to coworkers. I have also learned about a musician who uses his mobile phone to collect and share his musical ideas by, for example, recording a new rhythm with it for later reference. These observations and anecdotes suggest that certain occupational groups will develop multimedia cultures that differ from ordinary uses. Still, it can be argued that even these occupational uses transform small groups into self-reliant "telecocoons" (Matsuda 2005) rather than connect them to surrounding society, not to mention developing new forms of consciousness and taking stance to it. However, there seems to be at least one group that benefits from mobile video more than others. The deaf can use their mother tongue, sign language, with videophones (Kasesniemi et al. 2003).

Moblogs and Citizen Journalism: Mobile Multimedia and the Proactive Mittelbau

People can share images and other multimedia content with their phones not just by showing messages from the screen or by sending them to another phone. They can also send their messages via the Web. These contents even can be viewed with mobile phones provided they have proper software that parses the Internet contents into the small screen of the mobile phone. Software in newer phones makes it even possible to create Web sites with the phone. If one sends text to these sites to augment images, a photo album turns into a "moblog," a Web-based diary-like site from which readers can follow the writer's life and opinions. Perhaps moblogging could support the proactive Mittelbau?

As such, moblogging seems to be unusual even in Japan, where mobile Internet first became popular (for inside stories, see Matsunaga 2000; Natsuno 2003). In Japan, moblogging first took off around 2002. However, a Japanese survey from 2002 tells that only 0.6 percent of Web phone—mobile phones with an access to the Internet—owners had created a Web site. More prevalent uses of phones (more than 30 percent of users had tried at least once during the past year) were e-mail, music file downloads, image downloads, and visits to games or fortune-telling sites (Okada 2005, p. 49). When it comes to journalistic uses of moblogs in Japan, these seem to center on sharing newsworthy events in personal life and "stalking" celebrities with cam phone pictures rather than serious journalism. Even group diaries seem to be a rare occurrence, although Ito reports about the *"Sha-mail* Diary Confederation" (*sha-mail* refers to a popular handset) in which twenty-nine writers shared their diaries and *sha-mail* photos in 2004 (Ito 2004). The global situation today does not change this picture. In a recent study of moblogs, Döring and Gundolf (2005) observed that although by 2004 there were already hundreds of thousands of moblogs globally, only few were active after the first week. Most moblogs are personal in content, focusing on trite things of everyday life. Topical moblogs typically reflect interests in topics familiar from snapshot photography, such as children and pets.

In more institutional settings, moblogs may come to have more encompassing and long-term effects. For example, moblogs may change the way in which a whole community understands itself. A pertinent example comes from a community-development study of the Shibamata neighborhood, located at Tokyo's easternmost edge. This neighborhood had enjoyed a constant stream of tourism after the popular movie and TV series *Otoko wa tsuraiyo* (*It's Tough to Be a Man*) was located in the neighborhood in 1969. When the series ran out in 1996 after the death of Kiyoshi Atsumi, the actor who played the main character, tourism started to decline, and the local community had to redefine its charms. In 2004, a group of researchers led by Fumitoshi Kato (Kato and Shimizu 2005) sent a group of students for field work with camera phones in Shibamata. They set up a moblog site for the neighborhood and created a series of postcards from student pictures to display the attractive qualities of the neighborhood. Some of these pictures were linked to audio files that could be accessed on the Web by scanning a bar code with a camera phone. As Kato and Shimizu observe, whenever someone sends multimedia messages to a moblog, one gives clues about oneself. In Shibamata, moblogs are above all social things:

"Community-moblog" can be understood as a "place" for one's face-work. . . . In posting a photo, a member is constructing and maintaining the relationships with others. An individual's postings are not only displaying to other members what he/she has been, but also, he/she is displaying about him/herself, and his/her understandings about the relationships with other members. By sharing the "community-moblog," members define, redefine, the situation within which they are embedded. (Kato and Shimizu 2005)

This study points to one way in which multimedia phones might be a proactive force in social change. It has similarities to several critical art projects (see Dewdney and Lister 1988). Under certain conditions, community moblogs with a critical edge might sustain social movements and perhaps even smart mobs discussed by Rheingold (2003a).

Another way in which mobile multimedia may empower the Mittelbau is through citizen journalism. As soon as there are hundreds of millions of camera phones in society, things previously unseen by news media become increasingly photographed. Perhaps the best known recent example comes from London bombings by the terrorist organization Al Qaeda in July 2005. Pictures and videos from dark "tube" tunnels were sent immediately around the world. This footage mostly came from eyewitnesses' and victims' camera phones. It ended up on Weblogs in a couple of hours and was picked up by major news media like the BBC, CNN, and London's the *Sun* shortly afterward, raising issues of reliability and copyright of images sent on the fly, as well as privacy and excessive risk-taking by would-be citizen journalists (Noguchi 2005; Snoody 2005). It is also fairly easy to imagine moblog sites maintained by consumer groups where people could send photographs of dangerous goods or misleading advertisements they spot in shops. Such sites would collect evidence for consumer magazines, turning camera phones into instruments of consumption criticism. However, it is just as easy to imagine racist groups spreading propaganda with camera phones.

Thus, although early Japanese and emerging Western evidence suggests that camera-phone journalism largely concentrates on ordinary events and stalking celebrities (Ito 2004), this is not the whole picture. As Dunleavy (2005) notes, with London bombings, camera phone images became an accepted part of quality journalism: camera phone pictures were run on the front pages of the *New York Times* and the *Washington Post*. Another event that proved the value of camera phones (and amateur video) was the Asian tsunami at the end of 2004. Although media reacted quickly to the disaster, government response was slow and less reliable than the ad hoc responses from citizen journalists in Sri Lanka and Thailand. Although early fears about everyone becoming paparazzi may to some extent have been proven to be true, as in the case of rewards promised by tabloids for fresh cam phone photos, camera phones also have proved their social value in journalism. By the end of 2005, there were at least three Web-based photo agencies that sold pictures submitted by ordinary citizens to media (for example, www.celljournalist.com and www.scoopt.com). The first international conference on moblogs and journalism took place as early as 2002 in Tokyo (Rheingold 2003b), and at least one journalistic book based on observations on camera phone photos has been published (Margolis 2005).

With the possible exceptions of moblogging in Japan and citizen journalism, moblogs have not yet become a particularly popular form of sharing mobile multimedia content. Still, new forms of social action may be taking shape in the cross-section of wired and mobile digital technologies. However, it is just as likely that moblogs will simply function like photo albums: for monadic clusters, they serve as reservoirs of memories, filled with stories of the self, family, and friends, as well as trophies such as homes and cars. If anything, experience from personal and social uses points to the latter vision rather than to significant transformations in politics. Incentives to share multimedia with the wider society are today mostly provided by institutions of the caliber of *People* and the *National Enquirer*. Still, as the London bombings show, under some circumstances, multimedia messages may make a more valuable contribution to society. I am probably not much off the mark if I predict that the moblog culture will grow in two main directions, one increasingly personal and social, another increasingly tied to more significant institutions such as journalism. Personal content may occasionally become interesting enough to catch the public eye, but mostly under special circumstances like the tsunami disaster or the London bombings.

Mobile Multimedia as Mass Media

I first analyzed mobile multimedia mainly as a personal technology that, under certain circumstances at least, may also feed into national and even international news sources. However, even then, multimedia phones are firmly in the hands of ordinary people who create content with them. With the possible exception of moblogs, people share content with only a few of those people whose contact information they have in their phones. Now, there is another way in which significant institutions such as the media have an interest in multimedia phones. Multimedia phones can be used as terminals for receiving digital video and TV program stream, and an increasingly number of newspapers around the world can be accessed with multimedia phones. Such content is produced and edited by professional moviemakers, advertising agencies, journalists, and editors, for whom mobile phones provide another increasingly attractive channel.

Probably the first experiments in making movies for mobile phones were known as "micromovies" (see Boyd Davis 2002; Metso et al. 2004). This term describes small-scale movies that can be viewed with various mobile devices including mobile phones. In contrast to traditional cinema, which is made for large audiences and is watched in theaters or TV screens, micromovies are more personal and are shown in mobile terminals that usually have an input equipment that is sufficiently elaborate for playing computer games. Consequently, while most micromovies are just movies produced for a small screen, some movies have interactive features. For example, the viewer can select from several possible plots at predetermined spots, or use personal information stored in the phone to select from alternative courses of the story. There have even been several micromovie festivals at least in France, the United States, and Finland, and movie prizes even have been given to the best works in these festivals.

Of course, showcases in media art do not mean success in society. Micromovies have been popular within a small group of technology enthusiasts and digital artists rather than the public at large. However, in one country, South Korea, mobile movies has become a mass phenomenon. Ok (2005) describes how SK Telecom launched the mobile movie service "June" in 2002, and started to broadcast live TV feed, news broadcasts, sports shows, movies, TV drama, and animation through the service. At present, the movie/TV category produces 17 percent of June revenues. Ok distinguishes two content categories in this service. "Migrated cinematic imaginary" consists of reusing already existing material, while "original cinematic imaginary" consists of new content exclusively produced for the mobile phone. An example of a successful movie in the latter category is *Five Stars*, which, unlike traditional movies, has a game-like structure with an open ending. The narrative focuses on who will be loved by the main character; viewers can vote for their favorites on a Web site also accessible with a mobile phone. About 75,000 users ordered this drama following the first fifteen days of the launch, and in all, 400,000 users have watched it.

The billion-dollar question at the end of the first decade of the new century is how will mobile TV develop. In an interview for Finnish Broadcasting Company in December 2005, the head of Nokia's multimedia unit Anssi Vanjoki estimated that by the end of 2008 about 20 percent of all mobile phone users globally will have a handset with TV capacity. However, at the end of 2005, mobile TV was still in its infancy even in the most advanced mobile markets. For example, media companies, mobile carriers, and handset manufacturers have carried out studies in many European countries including Sweden, Finland, and the United Kingdom (see Södergård et al. 2003; BBC 2005). Also, the first field studies of technology were conducted in 2000–2001, but large-scale consumer studies are only currently underway. Some commercial services are already working, and more networks and content are coming to the market within a year or so for the most advanced mobile countries and cities. I suspect that TV will achieve some measure of success: forces pushing it—public broadcasters, media empires, and mobile phone operators—are powerful enough to withstand losses for years.

However, the commercial value of mobile TV is yet to be proven. An example comes from Finland, in which all major Finnish TV companies were recently involved in a study in which 487 consumers participated in a field trial with sixteen TV channels in three occasions (www.finnishmobiletv.com). In the first study, participants watched TV on average for about five minutes daily, mostly late in the evening. In two subsequent studies, the time increased to about twenty minutes. However, 58 percent of participants thought that mobile TV is going to be popular in the future. As expected, the heaviest users were young men with experience in mobile services, mobile Internet, smart phones, and camera phones, and had a habit of listening to music while on the move. Also unsurprisingly, the reasons for watching mobile TV were avoiding boredom (for example, while waiting for something and being stuck in traffic jams), staying updated (watching news), maintaining background entertainment when doing other things, and creating their own space.

Inevitably, a few marginal industries already have been taking advantage of multimedia technology—mostly MMS—with modest success. Just like ringtones and logos, downloading and sending background images (or *wallpapers*) for phones has gained a degree of popularity in several countries. The most reliable estimates are from Japan, where 34.8 percent of multimedia phone owners had downloaded images from the Internet at least once by 2002 (Okada 2005, p. 49). Of course, Japan may be an exceptional case, and we should be cautious of extrapolating from Japan to other countries. However, targeted mostly at teenagers and the young urban set, similar wallpaper industries exist in other countries as well. Finally, there is the marginal case of porn, which was probably the first industry to exploit the new channel, following the lead of sex lines and chatrooms (see Pertierra 2007). I received the first mobile multimedia porno ad from Switzerland as early as summer 2002.

We should remember that with the partial exception of civic journalism made possible by moblogs, mobile mass media is subject to the same limitations and social constraints as any journalism. All content in media is edited, designed, and dramatized. Governments do try to influence and sometimes, regulate, any program stream including one that takes place in mobile phones. Governments and other institutions try to control ordinary uses of camera phones too, for example, by banning them in military compounds and government agencies, and Saudi Arabia has banned camera phones altogether. However, it is far more difficult to control a ubiquitous technology embedded in millions of mobile phones than mass media. In its most ordinary uses, mobile mass media will probably be an individualizing technology, something designed to be followed and enjoyed alone. I find it difficult to imagine how the fall of New York's Twin Towers could have united hundreds of millions of people into a global, if temporary, community if they had seen these dramatic pictures from mobile phones. However, one should remember that most research on mobile mass media comes from some of the most stable societies in the world. It is difficult to get Finns on the street for any social or political cause. How would people in more labile societies with massive urban crowds accustomed to demonstrations react to mobile television?

Mobile Multimedia in Society

In this chapter, I analyze mobile multimedia in society by reviewing three dimensions of its *Apparatgeist*, as Katz and Aakhus (2002) have called interpretations people develop to make sense of mobile technologies. I also evaluate their significance by relating them to an analysis of mobile telephony Kenneth Gergen presents in this book. When it comes to the first Apparatgeist dimension, personal and social uses, the main message of my analysis is in line with Gergen's more somber picture. People capture ordinary things in immediate life and share them with their friend and acquaintances in monadic clusters that become even emotionally and relationally more self-reliant than before. It ties friends, acquaintances, and couples closer together by giving them means to share sensations as well as trivial and entertaining observations rather than information needed in solving problems in life. Though multimedia messages mainly focus on sensuous and emotional aspects of immediate life, it is too early so say whether messaging is just about harmless reproduction of ordinary things as Scifo (2005) and Koskinen (2007a) imply, or whether it turns the users to seek sensations and pleasure, as Rivière (2005) suggests. Probably both interpretations are right, given that quite often multimedia messages are about "pleasurable" things like food and having good times with friends and partners.

Since Japanese evidence largely concurs with early European and American evidence, we may see personal and social uses as the baseline for understanding mobile multimedia. Do other aspects of the Apparatgeist change this baseline?

Mobile multimedia phones appear to have brought about some changes to journalism, but convincing evidence is limited to two major news events, the Asian tsunami at the end the of 2004 and the London bombings in July 2005. Similar events will no doubt occur in the future, but if they have a lasting impact on journalism remains to be seen. It also remains to be seen whether mobile multimedia will erode journalistic standards, as some people have suspected, perhaps referring to the ever-increasing phenomenon of gossip columns that already have gotten a boost from camera phone photography (see Dunleavy 2005). With the exception of Korea, mobile TV and mobile movies have had even less success than mobile multimedia, which is also a marginal phenomenon in the bigger picture of mobile telephony, at least if compared to the huge success of text messaging. Mobile mass media function as any mass media. They give people means to follow edited media stream wherever they happen to be, splitting the audience into individuals.

What about how mobile multimedia relate to Gergen's more hopeful vision of the proactive Mittelbau? A precedent comes from text messaging, which may have significant consequences in society. In the Philippines, President Joseph Estrada reputedly fell victim to SMS in 2001. Although Pertierra and colleagues (2002, pp. 101–124) have shown that other media was as important in mobilizing the demonstrations, with the president's major opponents inviting people to demonstrations in church services, TV, and radio, mobile phones no doubt played their part in this process. There is no evidence of camera phones being used in organizing such activities, but there is no reason to think why they could not be used in this function either. I suspect that moblogs appear to be meaningful for small groups primarily. They function much like traditional photo albums that reserve memories, make the past reviewable, and make it possible to share significant moments in life with one's family and friends. Interest in such albums seldom goes beyond family members and friends, who know the background stories needed for understanding them.

This chapter places mobile multimedia in society by exploring its Apparatgeist. If I read the literature right, mobile multimedia phones primarily contribute to the grouping of society into small, monadic clusters rather than giving new means for organizing the proactive Mittelbau. These monadic clusters create microcultures around people, things, and events they face in immediate life. Civic affairs become little more than matters of passing commentary. Still, it is important to remember that most studies have focused on groups of friends. No studies of multimedia in the hands of social activists have yet been conducted, and studies of transitional societies are also still yet to be done. It remains to be seen whether mobile multimedia can become a technology that changes the consciousness of its users toward society and, following Katz and Aakhus's (2002) formulation, whether it alters minds and societies following the lead of text messages and mobile phones—or whether it just reinforces our existing habits and institutions. If we believe voices from the mobile industries and mass media, multimedia will change the way in which we experience our world and interact with it. After reviewing the Apparatgeist of mobile multimedia, I would like to argue to the contrary. People will surprise the mobile industries once again by defining mobile multimedia as yet another ordinary technology that is good primarily for personal and social activities.

References

Battarbee, K., and I. Koskinen. 2004. Co-experience—user experience as interaction. *CoDesign Journal* 1: 5–18.

BBC. 2005. Major UK mobile TV trial starts. http://news.bbc.co.uk/1/hi/technology/4271474.stm.

Boyd Davis, S. 2002. Interacting with pictures: Film, narrative and personalization. *Digital Creativity* 13: 71–82.

Chalfen, R. 1987. *Snapshot versions of life.* Bowling Green, Ohio: Bowling Green State University Press.

Dewdney, A., and M. Lister. 1988. *Youth, culture, and photography.* London: Macmillan.

Dunleavy, D. 2005. Camera phones prevail: Citizen shutterbugs and the London bombings. www.digitaljournalist.org/issue0507/dunleavy.html.

Döring, N., and A. Gundolf. 2005. Your life in snapshots: Mobile weblogs (mblogs). In *Thumb Culture: The Meaning of Mobile Phones for Society*, edited by P. Glotz, S. Bertschi, and C. Locke. Bielefeld: transcript Verlag.

Ito, M. 2004. A new set of social rules for a newly wireless society. *Japan Media Review.* Annenberg School for Communication, USC. http://www.japanmediareview.com/japan/wireless/1043770650 .php.

Ito, M. 2005. Intimate visual co-presence. Presented at the Workshop Pervasive Image Capture and Sharing: New Social Practices and Implications for Technology at Ubicomp'05, Tokyo. http:// ubicomp.org/ubicomp2005.

Kasesniemi, E.-L., A. Ahoren, T. Kymäläinen, and T. Virtanen. 2003. *Elävän mobiilikuvan ensi tallenteet. Käyttäjien kokemuksia videoviestinnästä.* Espoo: VTT Tiedotteita 2204. [*Moving Pictures. User Experiences about Video Messaging*, in Finnish].

Kato, F., and A. Shimizu. 2005. *Moblogging as face-work.* Presented at the Workshop Pervasive Image Capture and Sharing: New Social Practices and Implications for Technology at Ubicomp'05, Tokyo. http://ubicomp.org/ubicomp2005.

Katz, J. E., and M. Aakhus. 2002. Introduction. In *Perpetual Contact: Mobile Communication, Private Talk, Public Performance*, edited by J. E. Katz and M. Aakhus. Cambridge: Cambridge University Press.

Kindberg, T., M. Spasojevic, R. Fleck, and A. Sellen. 2004. How and why people use camera phones. http://www.hpl.hp.com/techreports/2004/HPL-2004-216.html.

Koskinen, I. 2005. Sound in mobile multimedia. Proceedings of Designing Pleasurable Products and Interfaces DPPI 2005, Eindhoven, the Netherlands.

Koskinen, I. 2007a. *Mobile Media in Action.* New Brunswick, NJ: Transaction Publishers.

Koskinen, I. 2007b. Managing banality in mobile multimedia. In *Place, Identity, and Media*, edited by R. Pertierra. Manila: University of the Philippines.

Koskinen, I., E. Kurvinen, and T. K. Lehtonen. 2002. *Mobile Image*. Helsinki: IT Press.

Kurvinen, E. 2002. Emotions in action: A case in mobile visual communication. Proceedings of the 3rd International Design and Emotion Conference, Loughborough, England.

Kurvinen, E. 2003. Only when Miss Universe snatches me. Teasing in MMS messaging. Proceeding of Designing Pleasurable Products and Interfaces, Pittsburgh, Pa.

Ling, R., and T. Julsrud. 2005. The development of grounded genres in multimedia messaging systems (MMS) among mobile professionals. In *A sense of place*, edited by K. Nyíri. Vienna: Passagen-Verlag.

Margolis, J. 2005. *MOB_LOG: Scenes from My Mobile*. London: Artnik Books.

Matsuda, M. 2005. Mobile communication and selective sociality. In *Personal, Portable, Pedestrian*, edited by M. Ito, D. Okabe, and M. Matsuda. Cambridge, Mass.: The MIT Press.

Matsunaga, M. 2000. *i-mode. The birth of i-mode*. Singapore: Chuang Yi Publishers.

Metso, A., M. Isomursu, P. Isomursu, and L. Tasajarvi. 2004. *Classification of mobile micromovies*. Proceedings of the 2004 IEEE International Conference on Multimedia and Expo, ICME 2004, Taipei, Taiwan.

Natsuno, T. 2003. *i-mode strategy*. Chichester: John Wiley & Sons.

Noguchi, Y. 2005. Camera phones lend immediacy to images of disaster. The *Washington Post*, July 8, p. A16. www.washingtonpost.com/wp-dyn/content/article/2005/07/07/AR2005070701522 .html.

Ok, H. R. 2005. Cinema in your hand, cinema on the street: The aesthetics of convergence in Korean Mobile (phone) Cinema. Proceedings of Seeing, Understanding. Learning in the Mobile Age. Communications in the 21st Century: The Mobile Information Society, Budapest, Hungary.

Okabe, D., and M. Ito. 2004. Camera phones changing the definition of picture-worthy. *Japan Media Review*. Annenberg School for Communication, USC. www.ojr.org/japan/wireless/1062208524 .php.

Okada, T. 2005. Youth culture and the shaping of Japanese mobile media: Personalization and the *Keitai* Internet as multimedia. In *Personal, Portable, Pedestrian*, edited by M. Ito, D. Okabe, and M. Matsuda. Cambridge, Mass.: The MIT Press.

Pertierra, R. 2007. The transformative capacities of technology: Computer mediated interactive communications in the Philippines—promises of the present future. In *Social Construction and Usages of Communications Technologies: Asian and European Experiences*, edited by R. Pertierra. Manila: University of Philippines Press.

Pertierra, R., E. F. Ugarte, A. Pingol, J. Herrandez, and N. L. Dacanay. 2002. *Txt-ing Selves. Cellphones and Philippine Modernity*. Manila: De La Salle University Press.

Rheingold, H. 2003a. *Smart Mobs: The Next Social Revolution.* Cambridge, Mass.: Perseus.

Rheingold, H. 2003b. Moblogs seen as a crystal ball for a new era in online journalism. *Online Journalism Review.* www.ojr.org/ojr/technology/1057780670.php.

Rivière, C. 2005. Seeing and writing on a mobile phone: New forms of sociability in interpersonal communications. Proceedings of Seeing, Understanding, Learning in the Mobile Age. Communications in the 21st Century: The Mobile Information Society, Budapest, Hungary.

Scifo, B. 2005. The domestication of the camera phone and MMS communications. The experience of young Italians. In *A sense of place*, edited by K. Nyíri. Vienna: Passagen-Verlag.

Simmel, G. 1949. The sociology of sociability. *American Journal of Sociology* 55: 254–261. E. C. Hughes, trans. (Original essay published in German in 1917.)

Snoody, R. 2005. How mobiles changed the face of news. *BBC Review of 2005.* http://news.bbc.co.uk/nolavconsole/ukfs_news/hi/newsid_4550000/nb_rm_4553802.stm.

Södergård, C., ed. 2003. *Mobile television. Technology and user experiences.* Espoo, Finland: VTT Publications 506. www.vtt.fi.

19 Mobile Communication and Sociopolitical Change in the Arab World

Mohammad Ibahrine

Just as in other regions, the Arab world has witnessed a rapid diffusion of the mobile phone in recent times. For Arabs, the mobile phone is not just for personal communication; it is also a multifunctional personal device. Mobile phones equipped with new trendy features such as Internet access, cameras, and MP3 players have become popular, particularly among Arab adolescents, who have played a very active role in adopting and appropriating multifunctional mobile communication services. Given the widespread adoption of the mobile phone within the Arab world, an important question is whether this diffusion is causing communication-related social and political changes. The primary objective of this chapter is to examine the role of the mobile phone relative to three topics.

The first topic probes the ramifications of mobile-based peer-to-peer production and distribution, focusing on its potential in changing entertainment configurations in the traditional media landscape. This has led to new conceptions and practices in the emerging mobile communication. In the Arab world, the masses have become capable of creating and distributing new content (texting, messaging, imaging, and videos, etc.).

The second topic examines mobile phones as used today by broader strata of the Arab population. Mobile phones have large impacts on social life. While mobiles are strengthening the traditional flow of communication between individuals and communities by perpetuating traditional (communication) relationships, they are weakening premodern patterns of controlling communication by disrupting traditional forms of hierarchy. Due to mobile communication, users interested in doing so are able to erode gender segregation in public spaces as well as in schools and private social gatherings.

The third topic looks at the potential of mobile phones when used as a political communication medium, focusing on coordinating political actions by nonstate actors. In countries where the authoritarian grip over the channels of political communication is tight, civil society groups and activists have turned to the mobile phone as an efficient tool for creating and distributing political messages to targeted audiences, especially to younger supporters, and for mobilizing followers and supporters for demonstrations.

In the last section, I examine the reactions of political and religious authorities in the Arab world. They are challenged not only by the recent popularity of the mobile phone, including texting, messaging, and imaging, but also by the consequences of the emerging mobile communication environments. Also included here is the analysis of how Arab political and religious authorities have sought to control the new mobile "technologies of freedom" along with their concomitant liberalizing and democratizing effects.

Diffusion of the Mobile Phone in the Arab World

Before proceeding in the analysis, it is worth noting that the Arab world is not a uniform political or social entity; however, the many countries and nations within this geographical region share cultural and religious patterns that make it a useful unit for analysis to examine the consequences of mobile communication. Significantly, the entire region is dominated by authoritarian regimes and engulfed in religious (Islamic) traditionalism and conservatism. This is the case even in socially liberal societies such as Lebanon and Tunisia and politically liberal countries such as Morocco and Jordan. In this regard, it seems appropriate to speak of the Arab world as a religiously and culturally closed society, especially when one wishes to examine how this region is being altered by the spread of mobile communication. So to navigate these similarities and differences I sometimes refer in my analysis to the Arab world generally and at other times to specific national and subcultural situations.

Al-Jawal, Arabic for the "one on the move" (mobile phone) boomed across the Arab world in the early 2000s, after privatization undermined the monopoly of the incumbent Arab telecom companies (Nield 2004, p. 26). In Arab countries, mobile penetration rates range from 3 percent (Sudan) to 85 percent (the United Arab Emirates). About 80 percent of mobile subscribers choose the prepaid phone cards instead of fixed-term contracts because prepaid service fits with the needs of those with lower income and education (ITU 2004).

Mobile penetration, not surprisingly given the profile of users in other regions, is much higher among Arab college students and high school students than the general population (Arab Advisory Group 2005). The mobile seems to be a nearly ubiquitous accessory for most social classes of teenagers, who favor SMS as an efficient and trendy form of personal communication.

In fact, it seems that 90 percent of Arab teenagers claim to text more than they talk on their mobile phones (Arab Advisory Group 2005). Youngsters and young adults, "Arab Generation Txt," use SMS for receiving notifications, chatting with friends, and voting in contests or participating in TV entertainment show polls. Since SMS was launched commercially for the first time in 1998 in Saudi Arabia, figures have gathered momentum year upon year and October 2005 marked a major text messaging milestone. Figures for October 2005, which coincided this year with the month of Ramadan, leapt to a remarkable sixty-five million messages, the highest ever total, according to semiofficial statistics (Fawaz 2005).

Although the camera phone has become popular only recently in the Gulf countries, camera phones are among the most preferred tools among Arab youth. In the UAE, within two months of its introduction in 2003, more than one hundred thousand people had subscribed to the advanced service of multimedia messaging service (MMS) (Kawach 2003). It was reported that in the wealthy Gulf countries even ten-year-old children possess the Bluetooth-enabled mobile phones (Aboud 2005).

While Dubai city supposedly has the highest mobile phone density in the Arab world, Al-Jawal is spreading rapidly, not only rivaling but surpassing that of the seemingly ubiquitous communication tool, the TV. Technological improvements are continuing to place mobile phones at the forefront of emerging communication tools. There is virtually no place where Arabs do not use their mobile phones, including public buses, restaurants, offices, clinics, and streets. It is common that Arabs use their mobile phones in places like mosques, despite a huge number of *fatwas* (religious edicts) prohibiting this kind of use in such places. Some *Ulama* (religious scholars) have issued fatwas declaring that musical ringtones on mobile phones are illicit. They argue that since music itself is un-Islamic, so too are musical ringtones. They advise mobile phone users to employ "neutral" ringtones. But it may be said that the ease and convenience of the mobile fits well with Arab culture, which stresses frequent and informal interaction among family and local groups.

Mobile Communication and Social Control and Social Tensions

The recent emergence of mobile phones has led some scholars to argue that the use of mobile phone and SMS to exchange visual images and verbal messages is bringing about social and cultural structural processes (Katz and Aakhus 2002; Ling and Yttri 2002; Rheingold 2002; Castells et al. 2004; Höflich and Gebhardt 2005). The common argument is that mobile phone and new communication technology have the potential to change people's social life. In Europe, the United States, and Asia such arguments and hypotheses were supported by empirical evidence. However, the impact of mobile phones on social behaviors of individuals and communities in environments characterized by high-tech sophistication and liberal cultural norms and values is less profound than in environments characterized by low-technological development.

The following section examines the use of text messaging (SMS) and imaging as it has developed into a form of mass communication among young people. It also shows how the rise of mobile communication environments supported by the peer-to-peer distribution of mobile content reveals the cultural tension between individualized pursuit of pleasure and the sustenance of traditional social and cultural values, especially those attached to religion.

Production and Distribution of Mobile Media Content by the Masses

In the Arab world, where private and public spheres are strictly divided and morally bounded, mobile communication is blurring the boundaries between the two spheres. Before the advent of the mobile phone, public spaces in the Gulf countries, for instance, were highly regulated through informal peer-based regulation and through institutional surveillance. But in an age of perpetual *visual* contact (see Koskinen's chapter 18 of this volume), with the use of camera-equipped mobile devices by just about everyone and everywhere, people can constantly take pictures in public spaces (Gillmor 2004, p. 48). Since the introduction of mobile phones, the traditional nature of private relationships within the public space is in the process of being altered.

In Saudi Arabia, for instance, where camera phones have quickly gained popularity, despite an official ban, young people use mobile phones not only to contact the opposite sex and thus avoid gender segregation, but also to snap pictures. They upload these visually shareable photos and clips. It was also reported that young males downloaded obscene pictures from Web sites and sent them to other mobile phones (Kawach 2003). As a result, people may suddenly receive pornographic images on their mobile phones from anonymous senders.

The equipped mobile camera phones with new Bluetooth technology marked the beginning of a new virtual and seamless flirting trend characterized by a transfer of phone numbers, songs, pictures, jokes, short video clips, and sometimes hardcore pornography (Aboud 2005). Young people in the Gulf countries have turned equipped mobile camera phones with new Bluetooth technology into a high-tech way to flirt. In Saudi Arabia, unrelated men and women caught talking to each other, driving in the same car, or sharing a meal risk being detained by the *mutawaeen* (Authority for the Promotion of Virtue and Prevention of Vice). Despite these risks and barriers, female and male teenagers have used the Bluetooth technology in public spaces because connecting by the Bluetooth technology is much better, easier, and safer than exchanging or throwing mobile phone numbers to the girls in shopping malls or even through car windows at traffic lights (Aboud 2005). It was reported that a number of young people use the Bluetooth-enabled mobile phones as their favorite method to arrange meetings without being intercepted and detained by the mutawaeen (Aboud 2005).

In the Gulf countries, young people are using their camera phones, for instance, in shopping malls, where the sex segregation is either not strictly imposed or laxly enforced. In such places, mutawaeen have often warned young men not to surreptitiously photograph female shoppers, and have sometimes arrested them for doing so. In some shopping malls, signals and announcements are made every few minutes specifying limitations on mobile phone use because it is illegal to photograph people in these shopping malls. For security agents, controlling people from taking pictures with their camera phones of female shoppers has become an additional security problem (Kawach 2003). To address the rising possibility of mixing, some segments of the Bahraini Islamist movements have called for segregated shopping malls (Doussary 2006).

In 2006, the Bahraini parliament passed the first law of its kind in the Gulf region outlawing the use of Bluetooth wireless technology. The new law includes lighter prison terms (three months in prison) and fines (100 dinar, or about USD 266). This law is primarily designed to prohibit young males from sending unsolicited e-mails or text messages to females on their mobile phones (Doussary 2006). The question remains open as to whether this new ban will affect mobile phone use among young men.

In 2004, three men produced a clip containing obscene images of a rape and posted it on the Internet. The mobile phone footage, which reportedly showed one of the three men raping a teenage girl, caused a scandal. The rape scandal broke out after the men reportedly circulated footage of the assault through mobile phones equipped with cameras. The young men accused of orchestrating and filming the rape were arrested and if convicted will face the death penalty.

While public places are highly regulated, private spaces have not been subjected to social control. Social gatherings among females in Saudi Arabia, for instance, was free from such regulation. But in the age of mobile-based peer-to-peer imaging and reporting, even private social gatherings have become subject to a new kind of censorship. Mobile phones are being confiscated from guests attending private weddings. A wedding party in July 2004 turned violent after a female guest was caught using her mobile phone to take digital photographs of other women at the segregated celebration and send them to people with sentimental captions. These photos were trafficked first among peers but posted later on the Internet for wide circulation. The woman was reportedly hospitalized after being badly beaten by other female guests.

Photos are newsworthy among friends and families and often serve as the topic of conversation. Sharing photos has become a popular practice among teenagers. Photos of the opposite sex are the subject of everyday talk: some of the photos are sent to others but because of economic considerations were generally posted on the Internet on personal blogs, making a visual archive of photos. There are few moblogs, which are personal, but their potential for building networks of people and disseminating news cannot be underestimated. Ito argues that "the moblog thus became a site of shared knowledge" (Ito 2005, p. 3). Arab blogs were full of private photos taken by mobile phone cameras, which could be viewed by anyone. In Morocco, "made-in-Morocco," a popular blog among young Moroccans, has enraged a number of people including human rights activists and liberal journalists, who are calling for a law to protect the private sphere. Similarly, in Bahrain there has been a discussion to issue a law that prohibits taking photos of private people and posting them on the Internet without their consent.

The easy availability of blogging tools can make the convergence of mobile devices and digital cameras more dramatic for entertainment. The emergence of "moblogging" appears to be growing as a hobby for the younger generation.

New Mobile Communication Patterns and Social Change

In the Arab world, interpersonal communication generally retains a strong traditional overtone and operates according to prescribed practices. It is typically assumed to be dominated by restricted flow of information, discreetness, and secrecy. Such features structure interpersonal communication patterns in private and public spaces alike. As a result, Arab people have limited access to novel or so-called modern forms of social life (Fandy 2000, p. 382). For instance, Arab traditional interpersonal communication between different sexes is socially and culturally defined. In accordance with their conservative notion of gender roles, females still adhere to traditional gender segregation (Wheeler 2000, p. 443).

Upon its introduction in the early 1970s, the fixed telephone facilitated social contact. However, the more dramatic change was brought about by the mobile phone because it makes communication much more convenient than the fixed phone does. Via Al-Jawal, a great number of adolescents in Arab countries are able to operate in virtual isolation and are also capable of escaping parental surveillance as well as the demands of existing social structures. Students use Al-Jawal to call each other during class and to photograph teachers and schoolmates (Geledi 2005). Since SMS messages and images cannot be intercepted by other friends or teachers, students send each other indecent text messages (Al-Qarani 2005). Its usage is highly deferential to institutions of learning such as schools, colleges, and universities.

El Baghdad found in a 2005 survey among one hundred undergraduate and graduate students from both the American University in Cairo and Cairo University that seventy-nine (forty-two females and thirty-seven males) believed that the mobile camera phone is a new medium that unravels their privacy. For them, the mobile phone can invade their privacy by other people's misuse as in the circulation of photos of people captured in embarrassing and inappropriate situations. Out of the seventy-nine participants who considered MMS threatening to their privacy, 96 percent believed that a new law is needed to regulate the mobile phone in public spaces. Respondents even suggested a ban of the use of camera phone in places such high schools, swimming pools, public bathrooms, hospitals, clinics, and beauty centers (El Baghdad 2006).

The Saudi Ministry of Education, for example, plans to ban mobile phone camera usage in schools, colleges, and universities. Mobile phone camera policies and regulations already have been issued warning students carrying mobile phone cameras on campus to a SR 500 fine and three-year suspension (Geledi 2005). To demonstrate the seriousness of the phenomenon, a woman was expelled from the university in March 2004 for taking pictures of unveiled colleagues with a camera-equipped mobile phone and posting them on the Internet (Geledi 2005).

The Lebanese channel LBC began in 2002 *Star Academy*, an Arab talent show. During *Star Academy*'s third season, in 2005, young Arab viewers voted by sending fifteen million text messages (*Economist* 2005). The use of SMS for voting purposes in entertainment is widely believed to be superior and effective compared with using the voice telephone service and e-mails. A young Saudi became a national hero (Hammond 2005) after winning the song contest; his victory seems attributable to his wealthy countrymen and women who generated vast numbers of supporting votes. In May 2005, the Ulama, not only in Saudi Arabia but also in other Arab countries, condemned the popular show and issued fatwas prohibiting the participation in the show either as an actor or as a voter. Despite these fatwas, the next year young Saudi males and females managed to vote via the Internet, bypassing the regime servers that not only control access but filter messages intended to block them (Hammond 2005).

The Saudi government banned the use of mobile phones to vote for *Star Academy*. Under combined constraints of the Saudi government and opinion leaders, Saudi and Emirati mobile firms operating in the Saudi mobile phone market announced that they will block customers and users from text-message voting for *Star Academy*'s fourth season in 2006. These companies feared a brand damage that the *Star Academy* may cause them given its infamous reputation among conservatives, traditionalists, and Islamists who constitute the wealthiest segments in the Arab world (Hammond 2005).

Though physically separate, females and males are able to exchange through mobile-based technologies popular sentimental images featuring babies, blowing kisses, animated cartoons doing belly dances, or dreamy and sappy Arabic songs. According to anecdotal evidence, Al-Jawal has become a hotbed for dating (Al-Qarani 2005).

Castells argues that mobile phones have significantly enhanced the autonomy of their users (Castells 2004, p. 237). As the "back door of personal communication" (Kasesniemi and Rautiainen 2002, p. 171), mobile phone and SMS help Arab females and males explore new forms of dating and flirting, one of the thorniest taboos in these religiously embedded societies. Of particular relevance to the following discussion here is the case of *Misyar* (easy and temporary marriage). While tantamount to a quasilegitimate liaison, Misyar is not widely practiced in the Islamic countries, except in Saudi Arabia and Egypt. It is a kind of marriage in which the partners agree to remain secret about it: this marriage allows the two partners to meet occasionally in a flat or a hotel room.

What is at stake is that before the arrival of the mobile phone, this kind of temporary marriage was risky and time- and energy-consuming for evading social control. With the mobile phone, the rules of the game have been made easier. It was reported that there are currently more than forty-five thousand Misyar marriages in Saudi Arabia (Al-Arabiya.net 2005). Mobile phones with SMS have fulfilled a function for lovers and friends, who are adopting Misyar as a more flexible manner of making (sexual) appointments.

Before the advent of Al-Jawal, this kind of marriage allowed a limited degree of social freedom. With the help of Al-Jawal, this marriage is becoming popular and a considerable number of young and adult people have been "emancipated" from social constraints and the surveillance of religious police. As with meetings for mutual interest, as in so many other areas, the observation of Katz and Aakhus remains apt: "whenever the mobile phone chirps, it alters the traditional nature of public space and the traditional dynamics of private relationships" (Katz and Aakhus 2002, p. 301).

Mobiles, Mobilization, and Arab Mobs

In an attempt to frame the potential of the mobile phones, Katz and Aakhus argue that communication scholars, political communication researchers, and social scientists should not underestimate the mobile phone's ability to help effect large-scale political change (Katz and Aakhus 2002, p. 2). Recent scholarship has demonstrated the role of the mobile phone in the mobilization of marginalized groups at critical political events in some newly established democracies such as the Philippines, South Korea, and Spain, where the outcome of the political use of the mobile phone was radical and revolutionary (Castells et al. 2004, p. 221; Suárez 2005).

It has been argued that key features of the mobile phone, particularly the flexibility and hyper-coordination, would nurture personal relationships that may not have otherwise developed and would allow greater freedom for collective expression. The mobile phone's ability to continually synchronize movements may turn out to be a challenge for authoritarian regimes and their repression arsenals: "everything is virtual until the parties, the places and the moments come together to make it real" (Plant n.d., p. 61). This increases the flexibility for the persons involved, and there is total control of the preparation and arrangement taking place before the demonstration.

Richard Ling distinguishes between two uses of the mobile, namely the instrumental and the expressive form. The instrumental use of mobile phones is referred to as "micro-coordination" while the expressive mobile phone use is termed "hyper-coordination" (Ling and Yttri 2002). Micro-coordination involves using a mobile phone for logistical purposes, such as confirming the place and time of a meeting or asking a family member to stop by the store on their way home. Hyper-coordination, on the other hand, involves using a mobile phone to communicate emotion or generally maintain social relations. Hyper-coordination has arguably augmented people's social interactions. While Ling points out that the mobile phone has contributed to the coordination of social interaction, I argue that the mobile phone has a potential in coordinating political action, and increases the functional capacity of collectivities, organizations, and individuals during demonstrations. In the political field, Ling's notion of hyper-coordination stressed the functional dimension as opposed to the affective dimension inherent in the expressive use of mobile phones.

Where the mass media is tightly controlled, human rights activists have added mobile phone functions (texting and imaging) to the technologies at their disposal, such as mass e-mails, the Internet, and blogs, to organize actions. True, the mobile phone performs the mobilization function much more efficiently than other communication channels, but not anytime anywhere. In current Arab political settings, fundamentally at odds with spontaneous grassroots mobilization, the potential of the mobile phone to mobilize smart mobs is extremely limited. Authoritarian regimes do no allow spontaneous demonstrations. Throughout the Arab world, freedom of assembly is virtually nonexistent and ad hoc political groupings and demonstrations are unlawful (Goldstein 1999). In such authoritarian contexts, it is the hyper-coordination rather than smart mobs that make political sense in mobilizing people into the street. The mobile phone is regarded as a means of coordinating the activities of protesters. While the notion of smart mobs puts emphasis on the ad hoc groupings dimension, the term *hyper-coordination* places premium on contacts within the preceding face-to-face interactions or within the networks of the organization.

While smart mobs can potentially have a strong impact on the political process in established democracies such as the United States or developing democracies such as the Philippines, South Korea, and Spain, established networks of organizations benefiting from "hyper-coordination via mobile phones" (Ling and Yttri 2002, p. 139) can have some political impact on authoritarian regimes.

Recently, instances of grassroots sociopolitical mobilization were identified in a number of Arab countries including Bahrain, Egypt, Kuwait, Lebanon, and Morocco (Coll 2005; Glaser 2005a, 2005b; Howeidy 2005). An illustrative case of the significant role played by mobile communication is the Egyptian presidential and parliamentary election of 2005. Mobile phones were empowering and mobilizing marginalized groups at this critical political moment by increasing the range of alternative actions available to individuals, opposition forces, and civil society groups, particularly those consisting of a strong network of students, activists, and young professionals.

Eid (Islamic new year) SMS greetings are enormously popular in Egypt. Since about 2002, many of these greetings have taken on an increasingly political color. At first they were often generally critical of the Mubarak family. Over time, however, they have become more directly mocking of the government, especially Hosny Mubarak's preparations to have his son succeed him (Howeidy 2006). It is difficult to assess whether these messages are a relief valve or, more likely, an attempt to spread and voice political dissatisfaction in preparation for more active expressions of political dissent.

The Kifaya Movement, a newly formed political force in Egypt, which says "enough" to despotism and monopoly of authority, turned to the mobile phone as an efficient tool for political communication for a number of reasons. It used text messaging to compose its own statements and deliver its political message to targeted audiences, especially to younger supporters. It also used mobile phones to mobilize followers, arrange for flash demonstrations, spread news about detained activists, distribute anonymous political digital graffiti, and circulate candidate slates. The Kifaya Movement's candidate for the 2005 Egyptian presidential election used text services to call supporters to the polls (Howeidy 2006).

Civil society groups and human rights activists used their mobile phones, including text messages, to call a rally on the day of the May 2005 referendum. Their aim was to demand political reform and open space for democratic political actions. The police attacked demonstrators who supported multiparty elections. While taking part in the demonstration in the streets, Egypt's human rights activists photographed incidents using their camera-equipped mobile phones, in particular of police attacking and beating the demonstrators (Glaser 2005b).

Being aware that their actions were being photographed and knowing the photographs could later cause alarm among the population and harm to Egypt's image abroad, the police attempted to confiscate these mobile phones to contain the photos. But in most circumstances, human rights activists managed to send and distribute their photos, which have also served as evidence in court against the police.

Afterward photos were posted on the Internet, which then reached a wide audience. In due course, activists have amassed a large database of photo and video galleries documenting the various abuses, including torture. In this way, they have revealed the regime's brutality to the whole world.

On Election Day, September 7, 2005, activists and independent elections monitors took photos of any harassment or election-rigging at polling places and posted them on their blogs, together with first-hand reports and detailed descriptions of the events during Election Day. By using phones, protesters became broadcasters, receiving and sending both news and gossip about the culpability of the regime. Being aware of such monitoring and publicity activities, security forces may have been more restrained in their behavior.

The political use of the mobile phone has allowed new flexible and creative applications. Political messages serve not only to inform receivers, but also to confirm the relationship and tie the sender and receiver together, since the message goes directly to the individual and at the same time to a wide audience (Howeidy 2006).

More generally speaking of the Arab context, it seems mobile phones are used primarily in the process of mobilizing existing networks rather than stimulating new ones. Most phone contacts originate only after there has been face-to-face interactions (Glaser 2005b). Likewise, it follows that the sharing of information is seldom used to explore and recruit potential activists. SMS messages may be effective in arousing people who already belong to an established network or in stimulating anger or discontent. But this is different from the wide-scale political mobilization that seems to have been seen in other cultures and contexts such as the Philippines or Spain (as discussed in other chapters in this volume). Mobile communication has the greatest chance for success when the mobilization process starts within the networks of the organization, reaching out to its mobile-equipped members, sympathizers, and activists. By itself, it appears that in the Arab world, the mobile phone and SMS impact on directly mobilizing political forces and coordinating electoral campaigns is limited. To date there is scant evidence this impact will be made. It certainly did not happen in the outcome of the 2005 presidential election in Egypt.

Religious and Political Reactions

Mobile technologies are increasing the range of alternative actions available to individuals or social groups. They are also increasing the margin of freedom and autonomy. However, these significant gains are counterbalanced by increased religious, political, and social control (Geser 2004, p. 16).

It is argued that mobile phones are in the process of undermining traditional mechanisms of social control, which have secured the segregation of the sexes, as well as the differentiation between private and public spaces. Instead, each individual now is armed with a mobile phone, which has radically transformed these boundaries, making them increasingly permeable. Mobile communication is blurring the boundaries between personal and public because the difference between personal behavior and social roles, and regulating the boundaries between different social relationships, groupings, organizations, or institutions, has become permeable. Therefore, the demand for social control will rise because social differentiation can no longer be based on spatial segregation and political authority cannot secure the flow of information.

After all, Arab regimes that have put so much effort into controlling the free flow of information were hardly going to ignore a publishing tool that will be easily accessible by what will soon be more than one hundred million mobile phone users in the region. Unlike the Internet, mobile phone diffusion has reached a large proportion of the Arab population. In addition, the mobile phone is more personal, more direct, and easier to use than the Internet, which is typically accessed via a personal computer from a fixed location. Unlike the Internet, mobile phone users are directly accessible at all times and locations. Compared to the Internet, the impact of the mobile phone is conceivably greater.

In response to these rapid developments, Arab regimes are finally turning their attention to the mobile phone and extending their authoritarian grip over mobile communication. In 2005, Saudi Arabia and the United Arab Emirates blocked flickr (a photo-sharing application and Web site) a number of times.

Arab authoritarian regimes are trying to respond to emerging mobile communication environments. Arab regimes were able to censor the information flow on the Internet to a certain extent but are now finding it more difficult to screen text messages sent via mobile phones. Perhaps even more problematic is the question of images transmitted via mobiles. In the past, strenuous efforts have been made to address the individually empowering (and also threatening) power of digital mobile imaging and distribution. The case of Saudi Arabia in this regard is illuminating.

Being aware that even high-tech censorship mechanisms remain ineffective in the face of mobile communication technologies, the political regime in Saudi Arabia initially made recourse to traditional control mechanisms of bans and physical enforcement, but also suffered widespread flouting of the rules. For instance, Saudi Arabia imported six million mobile phones in 2004 (Lettice 2004). Saudi media reported that camera phones are still freely and widely available in Saudi Arabia, despite the ban. They are even widely advertised in many Saudi cities (Harrison 2004). Even though Saudi authorities now permit camera phones, they arrived most reluctantly to this position after having banned camera phones in October 2002 (Mishkhas 2004). In fact, at one point, the Saudi regime urged the highest religious authority to issue a fatwa prohibiting the use of mobile phones and particularly camera phones in public spaces for the dissemination of visually shareable information (Mishkhas 2004; Geledi 2005). What is striking is that some of the fatwa's text was also distributed by SMS. A translation of the fatwa includes the following assertion:

The exchange of telephone or mobile messages between boys and girls is something that is shunned by the Islamic Sharia, and could lead to things that are forbidden and to untold problems...Some of the youths indulge in such practices only to cheat and receive the girls into committing moral blunders...Some of the girls have had their photos taken by the mobile cameras, and that such practices have led to extreme moral damage to the modesty and chastity of the girls that were involved. (International Islamic News Agency 2004)

Sheik Abdul Aziz bin Abdullah Al-Sheik, Saudi Arabia's Grand Mufti, explained that the exchange of mobile phone messages between boys and girls is un-Islamic. For him, some of the youths indulge in such practices only to commit moral blunders, adding that some of the girls have had their photos taken by mobile cameras, and that such practices have led to extreme moral damage to the modesty and chastity of the girls that were involved. The edict was a final resort after the ban in October 2002 on the importation and sale of camera phones failed to dent their popularity (Kawach 2003). To ensure that the ban was enforced, the mutawaeen, who regularly patrol the streets, confiscated mobile phones with cameras if their owners used them in public areas. But according to Saad El-Dosari, a Saudi sociologist, "camera phone devices exist everywhere in Saudi Arabia; retailers and wholesalers have all facilitations to import these devices. The ban is merely words on paper" (El Baghdad 2006, p. 11). The main objective of the political regime and religious authorities is to immobilize the mobile whenever they can.

In this context, four ministries (Finance, Commerce and Industry, Interior, and Communications and Information Technology) attempted to lift the ban on mobile phones with cameras (Mishkhas 2004). They devised new measures to stop camera phones being used to intrude on people's privacy (Mishkhas 2004). The four ministries submitted requests to the Royal Court, which were based on three arguments: 1) mobile phones with built-in cameras are a communication technology similar to television and the Internet; 2) despite the ban, Al-Jawal has remained widely available because the ban has led to the emergence of a flourishing black market; Saudis bought easily smuggled mobile phones with cameras from neighboring Bahrain or the Emirates; and 3) cameras have become and will remain a standard feature of mobile phones.

The ineffectiveness of the ban and the political weight of the four ministries have urged the regime to lift the ban. The use of mobile camera phones was legalized in December 2004 in Saudi Arabia. But their use is still not permissible everywhere. In places such as schools, colleges, and universities the ban is still in force (Geledi 2005).

Conclusion

Mobile communications point to a future that offers a wealth of knowledge and information to Arab communities. The production and distribution of mobile media content by and for the masses has transformed mobile phones into a remarkably effective channel for mass communication (Gillmor 2004; Koskinen 2004). These changes in communications lead, in turn, to subtle changes in the communication practices and expectations that accommodate the mobile phone in private and public life. The growth of mobile phones has created the space for a range of social interactions to occur beyond the place-specificity of the home or school. For women in Arab culture in particular, the mobile phone enlarges peripheral layers of social relationships; it also substitutes for face-to-face relationships between teenagers of different genders. The political use of the mobile phone is still in its infancy, but due to its flexibility and accessibility it is posing a serious threat to the existing Arab authoritarian regimes where information and dissent is strictly controlled. The mobile phone is empowering some at the expense of others: users at the expense of nonusers, younger generations at the expense of older, females at the expense of males, and the nonstate actors at the expense of state actors. As the mobile phone becomes part of the mainstream in Arab communities, it constitutes perhaps the greatest challenge yet to traditional forms of interpersonal, public, and political communication.

References

Al-Arabiya.Net. 2005. The mobile phone and the Misyar. http://www.alarabiya.net/Articles/2005/11/27/18993.htm#3 (in Arabic).

Aboud, G. 2005. Teenagers sinking their teeth into new technology. *Arab News*. http://www.arabnews.com/?page=1ion=0&article=58778&d=10&m=2&y=2005&pix=kingdom.jpg&category=Kingdom.

Al-Qarani, H. 2005. Bluetooth increases interaction of the sexes in Saudi Arabia. *Asharq Al-Awsat*. http://aawsat.com/english/news.asp?section=7&id=2896.

Arab Advisors Group. 2005. Only 56% of Arab cellular operators provide the MMS service, while Morocco and Lebanon have the highest SMS rates in the Arab World. http://www.arabadvisors.com/Pressers/presser-170105.htm.

Castells, M., M. Fernandez-Ardevol, J. L. Qiu, A. Sey. 2004. The mobile communication society: A cross-cultural analysis of available evidence on the social uses of wireless communication technology. Prepared for the International Workshop on Wireless Communication Policies and Prospects. http://annenberg.usc.edu/international_communication/WirelessWorkshop/MCS.pdf.

Coll, S. 2005. In the Gulf, dissidence goes digital: Text messaging is new tool of political underground. The *Washington Post*. http://www.washingtonpost.com/wp-dyn/articles/A8175-2005Mar28.html.

Doussary, S. 2006. The first law in the Gulf countries against the misuse of the Bluetooth wireless technology. *Asharq Al-Awsat*. http://www.asharqalawsat.com/default.asp (in Arabic).

Economist. 2005. Arab satellite television: The world through their eyes. *The Economist*. http://www.economist.com/PrinterFriendly.cfm?Story_ID=3690442.

El Baghdad, L. 2006. *Camera mobile phones: Between being a new powerful communication medium and a privacy menace*. Master's thesis at the American University in Cairo Journalism and Mass Communication Department.

Fandy, M. 2000. Information technology, trust, and social change in the Arab world. *Middle East Journal* 54(3): 378–394.

Fawaz, M. 2005. Ramadan dominates SMS in Saudi Arabia. *Islam-Online*. http://www.islam-online.net/English/News/2004-10/27/article06.shtml.

Geledi, S. 2005. Camera phones still banned at schools. *Arab News*. http://www.arabnews.com/?page=1&section=0&article=62707&d=25&m=4&y=2005&pix=kingdom.jpg&category=Kingdom.

Geser, H. 2004. Towards a sociological theory of the mobile phone. *University of Zurich Release 3.0.* http://socio.ch/mobile/t_geser1.pdf.

Gillmor, D. 2004. *We are the media: Grassroots journalism by the people, for the people.* Cambridge: O'Reilly.

Glaser, M. 2005a. Online forums, bloggers become vital media outlets in Bahrain. *USC Annenberg Online Journalism Review.* http://www.ojr.org/ojr/stories/050517glaser.

Glaser, M. 2005b. Blogs, SMS, e-mail: Egyptians organize protests as elections near. *USC Annenberg Online Journalism Review.* http://www.ojr.org/ojr/stories/050830glaser.

Goldstein, E. 1999. *The Internet in the Middle East and North Africa. Free expression and censorship.* New York: Human Rights Watch. http://hrw.org/advocacy/internet/mena/saudi.htm.

Hammond, A. 2005. Saudi telecom stops text vote for Arab talent show. *Reuters.* http://today.reuters.com/news/home.aspx.

Harrison, R. 2004. Camera phones freely available despite ban. *Arab News.* http://www.arabnews.com/?page=1&section=0&article=42472&d=4&m=4&y=2004.

Höflich, J., and J. Gebhardt. 2005. *Mobile kommunikation. Perspektiven und forschungsfelder.* Berlin: Peter Lang.

Howeidy, A. 2005. Lessons learned. *Al-Ahram Weekly.* http://weekly.ahram.org.eg/2005/771/eg8.htm.

Howeidy, A. 2006. E-mail interview, January 12. (Amira Howeidy is the assistant editor-in-chief at the authoritative Egyptian newspaper *Al-Ahram Weekly*.)

International Islamic News Agency. 2004. Daily news bulletin. *Rajab 18*, 1423. http://www.islamicnews.org/english/en_weekly.html.

Ito, M. 2005. *Intimate visual co-presence.* http://www.spasojevic.org/pics/PICS/ito.ubicomp05.pdf.

ITU (International Telecommunications Union). 2004. Mobile cellular subscriber. http://www.itu.int/ITU-D/ict/statistics/at_glance/cellular04.pdf.

Kasesniemi, E. L., and P. Rautiainen. 2002. Mobile culture of children and teenagers in Finland. In *Perpetual Contact: Mobile Communication, Private Talk, Public Performance*, edited by J. E. Katz and M. A. Aakhus. Cambridge: Cambridge University Press.

Katz, J., and M. Aakhus, eds. 2002. *Perpetual Contact: Mobile Communication, Private Talk, Public Performance.* Cambridge: Cambridge University Press.

Kawach, N. 2003. Camera mobile phone may be new security irritant. *Gulf News.* http://search.gulfnews.com/articles/03/09/06/96888.html.

Koskinen, I. 2004. Seeing with mobile images: Towards perpetual visual contact. Paper presented at the Mobile Information Society, Conference in Budapest. http://www.fil.hu/mobil/2004/Koskinen_webversion.pdf.

Lettice, R. 2004. Saudi ministers urge removal of camera phone ban. *The Register*. http://www.theregister.co.uk/2004/11/10/saudi_camera_phone_ban.

Ling, R., and B. Yttri. 2002. Hyper-coordination via mobile phones in Norway. In *Perpetual Contact: Mobile Communication, Private Talk, Public Performance*, edited by J. Katz and M. Aakhus. Cambridge: Cambridge University Press.

Mishkhas, A. 2004. Saudi Arabia to overturn ban on camera phones. *Arab News*. http://www.arabnews.com/?page=1&section=0&article=5618&d=17&m=12&y=2004.

Nield, R. 2004. Orbits of influence: A special report on telecoms. *Meed*.

Plant, S. (n.d.) On the mobile the effects of mobile telephones on social and individual life. http://www.motorola.com/mot/doc/0/234_MotDoc.pdf.

Rheingold, H. 2002. *Smart mobs: The next social revolution*. Cambridge, Mass.: Perseus.

Rheingold, H. 2004. Smart mobs. In *Network logic—who governs in an interconnected world?* edited by H. McCarthy, P. Miller, and P. Skidmore. London: Demos.

Suárez, S. 2005. Mobile democracy: Text messages, voter turnout, and the 2004 Spanish general election. Paper presented at the 2005 Annual Meeting of the American Political Science Association. http://convention2.allacademic.com/getfile.php?file=apsa05_proceeding/2005-08-22/41689/apsa05_proceeding_41689.pdf.

Wheeler, D. 2000. New media, globalization and Kuwait national identity. *Middle East Journal* 54(3): 432–444.

20 Locating the Missing Links of Mobile Communication in Japan: Sociocultural Influences on Usage by Children and the Elderly

On-Kwok Lai

Location-Navigation Technology in Mobile Communication

Enthusiasts of mobile communication technology promise borderless, flexible, and ubiquitous contacts—real-time, round-the-clock, and anywhere, making geospatial conditions all but irrelevant (MIC 2005, 2006a,b; Srivastava 2005; Wieser 2005). Yet, studies of mobile communication seem to underscore the importance of spatial and temporal dimensions, showing that physical location and social position in real life remain paramount (Katz 2003). For instance, Fortunati (2005, p. 36) highlights the regulation of space in the new media while Hutchby and Barnett (2005, pp. 162–167) show the importance of "where" and "the locational relevance" of communicators in caller-identity management. For Japan, Ito, Okabe, and Matsuda (2005) find the mobile phone is used for intensive territorial-bound social networking, and Kamibeppu and Sugiura (2005), Miyaki (2005), and Okada (2005) note the mobile phone's role in friendship-building among colocated students. This chapter continues a line of findings for Japan that emphasizes sociospatial (location, place, and space) relevance of mobile communication. It especially considers the elderly and children in the need for socially and geospatially fixed anchors despite the "mobile" in mobile communication.

Locating Mobile Communication in Demographic Transitions

First we examine links between mobile communication and the two ends of the population, that is, the children and the elderly. This examination draws inspiration from Ruth Benedict's characterization of Japanese traditional life-course, namely, that children and the elderly are privileged with the greatest freedom (or more aptly, least responsibility) (Benedict 1959; White 2002: pp. 154–179). These ends of the life-continuum are becoming even more prominent due to Japanese demographics: Japan is a low birth-rate, rapidly aging society (in late 2006, its total fertility rate was 1.25 children per family, well below replacement rate, and more than 22 percent of the population was 65 or older). This perhaps makes children even more precious.

Alongside this profile is the government's promotion of the so-called ubiquitous Japan project to make Japan a true information society (MIC 2006a). The "u-Japan" project comingles with sociocultural and policy matters at several points. A leading technology of u-Japan is global positioning systems (GPS), especially location-based navigation tools embedded in some Japanese mobile phones, the so-called SAFE Navi of the au-KDDI. Its multifunction can track the whereabouts and movements of the children via the device they are carrying, for example, mobile phone model G'zOne (au-KDDI 2006). This is especially handy if parents want to know where the child is, or confirm the child's location. In addition, it has an "area-directional track notification" capability. This capability allows an e-mail to be sent to parents confirming that their child is following the usual route home or is in the appropriate place. Parents are notified immediately, too, if a child strays from the usual route. The service is also quite helpful when a child is ready to be picked up from a prearranged location since the child's "ready" status is automatically forwarded to the parent when the child arrives at a designated area.

In addition, the built-in GPS mobile phone is not just trendy but also provides a potentially life-saving aid for outdoor survival in case the user gets lost in the wild. It also comes with an "electronic compass" directional sensor. The submenu displays the user's facing direction. Even more usefully, there is an EZ Navi Walk map with heads-up display. Voice route guidance offers a full-scale navigation system for pedestrians similar to car navigation systems. Navigation is integrated with barcode reading to allow easy setting of destinations or the inputting of other information.

The satellite-based global positioning system also has voice-activated locational search: a user says the names of departure and arrival stations into the mobile handset and, along with the dates and times, the requested schedule information will be displayed or announced through a synthesized voice. And, for the ultimate contingency, it can provide evacuation guidance in the event of a disaster.

In Japan the location-navigation services are quite popular. The map and geospatial information datalinks (about where to go) rank second among paid mobile phone services, and fifth among all services (MIC 2005, 2006a,b). Location-navigation services for mobile communication contribute to the process of further integration of mobile communication into everyday life. At the very least, the locational technology embedded in the mobile phone reduces for Japanese their sense of uncertainty. This achievement is highly prized since in Japan extensive efforts are made at every level to reduce uncertainty.

Mobile Communication for Children's Safety

Crime against young people is frequently reported in Japan: tragic cases of child kidnapping-murders are reported in the press in sustained and lurid detail. This in turn raises parental fears about danger to children. Hence, to ensure child safety, paren-

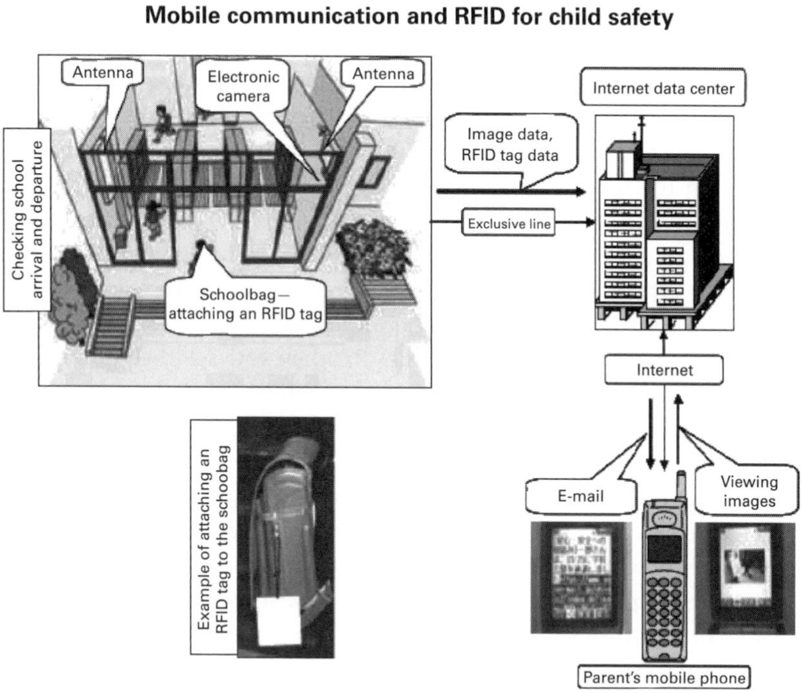

Mobile communication and RFID for child safety

Figure 20.1
Mobile phone network for school children's safety. Source MIC 2005.

tal monitoring of their children via mobile phone is an important service for which there is great demand.

In late 2004, the Kinki Bureau of Telecommunications of the Ministry of Internal Affairs and Communications (MIC) piloted a system that recorded the school arrival and departure times by radio-frequency identification (RFID) tags to notify parents of this information by e-mail or text message to mobile phone, under the cooperation of an elementary school in Wakayama prefecture (figure 20.1). It has been heavily subscribed: more than 81 percent of the participants checked daily on children's school arrival and departure, and user surveys found high satisfaction (MIC 2005, 2006a,b).

In response to a perceived wave of crime against youth, the Japanese government adopted in December 2005 a policy to further the use of mobile phones for child protection. As part of this initiative, the described pilot program is being expanded. Beyond the positioning of sensor for the RFID tag at the entrance and exit points of the school, the system will be expanded to include more location checkpoints and have tailor-made route monitoring. This way the safety of schoolchildren, say, from home to school or following a definite path, will be enhanced by the phone's existence, adapted to individual parent-children needs. Hence, there will be a network of surveillance for schoolchildren.

While few children are themselves are "owners" of mobile phones, they do use them frequently. Studies suggest that the overwhelming majority of school-age children are in regular mobile phone contact with parents (MIC 2005, 2006b; Miyaki 2005).

In another twist on the safety theme, the largest mobile phone network in Japan, NTT DoCoMo, has unveiled a new mobile phone model at USD 225. It features an alarm of 100 dB and an automatic emergency calling function to three preregistered numbers, all with the aim of keeping children (or other users) safe. Parents or guardians may also sign up for the "ima-doco" search (Where are you now?) location service. For a monthly payment of Yen 210 (USD 2), they can set their phones to automatically receive e-mail messages, updated at regular intervals, that track a child's location, in addition to having notification even when the handset is turned off (*Japan Times* 2005).

To better understand these issues and especially the utility of new services, in February 2006 we conducted focus groups with young mothers in their thirties and forties. Sixteen out of the twenty participants said they would like to have this location service as a way to provide extra protection for their children. All sixteen held this view despite a wide range of expertise with and exposure to using the mobile phone. They liked the reasonable price and extra assurance offered by the plan.

Not all uses are for safety, by any means, and the affective and expressive dimension must not be underestimated. This is reflected in one informant's comment:

For my kids, they are very excited about mobiles: they play with the new mobile phone by sending me text and photo-messages much more than previously—they even send me messages when they are at school! I feel good to be close with them! I reply to them more too! (Respondent Y)

In line with the idea that Japanese children like the mobile lifestyle, a Japanese publisher and network provider banded together in 2006 to market a "kids mobile." The device, priced at about USD 120, is a handset aimed at the younger-than-ten-year-old children's market. Built-in and mobile games are a key part of the offering, and are supported by a new range of mobile services (Fahey 2006). Hence, while the emphasis is on safety, once the device is available, its uses proliferate. Understanding that a phone in the hand is a great on-ramp, as it were, to the sale of persistent services, there is much joint enthusiasm on the part of the marketer and the customer to work together to buy and use mobile technology.

Reinforcing Sociospatial Nexus: From Familial Care to Community Support

Using mobile phone technology to collect and disseminate community alerts is becoming an important new venture in Japan. Its focus is the selective targeting for sourcing and dissemination of intelligence concerning the protection of potential crime victims. The initiative attempts to provide a safe environment for local residents and children. Against a nationwide backdrop of rising sex and violent incidents against children, a joint initiative by schools, local police, and parents in Hamakita City was undertaken. Based on mobile phone reportage of people in the area, local police disseminate real-time security alerts to preregistered schools as well as parents and teachers. These offer warnings and locations of suspicious strangers and any other incidents related to child safety. Similar schemes have been implemented in other localities in Japan, such as Ikeda, Osaka prefecture, and Tokyo's Arakawa Ward (*Asahi Shimbun* 2005).

All the discussion about personal safety and crime against children and women should not mislead the reader. Local communities in Japan remain quite safe. Yet Japanese society's collective search for certainty and safety, and desire to avoid risk, makes attractive various high-tech solutions such as those offered by mobile phones. And even though events are infrequent, they remain a source of concern and can be devastating in their consequences.

Based on our fieldwork, it appears that when such services are readily available, many citizens will use their mobiles to send reports to the police. Beyond the obvious public safety improvements such systems offer, people we have talked to say that the service enables positive community networking. They claim that using mobile phone technology at the local level gives them a greater sense of community security and mutuality. One respondent said, "We need both senses of human security, from the real community, as well as the communicative one—through mobile phones we have easy contacts with our family, friends and neighbors. [With mobiles], we feel we are all-together!"

Perhaps to readers of this chapter there will be an immediate and opposite concern, namely harmful and excessive surveillance and monitoring. Yet none of the people we spoke with, namely the so-called ordinary person in the street, brought up the issue. Apparently, when faced with real decisions, in the minds of most Japanese the benefits of safety and security far outweigh abstract calculations of losses to freedom.

Still the mobile phone is just one of the forces in redefining the sense of community, especially in terms of the territorial boundaries within which people live. Historically and still today, the local community's self-help and mutual-help societies have been vital to the maintenance of law and order and community integration. These include residents' self-governing community/neighborhood association (*ji-ji gai*), as well as the state machinery of the widespread and well-distributed neighborhood policy system of *Koban* (a small local police outpost at the street corners). Here, the novel offerings of mobile communication technologies, as shown in our fieldwork, can be understood as a reinforcement of social order and human security in an often alienated society facing rapid technological modernization. Hence, the mobile phone offers an additional safety and precautionary measure, aiming for social harmony and individual security. Interestingly, social norms are thereby reinforced rather than eroded. Thus Gergen's (2002) hope that the mobile phone could help re-establish a perceived loss of community and family grounding may actually be in the process of being realized.

Influencing the Sociospatial Sense by Mobile Communication

Extending Social Space of the Senior Adults by Mobile Communication

Japan is also concerned about the other end of the social dependency spectrum: an aging society and the demographic crisis it entails. Like other societies, one aspect of this is an age-specific digital divide in terms of mobile communication usage (OECD 2005; Wong et al. 2006). Although older people are increasingly becoming users of mobile phones (if compared with computers), from less than 10 percent in 2003 to 20 percent in 2005 for those over the age of seventy, there is a huge gap relative to younger users, as shown in figure 20.2, giving the results of 2003 and 2005 national surveys (MPHPT 2004; MIC 2006b). From the most updated *Communications Usage Trend Survey 2005* (MIC 2006b), digital divide still existed: the differences in the usage rates for people in their sixties or older and younger age groups are still remarkable (e.g., a difference of 20 percentage points between their fifties and their early sixties). Hence, age group had the greatest impact on ICT usage, for both mobile phone and computer usage rates, followed by household income. In general, demographic factors had a greater impact on computers than on mobile phones.

Despite low penetration rates, mobile communication technologies have been beneficial to Japan's elderly population and show even greater promise for the future. In its

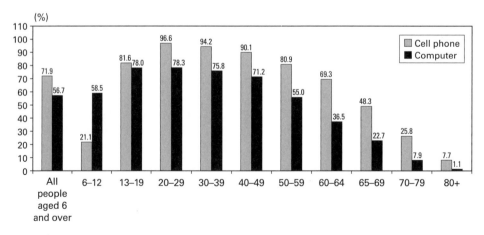

Figure 20.2
Age-specific digital divide: Mobile phone vs. computer usage, 2006. Source: MIC 2006.

early stages, the location-based service ima-doco was used to find senile elderly (and children, too). Mobile communication service oriented toward the health needs of the elderly is also reflected in au-KDDI's Helpnet and NTT's Life Support. The former is a one-button-push emergency service to signal the location of the caller; the latter connects volunteers with elderly people living alone (Srivastava 2004, p. 249).

To foster an active, healthy approach to aging, more initiatives from government and business have been launched to bring mobile communication access to seniors. These initiatives appear to be successful. According to one survey of seniors, all types of major participatory activities showed an upward growth compared to survey results of five years earlier due in part to the use of mobiles. These included health and sports (25 percent), hobbies (24 percent), and community festivals (20 percent). One new and important aspect of their community activism is their high use of information and communication technologies (ICT) and mobile communication. The survey found that 18 percent of the surveyed elderly were frequent users of mobile phones, 10 percent of faxes, and 5 percent of Internet and e-mail (DG-PCS 2005). Though Japanese elderly use of ICT is lower than that of younger generations, it appears to be much higher in comparison to the elderly in other Asian countries (Wong et al. 2006).

Recently, mobile communication service providers, particularly au-KDDI, have been targeting their business to the "silver-hair" consumers. They have begun providing a new line of simple-to-use mobile phones with big characters for both on-screen-display and panel operation. They also offer service plans with favorable pricing terms and direct-dial-up for the elderly to contact their chosen relatives. This product-cum-service line for the elderly-friendly mobile communication services has become quite popular in Japan, with a large following among elderly mobile phone users.

Compared with other Asian societies, the Japanese aging population has better social supports and is more likely to be active in community participation (Chi et al. 2001; Lai 2001; Wu 2004). And from our field observation, the elderly also have an open attitude toward embracing ICT in general and mobile communication in particular. Our recent fieldwork, with in-depth interviews and a focus group, also reveals the following characteristics of senior adults' adaptation to mobile communication. First, there is a differential adaptation path between the "young-old" (60–75) and the "old-old" (above 75 years old). As for men, the former group has generally had prior hands-on experience using ICTs in their work setting. As for women, they learned from peers and friends about using ICTs or mobile phones to coordinate their domestic household tasks such as shopping and the maintenance of sociofamilial networks. On the other hand, the latter group simply has not been fully aware of the availability of ICT gadgets and mobile communication nor how to use them. For the "old-old," then, learning mobile communication therefore requires more tailor-made handsets and the simplified telephony protocol (like single-button call). The "young-old" are frequent and active users of mobile phones and are far more likely to use both paid and free advanced mobile services.

Yet there is no guarantee that the systems, as currently designed, will yield the results expected. As one of our informants said,

I had high expectation for GPS location function, but it turns out only half good! I thought the GPS can help us in "tracking-down" where our parents-in-law [age early seventies], yet we can only know the location of the phones. Not about where they are—as my parents are, like many old people, absent-minded and they tend to leave their phones somewhere around the house or even not carry with them when going out. . . . It seems that the best way to track them down is stilling using the old way: the fixed line phone with its recording and answering functions. [We have to use] our previously agreed time of a calling appointment. (Respondent P)

Further, our focus group interviews with middle-aged (40s–50s) women found that, from their mobile phone use experience with their elders (parents and parents-in-law), the seniors prefer fixed-line communication:

My parent-in-law once told me that "my mobility is more important to me than carrying the mobile phone with me . . . and I find it stressful to carry [the phone] because I might forget where I left it." (Respondent Q)

In Japan, accessibility, popularity, and affordability of mobile phones in the market-place, coupled with the concern for personal safety against accidents and risks, are the key factors for elderly adoption of the Internet and mobile communications. Like other developed economies, in spite of the intrusiveness of mobile telephony that challenges social norms, elderly people are increasingly accepting that mobile phones can be a life-saving device, a call for assistance, in an emergency situation (DG-PCS 2006; MIC 2005, 2006a,b; MHLW 2004, 2005).

The emerging trend for more seniors to take up mobile communication can also been seen in, and reinforced by, the initiatives from the government's e-Japan/u-Japan project and the market-driven promotion for senior-friendly mobile phones. The redesigning process for elderly-friendly mobile phones is underway, too: simple and functional are the key concepts for the aging users. For instance, mobile phones are redesigned for seniors with bigger character-size(ing) for keypad and display, louder volume control, and preset phone numbers for their frequent calls. KDDI mobile phone is one example of a development in this direction: it is lightweight (less than 90 grams), has a long battery life (over one month without charging), and large buttons.

Bounded Mobile Communication Reshaping Japanese Life?

Our fieldwork further reveals some interesting observations of the differential socio-cultural influence of mobile phone (non)uses, with reference to gender and age (the bounded mobile communication). In an interview with a housewife in her late forties, the complexity of organizing family life with mobile phones can be highlighted, including the symbolic meaning of a mobile for the elders:

My parents-in-law accepted the mobile phones but only as a special gift! They sometimes use the phones with our presence when we visit them! But they love the phones as a present only and mostly place them with other precious souvenirs in the living room....Sometimes, they leave their phones unattended! (Respondent F)

Our case studies on Japanese using mobile phones for sociofamilial relationships and the protection of children and the elderly underscore such adaptation processes for sociofamilial needs. In actuality, it is the reinforcement of the existing social norm for searching out harmony and consensus (*wa*) and sense of human security.

In this chapter we examine only a small slice of the mainstream mobile activities, and there are enormous swaths of usage, both positive and negative, we have not addressed. These include the way mobile communication modes are transforming the formation of new social relationships (with strangers, for instance), and producing new deviant and dysfunctional social behaviors (Tomita 2005). But our focus is on the surprising areas of adoption and resistance at two ends of the Japanese demographic continuum, where we see many newly tailored uses emerging as well as reinforcement of and variation on deeply engrained proclivities.

In conclusion, the image of mobile communication used not just for personal gratification per se (as celebrated by media observers on the Japanese mobile miracle) is quite incomplete. Location is important and notions of unbounded movement and borderless activity do not seem to conform well to much behavioral data. The intergenerational uses of mobile communication exist alongside the modernization, personal interaction, and identity-formation projects; the latter have been heavily emphasized in prior research, the former much less so. Hence, the social aspect of mobile communication within a cultural milieu of obligation is one of the missing links for locating and understanding social processes of the rapidly emerging ubiquitous Japan.

References

Asahi Shimbun. 2005. Around Japan/Hamamatsu, Shizuoka Prefecture: Speedy e-mail warnings help keep children safe from molesters. http://www.asahi.com/english/Herald-asahi/TKY200511220129.html.

au-KDDI. 2006. au-KDDI G'zOne. http://www.au.kddi.com/english/product/lineup/gzone/index.html.

Benedict, R. 1959. *Patterns of culture*. New York: Mentor Books.

Chi, I., K. Metha, and A. Howe, eds. 2001. *Long-Term Care in the 21st Century: Perspectives from Around the Asia-Pacific Rim*. Binghamton, N.Y.: Haworth Press.

Director-General for Policies on Cohesive Society, Cabinet Office, Japanese Government (DG-PCS). 2005. Survey: Community participation of the elderly. Tokyo: DG-PCS.

Director-General for Policies on Cohesive Society, Cabinet Office, Japanese Government (DG-PCS). 2006. Annual report on the aging society. http://www8.cao.go.jp/kourei/whitepaper/w-2006/zenbun/18index.html.

Fahey, R. 2006. Bandai Namco may launch new mobile games device in Japan. *Gamesindustry.biz.* http://www.gamesindustry.biz/news.php?aid=15057.

Fortunati, L. 2005. Mediatization of the Net and Internetization of the mass media. *International Communication Gazette* 67(1): 27–44.

Gergen, K. J. 2002. The challenge of absent presence. In *Perpetual Contact: Mobile Communication, Private Talk, Public Performance*, edited by J. Katz and M. Aakhus. New York: Cambridge University Press.

Hutchby, I., and S. Barnett. 2005. Aspects of the sequential organization of mobile phone conversation. *Discourse Studies* 7(2): 147–171.

Ito, M., D. Okabe, and M. Matsuda, eds. 2005. *Personal, Portable, Pedestrian: Mobile Phones in Japanese Life*. Cambridge, Mass.: The MIT Press.

Japan Times. 2005. DoCoMo handset for kids boasts crime alarm, locater. http://search.japantimes.co.jp/cgi-bin/nb20051126a5.html.

Kamibeppu, K., and H. Sugiura. 2005. Impact of the mobile phone on junior high-school students' friendships in the Tokyo metropolitan area. *CyberPsychology & Behavior* 8(2): 121–130.

Katz, J. E., ed. 2003. *Machines That Become Us: The Social Context of Personal Communication Technology*. New Brunswick, N.J.: Transaction Publishers.

Lai, O. K. 2001. Long term care policy reform in Japan. *Journal of Aging and Social Policy* 13(2&3): 5–20.

Ministry of Health, Labour and Welfare (MHLW). 2004. White paper on health & welfare. Tokyo: MHLW.

Ministry of Health, Labour and Welfare (MHLW). 2005. White paper on the elderly. Tokyo: MHLW.

Ministry of Internal Affairs and Communications, Japan (MIC). 2005. Information and communications in Japan—white paper 2005. Tokyo: MIC.

Ministry of Internal Affairs and Communications, Japan (MIC). 2006a. Information and communications in Japan—white paper 2006. Tokyo: MIC. http://www.johotsusintokei.soumu.go.jp/whitepaper/eng/WP2006/2006-index.html.

Ministry of Internal Affairs and Communications, Japan (MIC). 2006b. Communications usage trend survey 2005. Tokyo: MIC. http://www.johotsusintokei.soumu.go.jp/tsusin_riyou/data/eng_tsusin_riyou02_2005.pdf.

Ministry of Public Management, Home Affairs, Posts and Telecommunications, Japan (MPHPT). 2004. Communications usage trend survey in 2003. Tokyo: MPHPT.

Miyaki, Y. 2005. *Keitai* use among Japanese elementary and junior high school students. In *Personal, Portable, Pedestrian: Mobile Phones in Japanese Life*, edited by M. Ito, D. Okabe, and M. Matsuda. Cambridge, Mass.: The MIT Press.

Monk, A., J. Carroll, S. Parker, and M. Blythe. 2004. Why are mobile phones annoying? *Behaviour & Information Technology* 23(1): 33–41.

Okada, T. 2005. Youth culture and the shaping of Japanese mobile media: Personalization and the *keitai* Internet as multimedia. In *Personal, Portable, Pedestrian: Mobile Phones in Japanese Life*, edited by M. Ito, D. Okabe, and M. Matsuda. Cambridge, Mass.: The MIT Press.

OECD. 2005. *OECD communication outlook, 2005*. Paris: OECD.

Srivastava, L. 2004. Japan's ubiquitous mobile information society. *Info* 6(4): 234–251.

Srivastava, L. 2005. Mobile phones and the evolution of social behaviour. *Behaviour and Information Technology* 24(2): 111–129.

Tomita, H. 2005. *Keitai* and the intimate stranger. In *Personal, Portable, Pedestrian: Mobile Phones in Japanese Life*, edited by M. Ito, D. Okabe, and M. Matsuda. Cambridge, Mass.: The MIT Press.

White, M. I. 2002. *Perfectly Japanese: Making Families in an Era of Upheaval*. Berkeley: University of California Press.

Wieser, B. 2005. Gadget-Generated growth: An overview of 3G for marketers. http://www.interpublic.com/read_file.php?did=302.

Wong, Y. C., C. K. Law, J. Fung, and J. C. Lam. 2006. Digital divide and social inclusion: Policy challenge for social development. Conference Paper delivered at East Asian Social Policy Network Conference, University of Bristol (U.K.). Unpublished paper.

Wu, Y. 2004. *The Care of the Elderly in Japan*. London: Routledge Curzon.

21 | The Effects of Mobile Telephony on Singaporean Society

Shahiraa Sahul Hameed

Singaporeans are almost constantly engaging with their mobile phones—making calls, sending text messages, listening to music, and playing games as they easily incorporate these activities into their daily lives, making it ubiquitous within this society. The mobile phone has become an essential part of life for the ordinary Singaporean. Once viewed as a luxury item, symbolizing status and power, it is now considered a necessity by most. Mobile phone penetration in October 2007 stood at 114 percent, with nearly 750 million text messages sent in that month, or the equivalent to 400 text messages per Singaporean above 15 years of age, according to Singapore's Telecommunication Regulations Authority (Infocomm Development Authority of Singapore 2005). The multiple conveniences afforded by text messages especially is not lost on tech-savvy Singaporeans; schools use text messages to inform parents of truancy; customers can use text messages to purchase drinks at discounted prices at bars, to order drinks from vending machines, and to receive news updates from the press; and by the government to update citizens on community events. If they wish, Singaporeans can have SMS alerts sent to themselves for e-government including renewal of road use tax and passports, medical examinations for domestic employees, as well as season parking reminders and notices concerning parliamentary action. Furthermore, the mobile phone almost perfectly complements the busy, intensively connected metropolitan lifestyle of Singaporeans.

Within the Singaporean context, high mobile phone penetration rates can be at least partially attributed to the Singapore government's aggressive campaigns aimed at promoting active use of information communication technologies (ICTs), along with competitive and relatively affordable pricing plans offered by the three main Telco operators (Singtel, Mobile One, and Starhub).

Alongside its aggressive promotion of the development of an e-society, the Singapore government also seems aware that widespread use of mobile phones carries risks for itself as well as the country's social fabric. As a result of these concerns, the active promotion of ICTs exists alongside a highly centralized system of government control of the uses of the very technologies it promotes.

When examining the impact of any new technology on a society, it is essential that the contextual, cultural, and social factors that shape the society it is being integrated into are taken into consideration (Kluver and Banarjee 2005). This chapter explores *Apparatgeist* theory proposed by Katz and Aakhus (2002), which advocates the use of a sociologic perspective in understanding the uses and the impact of the mobile phone on societies, in the context of Singapore. They argue that the mobile phone influences the lives of users, nonusers, and antiusers through its effects on their society. Although rejecting the idea of technological determinism, the underlying premise of this theory is that the use of ICTs shapes history and through that, society, as the technology constrains or facilitates the continuation of existing social practices based on a finite range of choices.

The *technosocial framework* posits that technology, the meanings assigned to it, and how it is used within a particular society is not technologically predetermined, but is the result of a two-way interaction between users and technology. As a result, both society and individuals do not passively receive the technology, but instead actively engage with it (Uotinen 2003). It is through this understanding that the technology can gradually transform lives formally and informally; as new patterns emerge, habits become ingrained in individuals and are transferred to and remain as part of the larger society. Therefore, the influence of mobile telephony on Singaporean society should not be seen as being purely external in nature, but rather as being embedded within the Singaporean culture, thereby making its impact culture- and technology-specific as the meaning of the technology and its uses are culturally modified to fit into the contemporary setting (Katz 2006).

Within the context of mobile phones in Singapore, it is suggested that how Singaporeans define the mobile phone will be related to the way they integrate the mobile phone into their daily lives, which are already structured around the transitional culture that is typically Singaporean. Therefore, the effects of mobile phones on Singaporean society are bidirectional, with culture influencing the way the technology is used just as much as the technology transforms the society, resulting in an interaction between the two. This chapter attempts to explore this proposition further, focusing specifically on the use of text messaging, as text messaging is one of the most popular features among Singaporeans, largely due to its asynchronous quality that provides a "window" for the receiver to decide if and when to respond, in addition to providing users with more convenient, simpler ways to send messages.

The mobile phone has also become more personalized in this society, as they are "Singaporeanized" to fit into the local culture. For example, ringtones are seen as an example of an identity marker that helps in the construction and maintenance of an image (McVeigh 2003). An excellent example of this within the Singaporean context are the polyphonic MP3 mobile phone ringtone versions of national songs that were made available online in 2003 by the Singapore government around the time Singaporeans celebrate National Day. Singaporeans are generally patriotic and the National Day celebration is a huge affair, a time when patriotism and national identity is at the forefront of government and public concerns. These ringtones quickly became popular due to the catchiness of the tune, and as news of the availability of these ringtones spread—mainly through word of mouth.

Drawing upon different sources—journalistic accounts, in-depth interviews, focus group and survey data—I seek to provide an overview of how the mobile phone and text messaging is being integrated into the Singaporean culture in three domains: youth and education, religion, and national security.

Youth and Education

Mobile phone culture and adoption in Singapore is largely driven by teens, and for them the mobile phone has become the preferred medium of communication (Ho, Tan, and Yeo 2003). Mobile phones are so widely deployed among Singaporean youth that no one is ever "inaccessible by virtue of being away from the phone" (Sherry and Salvador 2002, p. 115), and they are continuously connected to both family and friends. In fact, as early as 2002, 70 percent of Singaporean youth fifteen to nineteen years old owned their own mobile phones. Mobile phone operators here are very aware of this significant market trend, and as a result aggressively target their flexible pricing plans at the more tech-savvy, life-style conscious younger Singaporean who seems to have an unlimited supply of disposable income; for example, operators market by riding on the popularity of pop groups and by offering flexible price plans (Lim and Tan 2003).

Albeit changing, the educational setting in Singapore is largely based on a top-down framework and is hierarchical in nature. In today's schools, Singaporean students readily admit to using their mobiles to send text messages during class, though it is explicitly prohibited, and report feeling distressed when deprived of their phones (Ho, Tan, and Yeo 2003). As a by-product of the popularity of text messaging among this group, some teachers at secondary and tertiary institutions have complained that student have incorporated the abbreviations that are so popular in text messaging into their schoolwork, especially when the task is handwritten (Ng 2001). Some teachers are also worried that the casual tone used in text messages may exacerbate the already present problem of students using *Singlish* (Singlish is the colloquial form of English that is popular among many Singaporeans and incorporates words from Chinese dialects, Malay and Tamil) in their academic work.

The increasing use of mobile phones within this setting also has been pointed to as a cause for disruptions, limited attention spans, and the decline in teacher authority within the classroom. Although the traditional power distance (Hofstede 1991) inherent in teacher-student relationships within the Singaporean educational is slowly changing as the society moves toward adopting more liberal values (especially among the younger generation), strong Confucian ethics of teacher authority, respect, and control are still largely adhered to in everyday school life. Yet an important part of maintaining authority requires the separation of role from individual behavior since the latter is subject to much variability and may be more open to criticism.

These strains were brought to the surface when in mid-2003 a controversial video was widely distributed in Singapore through MMS (multimedia service) and the Internet. It showed a teacher at a top Singaporean college disciplining a student (which was captured by a fellow classmate using a PDA with a built-in camera). The three-minute video showed the teacher berating a student for his homework, telling him it was "outdated and irrelevant," calling him a "sly, crafty, old rat," before tearing up it up. The classmate who had recorded the incident later uploaded the file onto the Internet, with the purported intention of finding out "what other students thought" (Davie 2003). The student was later reprimanded for breaking school rules that permitted students to bring mobile phones and similar devices to school but not to use them during class without their teachers' permission. But later, the teacher was also asked by the principal to apologize to the student in the video. Though this request was both unprecedented and unconventional, it seems school authorities were pressured to make their request of the teacher due to overwhelming public reaction to the teacher's behavior.

As the example shows, ICTs can reduce centralized control in schools, since students now have a means of accessing the outside world, even when they are physically present in class, as well as disseminating private classroom information to the outside world (Katz 2005). One aspect of the public debate that arose as a result of this incident was related to what was termed the invasion of the sacred sphere that is the classroom, and an infringement of the teacher-student privilege (Lim 2003). This is an issue that Singaporeans have been grappling with for some time in different arenas, largely due to the rise in popularity of the mobile phone, the Internet, and the convergence between two media, as the technology highlights individual actions that may otherwise go unnoticed due to the ease of recording and transmitting information it affords. Perhaps in response to the need to better control the internal institutional environment, more schools tightened rules and raised penalties for mobile phone abuse in schools after this controversy.

As such incidents come to public attention due to the increasing convergence of the mobile phone and similar technology (such as the Internet) that can shed light on areas formerly hidden from public view, the Singaporean society is likely to grow more vocal about their personal opinions. This could result in individualism overtaking more established norms of collectivism (Geser 2004), and greater direct challenges to existing power structures (Katz 2006).

The following section focuses on an area of life that is often considered to be even more sacred and private—religion, looking specifically at how the mobile phone has been integrated into the religious lives of Singaporeans at both the institutional and individual level. The analysis is based on interviews conducted by researchers from the Singapore Internet Research Center who interviewed leaders from the five largest religious groups in Singapore (Taoists, Buddhists, Christians, Muslims, and Hindus) to find out their views on the role of ICTs in the religious sphere.

Religion

There is a growing body of literature on the incorporation of innovative information technology into religion (Katz 2006). Much of the discussion sees an inherent contradiction between the two at an intellectual level (Barzilai-Nahon and Barzilai 2005) and may also be the case on the operational level. Mobile phones, for example, are often described as agents that can empower microsystems, facilitating the continuation of primary bonds even during times of spatial separation (Gergen 2002), which at least intuitively stands in contrast to the promotion of cohesiveness and sense of hierarchy and order advocated by most religious institutions that promote centralized control. Singapore as a case study contradicts this latter hypothesis insofar as religion is concerned. This is all the more the case since although Singapore is one of the most globalized nations in the world, especially in terms of ICT penetration, religious faith remains an exquisitely important component in the lives of most Singaporeans (Kluver et al. 2005; Khun 1998).

Based on our interviews, religious leaders in Singapore appear quite aware of the vast potential offered by ICTs, and already are taking advantage of it. In fact, all of the religious communities, with the exception of the Taoists, already have integrated the mobile phone into their existing infrastructure to reach out to their followers from a central database. Incidentally, the Taoist are also the religious group that were least likely to use the Internet for religious purposes, suggesting that their lack of use of the mobile phone could be because overall they are technologically less savvy than the other Singaporean religious groups. One reason for this phenomenon could be that most of the Singaporean Taoists belong to the older generation, and are therefore less comfortable with using newer forms of ICTs.

Already, text messages are being used to disseminate information on events, daily religious greetings, and excerpts from Holy Scriptures and reminders, attesting to the fact that religious leaders in Singapore are very aware of the potential of mobile phones to bind their individual religious communities together, and are using it. In fact, during the interviews, interviewees themselves were constantly interrupted by text messages and phone calls to their mobile phone!

In addition to these benefits at a communal level, individual empowerment resulting from the use of text messages can also transform traditional practices and norms as the mobile phone changes the "traditional nature of private relationships" (Katz and Aakhus 2002, p. 301). For example, the use of text messaging for sending personal messages resulted in an uproar in the Muslim world. This occurred in 2001 when a man in Dubai divorced his wife through a text message (which was the first incident of its kind), and then when the divorce was declared legal by the religious authorities at the United Arab Emirates. This practice is not encouraged, however; by Muslim family law, which Singaporean Muslims are subject to, a man can divorce his wife by saying "I *talaq* you" (*talaq* means divorce) three times, as long as the statement was initiated and uttered by him, the phrasing is unmistakable, and his wife hears the statement.

A leading Singaporean Muslim cleric interviewed as part of the study mentioned earlier, referred to this text-message divorce, stating that it is how the technology is used that makes it good or bad: "the problem is the person, not the tool. So you need to educate the person, not blame to tool" (Aris 2004). He also said that although text messages may have simplified the divorce process, it is nonetheless considered unethical under Islamic teachings to initiate divorce proceedings in such a manner (a view that was shared by the four other Muslim opinion leaders we spoke with).

Initially, both the Muslim Religious Council of Singapore (MUIS) and the Registrar of Muslim Marriages (ROMM) in Singapore accepted this simplified divorce process. However, after strong protests from Singaporeans, and especially women's rights associations (such as Sisters in Islam, Association of Women for Action and Research, and the Singapore Council of Women's Organization), they retracted their initial statements a little over a month after this ruling was first issued in Singapore and issued a second statement indicating that in addition to the pre-existing conditions, the initiator (the husband) must also inform the *Syariah* court within seven days that such a declaration was made. Failing to do that, he can be fined SGD 500 (USD 300) or face a six-month jail sentence under Syariah law. (Syariah law is Muslim law that Singaporean Muslims are subjected to for matters related to family, including marriage and divorce.)

Although this new ruling does not explicitly forbid the use of text messages for divorce, it is more restrictive than that passed by religious authorities in the UAE. This new ruling was passed after the Muslim authorities in neighboring Malaysia disallowed a Muslim man to divorce his wife via a text message. As Singapore prides itself in being a secular state viewing Malaysia as a more fundamentalist Islamic country, it is likely that the change in the ruling was also at least partially influenced by regional geopolitical considerations. Another factor that could have influenced the authorities' decision is that as a multireligious, multiethnic society, racial and religious issues are deemed to be extremely sensitive and viewed as paramount for securing Singapore's political and economic stability. As such, the initial outcry from the public citing women's rights in a modern, multicultural, multiethnic society such as Singapore could have played an important role in the making of this decision.

Together, the domestic and external regional dimensions seemed to have influenced the Singaporean Muslim authorities' decision. If so, the incident lends support to the notion that local cultural patterns regulate the specific applications of technology, rather than there being something deterministic about the technology itself that requires certain specific patterns of adoption (Kluver and Banarjee 2005).

National Security

The third area is the antisocial uses of the mobile phone. These are of substantial concern to both Singaporeans and the Singapore government alike, especially with the rise of terrorist-related activities in the region that are seen as a threat to the political and economic stability of the small island-nation. The antisocial and opportunistic use of the mobile can be attributed partly to the ease of sending text messages to a large number of people at the same time (that can be likened to sending mass e-mails), which could encourage the propagation of ill-founded rumors and spam messages. In the Singapore context, text messages have been used to spread rumors and convey hoaxes periodically—with hoaxes promising rewards to individuals who choose to forward these messages.

The SMS problem is noteworthy in the context of heightened terror concerns in Singapore. In 2002, shortly after the first anniversary of the 9/11 attacks on the United States in 2001, a twenty-year-old Singaporean sent a message to his friends warning of a bomb threat in Holland Village (a popular expatriate meeting place in Singapore). This first text message, originating from his mobile phone, spread around Singapore rapidly, alerting Singaporeans of the potential (but nonexistent) threat. As a result, Holland Village and its surrounding areas, which are usually bustling with activity, were deserted for a few weeks after this message. In addition to the psychological burden, the prank text resulted in stinging financial losses for retailers in the area.

A twenty-two-year-old student who received a text message warning him of a bomb in Holland Village was quoted in the local daily as saying that he acted on his first instincts, to warn his friends as he thought "it was real, it urged (him) to warn (his) loved ones" (Dawson 2002). A Singaporean civil servant who intended to visit Holland Village changed his mind after receiving a text message warning him of the potential bomb threat, as he thought it would be unsafe. He was also quoted in the local newspaper stating "such acts are really terrible as it's so easy to create mass hysteria that way." Another Singaporean woman who received a similar text message from an unrecognized number said, "It's quite scary to get something like that on SMS. People really shouldn't spread rumors like this" (Dawson 2002).

With the help of phone companies, police were able to track down the culprit in only two days. He was not charged because he had no malicious intent and claimed only concern for his friends. He was let off with a firm warning, and Singaporeans at large were reminded that those found guilty of transmitting false text messages could be fined SGD 50,000 (approximately USD 30,000), jailed up to seven years, or both under the nation's Telecommunications Act.

A little over a year after this incident, and during the time that the Severe Acute Respiratory Syndrome (SARS) was at its peak in Singapore, similar unconfirmed messages were sent in Singapore, prompting the government to take a stronger stand on this issue as to not upset the economic stability of Singapore that was already severely affected by SARS. Most of these hoax messages contained unverified claims of how to protect oneself from SARS, often bordering on the obsessive, resulting in people thinking that SARS was more infectious than it actually was. At the same time, there were other Singaporeans who chose to use their mobile phones to send more light-hearted messages that contributed to the "screw SARS" culture that rapidly emerged among younger Singaporeans.

Currently, approximately 35 percent of mobile phone users in Singapore use prepaid mobile SIM cards that can be purchased easily, which makes it difficult for authorities to track down the perpetrators of such rumors. These and other similar incidents, which in combination with an increase in regional terrorism, led the Singapore government to take a stronger stance to enable them to better control and hold accountable those who use mobile phones for harmful or restricted activities: holding both individuals and Telco operators jointly responsible. In 2005, the Singapore government announced that all prepaid mobile phone subscribers would have to register their SIM cards with service providers by November 1 of that year. Failure to do so could lead to their phones being deactivated.

New regulations requiring subscribers to be at least fifteen years old and limiting each individual to no more than ten SIM cards are also in the process of being implemented. In addition, mobile phone operators will be required to submit user information to the Infocomm Development Authority (IDA) and the Ministry of Home Affairs. Operators who fail to furnish the authorities with this information can be fined, or even have their licenses revoked.

The Singapore government is also thinking of using text messages proactively during emergency situations. It was recently announced that the feasibility of using text messages to help protect citizens from terrorist attacks by sending alerts to mobile phone users from a base station during an emergency is currently being studied. According to the IDA, all three mobile operators will be obliged to cooperate if these plans are implemented. To help solve the problem of hoax messages and difficulties in determining the authenticity of broadcasted information, it is likely that cell broadcasts such as this will be limited to emergency situations and only government authorities will have the ability to send out such messages.

Conclusion

There are numerous cross-cultural consistencies in the use of mobile communication, but also enormous differences among groups and individuals depending on cultural practices—even within a single society. In the Singapore context, for example, non-Muslims are not likely to have to grapple with the issue of divorce through a text message directly, as this particular practice of allowing a man to divorce his wife by uttering the word *talaq* (divorce) is limited to the Islamic faith. However, as the incident occurred within the confines of the multiracial and multireligious society, the government reacted strongly and quickly because the impact of the ruling had some bearings on the issues of religious harmony across Singapore.

At the same time, it is unlikely that an unconfirmed message about a teacher scolding a student would stimulate such widespread concern and involvement were it not for prior cultural conditions, as is the case in Singapore. As explained earlier, Singaporean society is largely transitional in nature, with Singaporeans themselves grappling with integrating newer and more modern values into their existing cultural framework, which is largely hierarchical and traditional. It is also unlikely that unconfirmed rumors about potential terrorist attacks and SARS would have such a strong impact on a country's society and economy, leading the government to take such a strong stand against pranksters, if it were not for the geopolitical and economic concerns of Singapore, with its geographical situation and societal composition.

Although mobile culture enables and promotes more effective means for communication, it also leaves users, nonusers, and antiusers in a double bind. That is, although the potential usefulness of mobile phones is large, there exist equally valid concerns about privacy, sensitivity, and security since mobile phones have the capability to disrupt the integrity of situations and individual encounters (Katz 2006). These concerns exist at both institutional and individual levels. For example, the Singapore government seems to be aware of these negative potentialities that have and may continue to arise from widespread mobile phone use and is in the process of implementing more stringent mobile telephony laws. These new regulations seem to complement existing ones that allow stricter control over the use of other media including the Internet and foreign TV channels, for example. At an individual level, however, Singaporeans are still in the process of searching for their own social and moral stand with regard to ICT use within their unique culture and way of life, as they individually and collectively explore the new era of mobility that we as a global society are moving toward. However, due to the dynamism of both culture and technology, this search will remain a process rather than arriving at a resolution.

References

Anderson, P. 2005. Mobile technologies and their use in education: New privacy implications. http://www.jisc.ac.uk/uploaded_documents/Mobile%20tech%20-%20privacy%20-%20PaulAnderson.pdf.

Aris, M. 2004. Personal communication.

Barzilai-Nahon, K., and G. Barzilai. 2005. Cultured technology: Internet and religious fundamentalism. *The Information Society* 21(1): 25–40.

Davie, S. 2003. Student's ticking-off goes from RJC to Net; "Counseling" for teacher whose tirade in class is recorded secretly and posted online; college investigating incident. *The Strait Times*, July 12 (Singapore). Retrieved from Lexis Nexis database.

Dawson, S. 2002. Fear for safety fueled SMS bomb hoax; Recent news about security measures at Holland Village did not help either, aiding the rapid spread of the spurious message. *The Strait Times*, Nov. 30 (Singapore). Retrieved from Lexis Nexis database.

Gergen, K. J. 2002. The challenge of absent presence. In *Perpetual Contact: Mobile Communication, Private Talk, Public Performance*, edited by J. E. Katz and M. Aakhus. Cambridge: Cambridge University Press.

Geser, H. 2004. Towards a sociological theory of the mobile phone. In *E-Merging Media: Communication and the Media Economy of the Future*, edited by A. Zerdick, A. Picot, K. Schrape, J. C. Burgelman, R. Silverstone, V. Feldmann, C. Wernick, and C. Wolff. Germany: Springer-Verlag.

Ho, E., S. C. Tan, and K. Yeo. 2003. The place of the handphone in teenagers' lives: Understanding Singapore youth mobile culture by examining teenagers' use of the handphone for communication and co-ordination among peers. Unpublished final year project: Nanyang Technological University.

Hofstede, G. 1991. *Culture's Consequences: International Differences in Work Related Values*. London-New York: McGraw Hill.

Infocomm Development Authority of Singapore. 2005. Statistics on telecom services. http://www.ida.gov.sg/idaweb/factfigure/infopage.jsp?infopagecategory=factsheet:factfigure&versionid=1&infopageid=I3558.

Ito, M., and D. Okabe. 2004. Technosocial situations: Emergent structuring of mobile email use. In *Personal, Portable Intimate: Mobile Phones in Japanese Life*, edited by M. Ito, M. Matsuda, and D. Okabe. Cambridge, Mass.: The MIT Press.

Ito, M., and D. Okabe. 2005. Intimate connections: Contextualizing Japanese youth and mobile messaging. In *Inside the Text: Social Perspectives on SMS in the Mobile Age*, edited by R. Harper, L. Palen, and A. Taylor. Netherlands: Kluwer.

Katz, J. E. 2006. *Magic in the Air: Mobile Communication and the Transformation of Social Life*. New Brunswick, N.J.: Transaction Publishers.

Katz, J. E. 2005. Mobile phones in educational settings. In *A Sense of Place*, edited by K. Nyíri. Vienna: Passagen Verlag.

Katz, J. E., and M. Aakhus. 2002. Making meaning of mobiles—a theory of *Apparatgeist*. In *Perpetual Contact: Mobile Communication, Private Talk, Public Performance*, edited by J. E. Katz and M. Aakhus. Cambridge: Cambridge University Press.

Khun, E. K. 1998. Maintaining ethno-religious harmony in Singapore. *Journal of Contemporary Asia* 28(1): 103–121. http://ccbs.ntu.edu.tw/FULLTEXT/JR-EPT/khun.htm.

Kluver, R., and I. Banerjee. 2005. Political culture, regulation, and democratization: The Internet in nine Asian nations. *Information, Communication, and Society* 8(1): 1–17.

Kluver, R., B. H. Detenber, W. Lee, S. H. Shahiraa, and P. H. Cheong. 2005. Internet and religion in Singapore: A national survey. *Singapore Internet Research Center Report Series 02*. http://www.ntu.edu.sg/sci/sirc/workingpapers/IR%20report%20-26%20Sept%202005.pdf.

Lim, H. C. 2003. Strait times forum. *The Strait Times*, July 17 (Singapore). Retrieved from Lexis Nexis database.

Lim, S. S., and Y. L. Tan. 2003. Old people and new media in wired societies: An exploration of the socio-digital divide in Singapore. *Media Asia* 30(2): 95–102.

McVeigh, B. 2003. Individualization, individuality, interiority and the Internet. In *Japanese Cyber-cultures*, edited by N. Gottlieb and M. McClelland. London: Routledge.

Ng, D. 2001. SMS lingo creeps into school work. *The Strait Times*, May 26 (Singapore). Retrieved from Lexis Nexis database.

Rheingold, H. 2002. *Smart Mobs: The Next Social Revolution*. New York: Perseus Books Group.

Sherry, J., and T. Salvador. 2002. Running and grimacing: The struggle for balance in mobile work. In *Wireless World: Social and Interactional Aspects of the Mobile Age*, edited by B. Brown, N. Green, and R. Harper. London: Springer-Verlag.

Uotinen, J. 2003. Involvement in (the information) society—the Joensuu Community Resource Centre Netcafe. *New Media & Society* 5(3): 335–356.

Mobile Communication and the Transformation of the
Democratic Process

Kenneth J. Gergen

Lewis Mumford's imaginative and challenging work (1934) on technology and cultural
history opened a new scholarly agenda. As Mumford made clear, cultural history is not
simply shaped by humans in relationships. Rather, any responsible account of cultural
history must take account of technological developments. The inventions of the clock,
the printing press, and the automobile, for example, have played an enormously important role in shaping the contours of Western history. For communication scholars,
this early work has stimulated a rich tradition of scholarship concerned with communication technologies and their roles in social change. Radio, telephones, TV, newspapers, and Web communication have all been the subject of intense interest. In this
respect, inquiry into the place of mobile communication in cultural transformation is
but in its infancy. The challenge is clear: mobile communication is the most rapidly
expanding communication technology on the globe. The door has been opened (Katz
and Aakhus 2002; Nyíri 2003), but it is imperative to sustain deliberation about its
place in cultural life, to sensitize ourselves to possible futures, and to open dialogue
on ways in which such trajectories might possibly be altered.

In the present offering I wish to open discussion on mobile communication in the
context of democratic society. My specific concern here is with the democratic process
and the critical place of communication within this context. There is now a substantial
literature on the implications of communication technology for the democratic process (see, for example, Pool 1984; Sclove 1995; Hewitt 2005). However, there are but
scant offerings on the contribution of mobile communication to political life. Fortunati (2003) sees the cell phone as making a strong contribution to democracy, as it
furnishes the individual with expanded possibilities for voicing political opinion. As
Kim (2003) sees it, this expanded form of expression is quite popular in the younger
generation. Of particular importance, these forms of political expression are essentially
horizontal, circulating across the populace and without monitors (Danyi and Sukosd
2003; Shapiro 1999). If the individual voter is informed about the nature of the world
only by those in power, democracy is nonexistent; the vote will effectively affirm that
which has been promulgated by those in power. With an expansion of horizontal

communication—among the populace—independent deliberation on government is enhanced. My primary concern, then, is with the significance of mobile communication within the democratic process.

In what follows I propose that on the national level we have witnessed during the past half century three highly significant transformations in the structure of political communication. Within this transformation, mobile technology has played, and continues to play, an increasingly significant role. The picture that emerges is one in which public deliberation on political issues is greatly enhanced, while simultaneously the potential for both extremism and disinterest in political issues is increased. In effect, while there have been substantial gains, there are offsetting developments that bear continuing attention.

Democratic Structure: The Visible Tradition

To appreciate these changes it is first essential to characterize the national democratic structure of communication in the midtwentieth century. Although much could be said about these complex conditions, for present purposes a rudimentary sketch will suffice. Let us focus first on the most visible structure prevailing for the greater part of the century. At the outset we have an overarching governmental establishment, composed of the various offices and institutions of the national government. As we have generally understood the democratic process, the governmental establishment should reflect and incorporate the opinions and values of the voting public. At the other end of the spectrum, then, we have the individual voter, endowed with the right and duty to observe the world, deliberate on issues affecting the society, participate in political discussions, vote on representatives and issues, and support the governing institutions.

As political scholars have long maintained, between this macrostructure and the individual there is (or should be) a domain of face-to-face relationships in which issues of common political concern can be debated. This is the domain of civil society, lodged within the microsocial processes of communication (see, for example, Seligman 1995; Goldsmith 2002; Ehrenberg 1999). Participation in civil society is not only important in generating independent deliberation about political issues, enabling expressions of resistance, inviting independent initiatives, and mobilizing organized expression. In addition, civil society is to serve as a bulwark against the raw pursuit of individual self-interest. Participation in the dialogues of the common good should balance the desire for individual benefit. In contemporary terms, the pursuit of individual rights would be tempered by a concern with duties to community.

To this mix we must add a fourth governing sector, one not envisioned by the founding fathers of constitutional democracy, but one which came to play an increasingly predominant role over the century. This is the sector of mediated communication. Through the 1950s, both the newspaper and the radio served as major vehicles for communicating information and opinion relevant to political deliberation. Communication of this form was largely monologic. The public was informed but did not contribute directly or significantly to the media themselves. The media were in a position to influence public dialogue, but did not significantly communicate public opinion to those in power. In effect, the process of democratic dialogue was truncated. Moreover, in important degree it may be said that the media primarily served the interests of the overarching institution of government. To be sure, critical deliberation on existing office holders and policies was common. However, because news reporting was held to standards of objectivity, and objectivity is defined as freedom from value bias, the media largely disseminated news issuing from, and largely controlled by, those who either occupied public offices or were candidates for these positions. In effect, the news media served largely to provide the public with information about the views and actions of those in governing positions, which information was largely shaped by these same individuals. In lesser degree did the media serve a critical function of government itself.

In brief, then, at midcentury we find roughly a four-tier structure of democratic process. As summarized in figure 22.1, the governing structure is a major source of politically relevant information. The news media disseminate this information to individual citizens, while providing a modicum of counter-opinion. Individual consumers thus enter active deliberation within the civil sphere. This is not at all to say that under these conditions political debate was tepid. While intense debate was possible, it was largely generated within the parameters set in the governmental sphere itself. That is, the major proposals and policies were created within this sphere, and these were often contested by an opposing political party. However, both the offerings and their

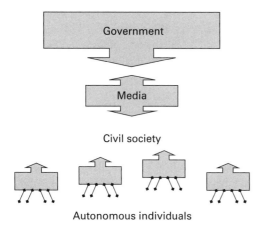

Figure 22.1
Structure of democracy: mid-twentieth century.

opposition were formulated within the governmental sphere itself. In effect, the major agenda or topoi of political deliberation did not originate within the populace itself.

Mobile Communication and Structural Transformation

While the political process of midcentury seemed comfortable enough for the majority of the public, it also shielded widespread oppression. Through various means, vast sectors of the African American minority were either discouraged from or actively prevented from voting. Many Japanese Americans had been incarcerated during World War II, with little national protest. Although women had painfully acquired the right to vote in 1920, little was done either to encourage their voting or to provide sufficient education for full political participation. Laws governing the Native American right to vote remain equivocal even today. Both by virtue of law and psychiatric prejudice, the homosexual voice was essentially absent from the political sphere. In effect, in numerous ways the prevailing structure functioned to suppress the flows of communication essential to full democratic process. Ironically, one might even suppose that the prevailing sense of satisfaction with democracy was dependent upon a vast suppression of difference.

The Emergence of a Proactive Mittelbau

Beginning in the mid-1950s, with the civil rights protest marches and sit-ins in the southern United States, a new communication structure began to develop, one with enormous implications for democratic process. In political terms one of the most important synchronies of the century was between the civil rights protests and the emergence of TV. No longer were minority claims of injustice and inequitable treatment taking place in the distant and abstract media of print and radio. Rather, the sometimes touching, sometimes horrific events in Selma or Tuscaloosa, Alabama, were suddenly there in the living room. With the advent of TV, the political passions of local or regional groups increasingly became national issues.

With the waves of antiwar protests and feminist demonstrations in the 1960s and 1970s, one may locate the solidification of what I shall call a proactive Mittelbau that is a structure of political communication lodged between the national government and the local or civil society, capable of both drawing participation from the local culture and speaking to government. In effect, we find the emergence of a political body capable of stimulating political engagement that, because of its ability to mobilize public support, stands in a more dialogically direct relationship to government. While early protests were relatively naïve with respect to national influence, more expressive than instrumental, the success of the civil rights movement in mobilizing national sympathy altered the character of public demonstration. Rather than simply expressing a point of view, such demonstrations could be staged for the specific purposes of activating support. Frustrated by, and largely independent of, official government proceedings, public remonstrance has thus served the interests of vast sectors of the population, supporting numerous rights groups (from life to choice, to gay and lesbian, animal life, and so on) along with groups concerned with the environment, Supreme Court nominations, globalization, and more.

While television is the technology primarily responsible for the politically proactive Mittelbau, and the Internet its major means of organizing across large geographical domains, mobile communication has served as its chief instrument of refinement. For example, Paragas (2003) has documented the use of text messages for purposes of mutually affirming political commitment, and coordinating political protest. Lai (2003) sees potential in mobile phone technology to support democracy across territorial boundaries. The most extensive account of the ways in which mobile communication facilitates political activism is contained in Howard Rheingold's *Smart Mobs*. Like others, Rheingold properly sees mobile communication within the context of technologies that facilitate horizontal communication across broad spectra of society. However, mobile technology is particularly effective in creating ad hoc social networks for achieving effective demonstrations. In this vein, a colleague recently recounted for me his experience with an urban demonstration that had been outlawed by the police: Small groups of demonstrators were scattered throughout the city. As one group executed its protest, snarling traffic and creating public consciousness, the police began to converge. Lookouts posted some blocks from the protest and armed with cell phones quickly communicated to the group the impending threat. The demonstrators immediately disbanded and melted into the crowd. At the same time, they phoned their colleagues in other parts of the city. Soon another demonstration began, and the scenario was repeated. In effect, the protestors were able to sustain a full day of demonstrations without a single arrest.

From Civil Society to Monadic Clusters

While communication technology has been instrumental in bringing a politically effective Mittelbau into being, a second transformation in politically relevant communication has also been under way. In this case we do not have a new arena of dialogue so much as a significant transformation in structure, a transformation in which the civil society is being slowly replaced by small, intensely interdependent communication clusters. Consider first the case of civil society. In recent decades, cultural commentators have become increasingly concerned with the erosion of the civil society. Sennett (1974) bemoans the loss of those bonds of association and mutual commitment out of which community is forged. For Bellah and his colleagues (1985), individualist ideology promotes a me-first orientation to social life, with a resulting lack of interest in community participation. Or, as one might say, the grounding of democracy in the freedom of the individual mind is set against the very kind of civic engagement necessary for effective democracy. With the publication of Robert Putnam's *Bowling Alone*, the loss of civic participation became a matter of broad debate. As Putnam demonstrates, over a broad range of indicators, voluntary communal participation has undergone decline. It is in the face of just such decline that the communitarian movement has sprung to life. In effect, communitarianism is emblematic of a disappearing tradition.

In an earlier work, *The Saturated Self*, I proposed that the undermining of face-to-face communities is largely the result of twentieth-century developments in communication technology. Entertainment technologies, such as radio, TV, and magazines offered entertainment within the confines of the private dwelling, thus replacing the more strenuous demands of face-to-face conversation in public venues. With the advent of the Internet and e-mail, even those within the home were split apart from one another for extended periods. With the development of transportation technologies (e.g., the automobile, jet planes, mass transportation systems), people were physically propelled outward from the local community—whether in terms of jobs, schools, shopping, or entertainment on the one hand, or long-distance travel (e.g., business, vacations, summer homes) on the other. In effect, the development and broad proliferation of mundane technologies of communication played an important role in the evisceration of the local, face-to-face communal life.

At the same time that communication technology has hastened the erosion of civil culture, mobile communication in particular has played a critical role in bringing about transformation. Essentially we are witnessing a shift from civil society to monadic clusters of close relationships. Cell phone technology favors withdrawal from participation in face-to-face communal participation. Indeed, as many commentators demonstrate, public cell phone use invites antagonism and scorn. Simultaneously, however, the cell phone favors intense participation in small enclaves—typically of friends and family. These "floating worlds" of communication (Gergen 2003) enable such groups to remain in virtually continuous contact. The individual may move through the day relatively disengaged from those about him or her, as physically absent participants in the favored cluster are immanently present. In my view, this creation of monadic clusters is having two substantial effects on democratic process. To appreciate these trajectories, consider the social implications of perpetual contact: As people coordinate words and actions together, so do they come to create meaningful worlds. Realities are constructed, values developed, and "good reasons" come into being (Gergen 1994). Locally fashioned assumptions are transformed into "obvious realities," universal in implication. While the process of world construction is embedded within all social interchange, it is most effective in small, dialogically engaged relationships such as those invited by mobile communication. It is here that we may locate a process of *circular affirmation*, that is, a form of interchange in which participants continuously affirm the views and values of each other. With the increased sophistication of camera phones, the co-present affirmation is further intensified (Scifo 2005). Under these conditions a univocal and compelling world is constructed. In effect, these atomic clusters of communication are powerful implements for creating and sustaining circumscribed realities, values, and logics.

In this context two major trajectories in political participation are especially invited. First there is *political detachment.* In many monads the dominant issues concern the immediate lives of the participants themselves. The cell phone is used primarily for the micro-coordination of social or family life, for social and emotional support, for enhancing the participants' safety, and for sharing experiences. In effect, communication functions to sustain the life of the group itself. Under these conditions, life outside the group recedes in significance. Issues of political concern, unless they immediately affect the lives of the participants, dwindle in importance. Supporting this view, Sugiyama and Katz (2003) explored the relationship between mobile phone use among university students and participation in civil society. Their data indicated participation in volunteer work and in political activities both receded with increased use of the mobile phone. Those who never used these technologies were most engaged in civil society. Ancillary data also showed that increased reliance on the mobile phone was associated with high frequency of socializing with friends. In effect, when friendship is central, issues unrelated to friendship recede in importance.

One could scarcely suppose that mobile communication is substantially subtracting from political participation. Cultural patterns are always complex and varied and we must consider a second trajectory invited by mobile communication, *dialogic disruption.* In particular, let us consider monadic groups in which political concern is intense. The participants are actively engaged in sharing opinions and information concerning political issues. Here again we must consider the tendency toward circular affirmation. To the extent that participants in these monadic groups tend toward mutual affirmation, there will be a resulting resistance to interfering or opposing ideas. One is rewarded for bringing to the group news and information that supports the dominant opinion. Deliberation on opposing ideas is replaced by tendencies toward consensus. When there is opposition, the tendency toward internal affirmation is only intensified. Those outside the group are viewed with disregard or contempt. In effect, the flows of political communication essential for viable democracy are interrupted. Dialogue communication among groups gives way to monologue within groups. The animosity so pervasive in the 2004 U.S. presidential election may stand as a case in point.

From Autonomy to Relational Life

A final shift in the structure of democratic process requires attention. Although more fully conjectural, here we confront the possible transformation in our very conception of democratic process. At the heart of the conception of democracy is the autonomous agent, the individual endowed with the capacities for choice independent from social influence. Although the social sciences, armed with deterministic theories of voting behavior have long inveighed against the voluntaristic assumption, everyday phenomenology has provided strong resistance to such threats. In spite of highly successful predictions of voting patterns, we must suppose that most people continue to see themselves as the originating agents of their own voting preferences.

Yet, in spite of the phenomenology of daily life, there is an important sense in which mobile communication invites a transformation in the functioning of the individual, one that represents a collapsing of agency into relational process. As I have outlined elsewhere (Gergen 2003), as the mobile phone brings small enclaves into continuous co-presence, so do actions that might otherwise be attributed to the autonomous self become embedded in dialogue. Private thought and public deliberation converge; isolated emotionality is replaced by emotional sharing (Rivière 2002). We increasingly locate practical reasoning in relationships as opposed to independent minds (Katz and Aakhus 2002). Personal significance is acquired through one's place within the network. The mobile phone functions symbolically, then, as an umbilical cord through which one draws vital nurturance from a larger, protective force. It is not one's individual thought and personal desire that is now central to the democratic process but relational interchange.

The Future of Democratic Process

The present analysis brings into focus several significant issues relevant to our understanding of contemporary political life. At the outset it underscores the substantial significance of communication technology in general, and mobile communication in particular, in reshaping major institutions and concepts. Our particular concern in this chapter is with what I suggest is a substantial shift in the landscape of democratic deliberation. It is helpful here to contrast the simplified structure of the democratic process prevailing in midtwentieth century (figure 22.1) with what now appears to be the case. As I have argued, communication technologies have first added enormous strength to the media as stimulants to the democratic process. No longer do they primarily serve as one-way communicating devices (from the top down), but as two-way influence structures with a strong horizontal communication thrust within the populace, and in the case of the Internet, horizontal deliberating devices. The power of the media in circulating expressions of opinion has fostered what I have

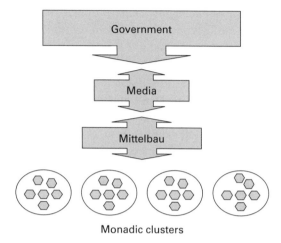

Figure 22.2
Structure of democracy: conditions of high technology.

called a Mittelbau of democratic expression. As indicated in figure 22.2, this domain of public remonstration is inserted between the local community and the national governing bodies. Here mobile communication is making an increasingly powerful contribution to the efficacy of political activism.

On the local level communication technologies are bringing about the decay of civil society in the traditional sense. Largely as a result of mobile communication technology, community and town participation is being replaced by atomistic communication cells. For many, these monadic groups become self-serving, and political issues recede in importance. Participation in the democratic process erodes. For others, there is reason to suspect a high degree of political deliberation, but of the sort that lends itself to a hardening of political opinion. Debate is replaced by diatribe.

To be sure, I am offering neither a definitive nor an empirically fortified account of political communication in the United States. Rather, my hope is that the present observations and speculations may provide both a stimulus to dialogue and a framework for continuing study. As I have indicated, there is extant support for much of what I propose. However, of particular importance to future research in this domain are three issues:

Developmental Transformation

With notable exceptions, existing research on mobile communication is synchronic. That is, it provides us with snapshots, frozen in time, of processes that quite likely undergo historical change. Two of these diachronic processes are particularly demanding of study. In the first case, we know very little about changing habits of mobile communication across the lifespan. At the present time, cell phone usage often begins during early adolescence and may presumably be extended for a lifetime. However, the particular uses to which it is put may vary considerably across the lifespan. The picture painted by much existing research is that as one enters the world of mobile communication there is an intensification of microrelationships. While much data from the youth sector surely supports this view, it is not clear that the demands of the microsphere remain significant as people move into professional roles. And, as they become parents, it seems likely that the particular microsphere sustained by mobile communication will be the family as opposed to friendship clusters. In the elderly community the function of mobile communication may shift again, with relational intensification replaced by more pragmatically functional communication. In each case there are different ramifications for democratic process.

There is a second diachronic concern that is far more general, and which necessarily renders all study of mobile communication speculative. Mobile communication technology is in a state of continuous transformation. Not only does the vastness of the market ensure that manufacturers continue their active search for means of "improving" the technology. In addition the microchip technologies enable small devices such as the cell phone to be "loaded" with additional technologies. So adept has been the ability of manufacturers to install such technologies as text messaging, photography, Internet access, music, TV, geographical locating, and the like, that the cell phone has been called "the technology that eats everything." As the technology shifts over time, so will its political usages. If, for example, the cell phone increasingly becomes a means by which the individual can sustain a privatized existence, then the monadic cells will ultimately go the way of the civil society. People will be able to carry with them their own music, entertainment, news, Web resources, and so on in such a way that organically based relations will become outmoded. We confront the daunting possibility that relations with actual persons, with all their foibles, will be replaced by mediated idealizations.

Niche Variation

At present most research on mobile communication is appropriately addressed to general patterns of usage and their implications. As a second level of inquiry, populations may be broken down by various demographic criteria (e.g., gender, age, income). Ultimately, however, an appreciation of the political implications of mobile communication will depend on a nuanced examination of subpopulations characterized by differing lifestyles. This form of niche analysis, more common in political and advertising research, may shed important light on qualitatively differing uses of mobile communication. Rather than issues of more or less usage, we turn to the functions played by mobile communication in differing styles of life. It is quite possible, for example, that the groups characterized above as monadic are not highly general, but represent a particular niche. For others, issues of friendship and in-group solidarity may be less important in their mobile communication practices than business or political concerns. Niche-based research would do much to illuminate these possibilities.

Localizing the Mittelbau

In terms of historical change, there is a recent shift in political communication that promises to be of significant consequence. As I have outlined, the proactive Mittelbau has largely relied upon technologies of TV, the Internet, and mobile communication to achieve efficacy. Yet, in recent years the Internet has come to play a new and potentially dramatic role in politics. It is not simply Web sites, Listserves, and e-mail that have enabled various political interest groups not only to announce plans for political protest (thus swelling their numbers), provide extended rationale to support their positions, facilitate entry to networks of communication, and furnish continuous news of developments. Rather, the Internet is increasingly used as a means of organizing and directing political activity in much the same way that political parties have functioned; for example, national grassroots movements have become more fully organized as they have become increasingly able to generate local networks. The efficacy of these networks, in turn, is enhanced by mobile communication technology. As but one example, during the Bush-Kerry election, the national venture, Move On, spawned an enormous array of local get-out-the-vote activities. In thousands of local communities, canvassers were enlisted to go from door to door to stimulate voting interest among Kerry supporters. Not only were local Move On representatives coordinating their activities through mobile communication, but they were also able to maintain close contact with regional and national organizing efforts.

Looking into the future, we may thus see the emerging Mittelbau come into being as the new domain of civil society, with small monadic enclaves replacing single individuals as the base unit of democratic society. On the grassroots level this could mean both an increase and decrease in political activism, depending on the character of the monad. Within the more active monads we may anticipate greater extremism. On the national level, this would also mean a vast increment of "power to the people." In an age in which the checks and balances within the central government seem to be eroding, and a strong, active, and partisan-based gridlock often stands in the way of progress, there is much to be welcomed in such a movement. The emergence of a proactive and politically engaged populace may represent a major advance in the democratic process.

References

Bellah, R. N., R. Madsen, W. M. Sullivan, A. Swidler, and S. M. Tipton. 1985. *Habits of the Heart: Individualism and Commitment in American Life*. Berkeley: University of California Press.

Danyi, E., and M. Sukosd. 2003. Who's in control? Viral politic and control crisis in mobile election campaigns. In *Mobile Democracy, Essays on Society, Self and Politics*, edited by K. Nyíri. Vienna: Passagen Verlag.

Ehrenberg, J. 1999. *Civil Society: The Critical History of an Idea*. New York: New York University Press.

Fortunati, L. 2003. The mobile phone and democracy: An ambivalent relationship. In *Mobile Democracy, Essays on Society, Self and Politics*, edited by K. Nyíri. Vienna: Passagen Verlag.

Gergen, K. J. 1994. *Realities and Relationships*. Cambridge: Harvard University Press.

Gergen, K. J. 2002. The challenge of absent presence. In *Perpetual Contact*, edited by J. E. Katz and M. A. Aakhus. New York: Cambridge University Press.

Gergen, K. J. 2003. Self and community in the new floating worlds. In *Mobile Democracy, Essays on Society, Self and Politics*, edited by K. Nyíri. Vienna: Passagen Verlag.

Goldsmith, S. 2002. *Putting Faith in Neighborhoods: Making Cities Work through Grassroots Citizenship*. Nobelsville, Ind.: Hudson Institute Publications.

Hewitt, H. 2005. *Blog: Understanding the Information Reformation That's Changing Your World*. Nashville: Thomas Nelson.

Katz, J. E., and M. A. Aakhus. 2002. Conclusion: making meaning of mobiles—a theory of Apparatgeist. In *Perpetual Contact*, edited by J. E. Katz and M. A. Aakhus. New York: Cambridge University Press.

Kim, S. D. 2003. The shaping of new politics in the era of mobile and cyber communication. In *Mobile Democracy, Essays on Society, Self and Politics*, edited by K. Nyíri. Vienna: Passagen Verlag.

Lai, O. K. 2003. Mobile communicating for (E-) democracy beyond sovereign territorial boundaries. In *Mobile Democracy, Essays on Society, Self and Politics*, edited by K. Nyíri. Vienna: Passagen Verlag.

Mumford, L. 1934. *Technics and Civilization*. New York: Harcourt Brace.

K. Nyíri, ed. 2003. *Mobile Democracy: Essays on Society, Self, and Politics*. Vienna: Passen Verlag.

Paragas, F. 2003. Drama*text*ism, mobile telephony and people power in the Philippines. In *Mobile Democracy, Essays on Society, Self and Politics*, edited by K. Nyíri. Vienna: Passagen Verlag.

Pool, I. de S. 1984. *Technologies of Freedom*. Cambridge, Mass.: Harvard University Press.

Putnam, R. 2000. *Bowling Alone*. New York: Simon and Schuster.

Rheingold, H. 2002. *Smart Mobs, the Next Social Revolution*. New York: Perseus.

Rivière, C. 2002. La pratique du mini-message, une double stratagie d'exterioisation et de retrait de l'intimite dans les interactions quotidiennes. *Réseaux* 20(112–113): 139–168.

Scifo, B. 2005. The domestication of camera-phone and MMS communication. In *A Sense of Place, the Global and the Local in Mobile Communication*, edited by K. Nyíri. Vienna: Passagen Verlag.

Sclove, R. E. 1995. *Democracy and Technology*. New York: Guildford.

Seligman, A. B. 1995. *The Idea of Civil Society*. Princeton, N.J.: Princeton University Press.

Sennett, R. 1974. *The Fall of Public Man*. New York: Norton.

Shapiro, A. L. 1999. *The Control Revolution—How Internet Is Putting Individuals in Charge and Changing the World We Know*. New York: Public Affairs.

Sugiyama, S., and J. E. Katz. 2003. Social conduct, social capital and the mobile phone in the U.S. and Japan, A preliminary exploration via student surveys. In *Mobile Democracy, Essays on Society, Self and Politics*, edited by K. Nyíri. Vienna: Passagen Verlag.

Culture and Imagination

23 Cultural Differences in Communication Technology Use: Adolescent Jews and Arabs in Israel

Gustavo Mesch and Ilan Talmud

Our purpose is to investigate how culture modulates the use of information and communication technologies (ICTs) by looking at Israeli teens from Jewish and Arab ethnic groups. This is a vital topic from both intellectual and policy viewpoints. In the first case, there are vigorous debates over technological determinism and the role of social structure and culture. In the second case, there are serious questions concerning equity, digital divides, and social capital (Katz and Rice 2002). The interaction between innovative information and communication technologies and traditional communities is often deemed problematic as these communities are perceived as resisting policies that promote expansion in the use of ICT because they are held to be a threat to their local and cultural practices.

Technology poses a challenge to four central dimensions of the tension that characterizes these groups. First, traditional communities are often characterized by a tight hierarchy, based on the subordination of large groups of individuals to authoritarian elites. Censorship is a major means by which the communal elite controls information flows. A second dimension of tension is patriarchy, which promotes a gender-based division of labor, in which females are relegated to unequal treatment. Communication technologies may be used to air disputes with the patriarchal hierarchy and promote desegregation according to gender lines. Discipline is a third dimension of tension, where traditional communities are highly disciplined and social behavior is based on tradition. While tradition directs the members to look inward into the community, technological innovations enable intensive interactions that cross rigid social boundaries such as age, gender, and community. The fourth dimension of tension is seclusion: the boundaries of the traditional community are highlighted as they facilitate the formation of a collective identity (Barzilai-Nahon and Barzilai 2005). ICTs are perceived as a symbol of individual freedom, inclusiveness, egalitarianism, and multifaceted interactions among individuals and groups (Haythornthwaite and Wellman 2002). For this reason, traditional communities are suspicious and perceive new communication technologies as a threat to cultural preservation. Clearly, this situation is not free of tensions, as the young and more secular attempt to adopt and take advantage of new technologies and may challenge the existing traditional social order.

Studies of Muslim populations in Western countries show a similar pattern of relatively lower access by these populations to information and communication technologies. A recent study in Australia traced a major networking initiative in a community including Turkish residents. The initiative provided free access to home computers, along with software, cabling, and intranet with multilingual (including Turkish) content and training in the use of these technologies. While the take-up rate in the community was relatively high, with over 75 percent of the households completing training and receiving a computer, only 20 percent of the Turkish residents elected to participate in the program that was implemented in 2001–2003. Their most common rationale for their lack of interest was fear that women and especially children could access inappropriate material and communicate with individuals outside the community (Meredyth et al. 2002, 2004).

Dutch studies have shown that although their Turkish Muslim population is underrepresented among Internet users, information technologies are used as a direct resource for young Muslims constructing and reconstructing their identities in a Western country (Mandaville 2000). The Internet is used to create social communication networks within the religious community and across gender lines. A study of Muslim mailing lists showed that females were active participants and questioned religious rules. The anonymity of the Internet makes it possible for young Muslim females to raise sensitive topics with males for discussion (Brouwer 2004).

Positions on the social impact of the communication technologies are dominated by two contrasting views, technological determinism and social constructivism. Early conceptualizations, assuming technological determinism, described the weakness of electronic media in supporting social ties. The "reduced social cues perspective" is based on the observation that electronic communication allows the exchange of fewer cues than does the face-to-face environment, and suggests that it is less appropriate for the support of emotional exchanges or the conveying of complex information and a sense of social presence. This early perspective, assuming technological determinism, as noted, was quite skeptical of the ability of electronic communication technologies to create and support strong ties.

Social constructivists, by contrast, argue that some features of electronic communication, such as anonymity, isolation, lack of "gating features," and ease of finding others with the same interests, make it easier for individuals to create and sustain strong ties (McKenna, Green, and Gleason 2002; Katz and Rice 2002). The formation of close interpersonal relationships requires trust, that is, a sense that intimate information disclosed in interpersonal exchanges is not widely disseminated and is not used to ridicule friends. The relative anonymity of the Internet reduces the risks of such disclosure, especially of intimate information, because such information can be shared with little fear of embarrassment resulting from disclosure to members of the close-knit, often transitive, face-to-face social circle (McKenna, Green, and Gleason 2002).

Although diametrically opposed, these two perspectives focus on technology as a major determinant of social relations. We join this debate by arguing that the social standing and cultural repertoires are associated with particular uses of cell phones and computer-mediated communication. Differences in individual choices of communication channels and their purposes reflect social and cultural differences among social groups. Using in-depth interviews with Jewish and Arab adolescents in Israel, we attempt to illustrate how social conditions and not the mere features of communication technologies influence their use.

Use of Communication Technologies among Adolescents in Israel

Israel is cleaved according to nationality: Israeli Jews (79 percent of the population) and native Israeli Arabs (18 percent). Israeli Jews and Arabs differ in culture, as reflected in different languages, religions, nationalities, cultural heritage, and family structure and values (Al Haj 1987). This cleavage is also apparent in the high residential segregation of the communities. Only 9 percent of the Arab population resides in mixed cities, the rest in communities that are entirely Arab. Even when Arabs and Jews share the same city they tend to reside in different neighborhoods and attend different schools (Smooha 1997).

The culture of Israeli Arabs is deeply rooted in the more general Arab culture and is more traditional and collectivistic than Jewish society. For example, almost half of the Arab population report going to pray frequently, compared to only 23 percent of the Jewish population (Smooha 1997). In both groups the family is central; the extended family is pivotal among the Arab—though hardly exists among the Jewish—ethnic clusters. In the Arab population, marriage is between men and women within the extended family, being first or second cousins (Al-Haj 1987). The Israeli Jewish family system is mainly of single nuclear units, run along democratic lines, similar in many ways to the situation in other Western countries. By contrast, the Arabs in Israel are a relatively collectivistic, communal, and traditional cultural group. The precise roles of family members are clearly laid out, and Arab females have a lower social status than their male counterparts and their parents. Arab adolescents perceive relations in the family as based on punishment, negative criticism, and restrictive control. Despite the process of modernization, many traditional notions of family life continue to exist, including a patriarchal pattern of authority, sharply delineated gender roles, conservative sexual standards, self-sacrifice for the greater common good, and honor and shame as regulators of moral norms. The Israeli Jewish family is a nuclear system characterized by democratic family relations, relatively permissive parental controls, similar in many ways to the case in other Western cultures. Israeli Jews tend to be egalitarian, with little respect for authority and with unusually high social network density (Fisher and Shavit 1995). Arab adolescents tend to report being closer to both their parents and having a more cohesive family structure than Jewish adolescents (Mikulincer, Weller, and Florian 1993). The higher levels of closeness to parents and family cohesion reported by Arab youth have been interpreted as reflecting family ties characteristic of a collectivistic culture, as compared with the more individualistic orientation of Israeli Jewish youth (Mikulincer, Weller, and Florian 1993).

The use of ICTs has grown rapidly but has yielded a digital divide in access to and use of communication technologies by nationality. Data for families with adolescent children show that 86 percent of Jewish and 62 percent of Arab households report computer ownership at home. The percentage of households with adolescents reporting an Internet connection is 72 percent for the Jewish population but only 45 percent for Israeli Arabs. As for cell phones, 89 percent of the Jewish households report owning cell phones versus 61 percent of the Israeli Arab population.

While in 2001 only 35 percent of the adolescent population had access to the Internet; by 2004 available access was for 65 percent. As to purpose, the overwhelming majority of adolescent Internet users reported mainly social purposes. Almost 70 percent of the Arab and 80 of the Jewish respondents said that they liked to meet new people through the Net.

In studies examining the relationship of Internet use and social involvement, 14 percent of Israeli adolescents reported having friends they met online (see Mesch and Talmud 2006). These adolescents had a more dispersed and heterogeneous network in terms of gender and age than did those without online friends (Mesch and Talmud in press).

Approach

To further inquire into the impact of Internet use on adolescents' identity, experience, and relational style, we composed a semistructured, open-ended questionnaire, which was administered to eighteen Arab and nineteen Jewish Internet users. The interviewees were selected by a snowball procedure, beginning with our students. We attempted to interview adolescent Internet users who were twelve to thirteen and sixteen to seventeen years old to maximize variability. We also attempted to represent equally both sexes, although we ended up with slightly more females than males. The interviews were held mostly at the interviewees' parental homes, but in a few cases, where the household was packed, in cafés. We trained graduate students, most of them with some experience in in-depth interviews and qualitative research, to probe the interviewees by leading questions, inquiring into their communication habits, hobbies, and particularities of communication style, friendship formation, and acquaintances. To minimize social desirability and to maximize comfort and trust, the interviews were held in the interviewee's native language with an interviewer from his or her own national group. The interviews were recorded fully and transcribed verbatim by the interviewers.

What We Learned

Choice of Communication Channel

From our examination of the responses of adolescents it emerges that the use of e-mail, Instant Messenger, and cell phone is very frequent and part of the daily lives of both Jewish and Arab adolescents. Naama, Jewish, 18 years old, said:

Ahh...not really...you know using the Messenger today is not like "WOW! I chatted today!!!!" It's something you do daily, something most people my age do, the ones I know...my friends do it. The Internet, e-mail, Messenger, cell phone is for us part of our daily life. We do homework, play games, and chat in the Instant Messenger at the same time. It's a regular, normal daily experience, like brushing your teeth every morning.

Adolescents in our study appeared to be aware of cultural differences; Ranin, a 17-year-old Arab girl, commented:

For me, chatting with Jews and Arabs is different. The Jews have a more open mentality and Arabs have a different mentality....I can't tell an Arab that I went out with a boy today...but with Jews I can share more the experiences I had when I went out with a boy...

Still, the choice of the communication channel depends on various factors, as can be seen in the following responses. Liron, a 16-year-old Jewish girl, said:

The most frequent way I communicate with my friends is by instant messaging. I use the phone less. Only when I have a friend that I want to talk to, and he doesn't have IM or isn't online. When I'm home I never use the cell phone. It's difficult to use your fingers to instant message friends and at the same time to hold the cell phone on your ear.

But besides being inconvenient, the reason for using instant messaging and the cell phone differs according to national lines. Liron continues:

Using the cell phone is much more expensive than using instant messaging. Most of my friends are on my buddies list. I know when they are online and when they are free to chat. The cell phone is expensive and I try to save by using it only when I am not close to a computer.

Liraz is a 17-year-old Jewish boy, for whom cost is an important factor in his choice of communication channel, but crossing gender lines is even more relevant:

I don't use the phone very much to communicate with friends. When I'm at home, making a cell phone or landline call is more expensive than using instant messaging. My parents watch the cell phone bill and try to restrict me. For me, it's much easier to communicate by IM, specially with girls...with a girl it's not like I choose to call, but it's different. It's not that you say I want to talk to her and phone her, and I need to start a conversation ... on the Net is like you see if she's there, so why not to send a message? If she answers then we just chat and if not well ... not.

Nuhas is a 14-year-old Christian boy:

I met one of my friends through the Net. She is a girl and lives in my village and she is also Christian. I got her e-mail from a friend of a friend. We chat every day about homework and school, about teachers and gossip. She studies at my school and I see her every day, but I don't speak with her there ... it's embarrassing ... you know ... it's not acceptable ... but chatting is different because only she and I know we chat ...

A better explanation for the advantage of the instant messenger over the cell phone is giver by Maha, a 13-year-old Muslim girl:

I met a boy through the Net. A friend of my friend gave to him my e-mail address and he added me to the list. We chat and I like the messenger more than the cell phone. Instant messenger is cheaper and my parents allow me to use the computer as much as I want. With the messenger I have privacy; nobody knows who I'm chatting with. With the cell phone I don't have privacy, my sisters and my parents ask me all the time who's on the phone.

Ibtisam is a Muslim girl, 17 years old:

I met two boys through the Net from a different Arab town in Israel. They added me [to their address book] saying that they have my e-mail. Both are boys. One is seventeen and the other eighteen and we chat about families and school. We haven't spoken by phone—there's no privacy and we don't want our parents to know that we chat. It is not accepted in our society.

Aisha is an 18-year-old Christian girl:

like IM ... Cell phones are more expensive ... although I like more to talk than chat because I don't type fast. But I have to compromise. Cell phones are expensive and my parents check the phone numbers that I call and ask questions. IM is more private. My friends have nicknames and my parents don't know who is male and who is female. A boy from another Arab town got my e-mail and added me to his list. We chat every day. He is my age. We only chat, don't talk by phone or meet, but we are good friends and is our little secret.

Marlene is a 15-year-old Christian girl:

I met people on the Net usually by mistake, you know [laughs]. They got my e-mail and they added me to their list. We speak about a lot of things, school, other friends and things that happen in our lives. I like to chat because it's more private than the phone. When I chat my parents don't know that I'm chatting and who I'm chatting with. When I speak over the phone my parents ask who it is and they ask me to shorten the talk.

In the choice of communication channel, an emerging theme, apparent in both Arabs and Jews, is the cost. Cell phones are relatively expensive and their use is more restricted to family communication, with parents and siblings. Instant Messenger is considered cheaper, and it allows multitasking—chatting and doing homework at the same time. The cost consideration is common to both groups. But in two themes, differences emerge as well. Reflecting the centrality of the family, the patriarchal structure of the Arab family and the restriction on cross-gender friendships, Arab adolescents made a choice of using IM for communication with friends of the opposite sex. Instant messaging provides them with privacy from the controlling eyes of parents and siblings, and the freedom to communicate with strangers. The latter are not total strangers but members of an extended friendship network. Internet friends are introduced to adolescent Arabs by other friends or distant family members. Chatting with them is easier as Arab adolescents socially construct chatting as a less serious transgression of social norms than phone calls and meetings in person.

Certainly, this is not the case of all the adolescents. Our interviewees included Arab adolescents who were not willing to communicate with strangers by any means. Usually these adolescents lived in small villages and close-knit neighborhoods and less in the large mixed cities (Haifa) and relatively large Arab cities (Nazareth). An example is Tigris, a 14-year-old Muslim girl:

I only chat with friends and people I know. I do not allow to anybody that I do not know to add me to their IM list. All of my friends study in my school or live in my neighborhood.

Language as a Communication Barrier

Language can be a barrier to access and use of communication technologies in a multicultural society like Israel. The Jewish majority speak Hebrew, have a good command of English, and study Arabic at high school. Arab adolescents speak Arabic as their mother tongue and learn Hebrew and English at school. However, the high residential segregation between the two communities means that until they finish high school, Arab and Jewish young people do not have many opportunities to use and practice the language of the other group. This social position is reflected in the use of a variety of languages for operating communication technologies. With the cell phone, the preference is the mother tongue, so no differences were found in language use for phone communication, and Jewish and Arab adolescents preferred to speak in their own language. But with the computer some interesting differences are noteworthy. The Jewish adolescents attested that the languages they used were English and Hebrew. When surfing for information they visited sites in English and Hebrew, and even saw this use as a way to improve their command of the foreign language. The Arab adolescents reported difficulties in the use of language. Their preference was to search for information in Arabic, but the sites that attracted their attention in this language were limited in number. They barely visited English and Hebrew sites. For the purpose of social interaction we were surprised by the most common response, which was that they invented a new language: Arabic in Latin script. The advantage of this use was that they could express their thoughts without having to change from language to language or to use a keyboard with three scripts (Arabic, Hebrew, and English), which would complicate typing.

Distance

Another central theme repeatedly emerging from our in-depth interviews was the problem of distance and the use of communication technologies to maintain social ties over short and long distances. Here is Nitza, an 18-year-old Jewish girl:

Communication is not a problem with friends who live nearby. I have a choice between a landline and cell phone. But with friends that live far away (another city or temporarily in another country) electronic mail helps us to remain connected. When friends go on vacation overseas we use electronic mail much more.

Lavie is a 17-year-old Jewish boy:

I met some of my friends when I was on vacation overseas. We keep in touch through e-mail. Instant messaging is difficult because of time differences, and cell phone is too expensive.

Ronit, 17, is a Jewish girl from Haifa:

I keep in touch with friends who live nearby more by cell phone and we meet face to face as well. But I have friends who live in other cities we are in touch more by e-mail.

Many Arab families in Israel have extended families abroad; nevertheless, the possibility of communication or maintaining ties with family or friends overseas was not mentioned by the Arab adolescents. The focus in their interviews was heavily on the use of communication technologies for the formation and maintenance of local and regional social ties that mostly were sustained virtually, without moving to meetings in person. Here is Faisal, 16, an Arab boy:

I started talking with a girl from another city. She is a Muslim like me. I got the e-mail from a friend; they are family and chat from time to time. We chat by computer and once or twice we talk over the phone. She is a year younger than me. We never met face to face.

Siham is an 18-year-old Arab girl:

Yes, I met on the Net a male from another village. He said that my e-mail address got onto his Buddy List by mistake...and we started chatting. I know that he is older than me and Christian like me. Our relationship is limited to IM. We have been good friends a year already and we have an agreement that if somebody gets onto my list by mistake and bothers me, he will take the e-mail and write to that person to stop...

This is what Lutfi, 17, an Arab, had to say:

I chat on the Internet with some people who are not at my school. They are from other villages in the north of Israel. I met them through family and friends. We tell jokes, what is going on in each other's school, songs, and we exchange the lyrics of songs. This is good. Because of the distance we cannot meet during the year...we think about meeting during the vacation...but we'll see.

These interviews show a different pattern of overcoming distance barriers. Jewish adolescents, conforming to the national disposition, were oriented to making ties overseas and keeping them up through e-mail. As we see in the next section, for Jewish adolescents, virtuality is a temporary and transitional state, soon to be transformed into meetings in person, thereafter followed by IM and cell phones as maintenance devices. By contrast, for the Arab adolescents, IM is a critical element not only in forming new relationships but also in maintaining them. The orientation is much less overseas, probably reflecting their complicated status as an Arab Palestinian minority holding Israeli citizenship. Social ties are created regionally, which overcomes the residential segregation and the friction of distance from other members of the Arab minority in Israel.

From Virtuality to Reality

The narrative of Israeli Jews defines an Internet friend in unstable status terms. An Internet friend is one who was met during an Internet-based activity, and has an unstable status, as the goal of the new acquaintance is to move from virtuality to a face-to-face relationship. Liron, the 16-year-old Jewish boy, tells us:

I like to play a game through the Internet. There is an entire group that plays…we start knowing each other by placing messages in the game bulletin board. We leave a message for each other and when there is a common topic ties are created. We started that way but very quickly arranged to meet face-to-face…We arrange meetings during the holidays because there is no school and it is easy to travel to another city. At the last meeting I met eight members. We became like friends.

When asked "What does it mean to be like friends?" he replied:

Is like schoolmates. We meet and do the same things that I do with my friends at school, we go together to eat something, hang around the shopping mall, chat about school, and spend the day together.

Cell phones play a key role in the transition from virtuality to a face-to-face relationship. It becomes the intermediary channel of communication, and is used to plan meetings in person of two people or more.

Liron said:

This week I met a girl. I got her e-mail from a friend. I added her to my messenger. We chatted a number of times, then I asked for the number of her cell phone and we arranged to meet.

Narkis, 18, is a Jewish girl from Haifa:

Yes, once I met a boy on a bulletin board. We had a number of chats by messenger. Then we exchanged cell phone numbers, had a conversation by phone, and arranged to meet. We are still friends and we talk from time to time by phone.

Here is Sharon, a 17-year-old Jewish girl from Haifa:

Yes, once I met on the Internet a boy who was a friend of a friend. We met and went out a couple of times and that's it...I met another one also met on the Internet, and went out a couple of times and we still meet because he is a friend of one of my friends. It started with chatting and then moved to the phone, and we still call each other from time to time.

Rushdi is a 17-year-old Muslim girl:

I met new people on the Internet from other schools and towns...You know—a friend brings a friend, so friends started adding me to the Buddy List...[Interviewer's probe: Did you ever go out with one of them?] No, never. We chat about school, about fights with siblings, problems with friends...but we did not met.

Latifa, 16, is an Arab girl:

I met some friends on the Net from other Arab cities. I never met them [personally] but we have been chatting for quite for a long time. I have been chatting with two girls for two years. I met them through a friend of mine—she introduced them to me. Maybe in the future we will meet [personally].

It seems that while both Jewish and Arab Israeli adolescents make new ties through the Internet, they perceive them in different terms. The Jewish adolescents see the virtual period as transitional. They are typically interested in converting the ties from virtual to real, and they see the process as one that develops from IM to phone to face-to-face. As for the Arab adolescents, their ties remain virtual, at least for now, and their position on whether to move to face-to-face ties in the future is vague. This finding is somehow paradoxical as Arab society is very closed, with hardly any outsiders. But apparently tradition plays a role regarding the desire for face-to-face ties. Personal connections are preferred in the context of the extended family and the town. Culture once again determines the creation of real ties, and sets limits on the ability of members of the Arab minority to capitalize on virtual ties and make them face-to-face, thereby increasing their social capital.

Implications

Studies on mobile communication technologies have focused on the differences between technology-mediated communication and in-person communication. The emphasis has been on how aspects of technology shape social interaction, disregarding the role of culture and social status as exogenous variables and determining the choice of communication channels and their specific use.

Attempting to overcome this limitation, we took advantage of Israeli society, with its deep divisions along national, cultural, and religious lines, to explore cultural differences in the adoption and domestication of communication technologies. We conducted an explorative study, using a sample of thirty-seven Jewish and Arab adolescents for a preliminary identification of their main themes, which are reported verbatim, regarding the choice of different communication technologies and their use.

We showed some indications of similarities and differences in the choice of communication channel and in the subjective meaning attached to the technology. The Jewish adolescents, reflecting a Western type of nuclear family, a lower network density (Talmud and Mesch 2006) and more individualistic orientation, tended to use communication technologies in a way that does not differ greatly from patterns observed in other Western societies. Their choice among cell phone, instant messaging, and electronic mail, mainly depending on cost considerations, was meant to expand and maintain their social ties. Network expansion was achieved through meeting new buddies online, but with the goal of moving on as fast as possible to cell phone calls and face-to-face meetings. The final purpose was to expand the face-to-face network with others who shared interests and hobbies, and to include them as permanent members of the social network. This trend aimed at converting weak virtual ties into strong face-to-face ties.

By contrast, the Arab adolescents used cell phones to maintain local ties, at times or in situations where face-to-face communication is not possible. In addition, their use of communication technologies reflected a traditional enclave, surrounded by Jewish society, showing higher network density with more transitive ties (Talmud and Mesch 2006). This is more disposed to social control. In such an environment the adolescents used communication technologies to overcome the limitations of residential segregation and ties that cross gender lines and are not accepted in their collectivistic and traditional society. Ties were created by "mistake": most of the Arab interviewees gave an account whereby they were added by mistake to the Buddy Lists of others. From their reports, ties with the opposite sex are apparently common, but are kept secret and tend to remain virtual to avoid breaking the rules of their society.

It is difficult to assess in this exploratory study to what extent these differences are a part of cultural repertoires per se or stem from properties of the social and residential structures of the two groups, yet it seems that price barrier is a function of social position in the stratification system, while gender and parental control issues are components of cultural practices. We also believe the future integration of cellular phones with Internet will have a separate effect on Jewish and Arab adolescents. While Jews will surf on cellular phones, Arab adolescents will continue using IM as a separate device to avoid family and parental monitoring. In conclusion, cultural dimensions remain an important factor in teens' choice of communication media.

References

Al Haj, M. 1987. *Social Change and Family Processes: Arab Communities in Shefar Am/Majid Al-Haj.* Brown University studies in population and development. Boulder: Westview Press.

Barzilai-Nahon, K., and G. Barzilai. 2005. Cultured technology: Internet and religious fundamentalism. *The Information Society* 21(1): 25–40.

Brouwer, L. 2004. Dutch-Muslims on the Internet: A new discussion platform. *Journal of Muslim Affairs* 24: 47–55.

Fischer, C. S., and Y. Shavit. 1995. "National differences in network density: Israel and the United States," *Social Networks* 17(2): 129–145.

Haythornthwaite, C., and B. Wellman. 2002. *The Internet in Everyday Life*. Oxford: Blackwell.

Katz, J. E., and R. Rice. 2002. *Social Consequences of Internet Use: Access, Involvement and Interaction*. Cambridge, Mass.: The MIT Press.

Mandaville, P. 2000. Digital Islam: Changing boundaries of religious knowledge. *ISIM Newsletter* 2: 23.

McKenna, K., A. S. Green, and M. E. J. Gleason. 2002. Relationship formation on the Internet: What is the big attraction. *Journal of Social Issues* 58(1): 9–31.

Meredyth, D., L. Hopkins, S. Ewing, and J. Thomas. 2002. Measuring social capital in a networked housing estate. *First Monday* 7(10): 16.

Meredyth, D., L. Hopkins, S. Ewing, and J. Thomas. 2004. Wired high rise: Using technology to combat social isolation on an inner city public housing state. In *Using Community Informatics to Transform Regions*, edited by S. Marshal, W. Taylor, and X. Yu. Hershey Pa.: Idea Group.

Mesch, G., and I. Talmud. 2006. "The quality of online and offline relationships: The role of multiplexity and duration." *The Information Society*, 22(3): 137–148.

Mesch, G., and I. Talmud. In press. Homophily and quality of online and offline social relationships among adolescents. *Journal of Research in Adolescence*.

Mikulincer, M., A. Weller, and V. Florian. 1993. Sense of closeness to parents and family rules: A study of Arab and Jewish youth in Israel. *International Journal of Psychology* 28: 323–335.

Shavit, Y., C. S. Fisher, and Y. Koresh. 1994. Kin and Nonkin under collective threat: Israelis during the Gulf War. *Social Forces* 72 (June): 1197–1215.

Smooha, S. 1997. Ethnic democracy: Israel as an archetype. *Israel Studies* 2(2): 198–241.

Talmud, I., and G. S. Mesch. 2006. Network density and virtual social capital: The case of Israeli adolescents. Unpublished paper, University of Haifa.

24 "Express Yourself" and "Stay Together": The Middle-Class Indian Family

Jonathan Donner, Nimmi Rangaswamy, Molly Wright Steenson, and Carolyn Wei

Across millions of households in India, amid normal conversations about finances, education, dating, relatives, and the home, a new topic is emerging: the proper role of the mobile phone in the family. Mobile use is rapidly growing in India. Subscriptions grew 73 percent between March 2005 and March 2006 alone (Telecom Regulatory Authority of India 2006). Though overall penetration is still a modest ninety million lines (8.2 percent of the population), the flourishing Indian middle class is driving much of the current growth. Some families are purchasing their first mobile, others are adding a second line for the spouse, and still others are adding lines as their children reach certain milestones. The handsets might be new, but the conversations about the mobiles are not. Instead, the questions about when to purchase them (or not) and how to use them (or not) are closely related to all the traditional conversations mentioned above. When Indian families talk about mobiles, they are also talking about money, about dating, about the home, and so on.

Drawing on three related studies of middle-class Indian families, this chapter considers how the mobile phone reshapes and reflects existing tensions within families. A wide body of research exists on how personal and mediated-communication technologies affect and reflect family dynamics. However, this chapter breaks new ground by viewing these processes in the context of urban family structures that are being renegotiated in response to rapidly changing social and economic conditions. Thus we argue that mobile use is central to our understanding of the tensions facing the new and expanding Indian middle class; it is not only a symbol of middle-class consumption but also a lens through which to see the family dynamic itself.

Mobiles and Families

The complex dynamic around family adoption and use of mobile telephones—similar to landlines (Betteridge 1997), TVs (Simpson 1987), and PCs (Lindlof 1992) before them—has fascinated researchers. Following Haddon (2003), we draw particularly on the domestication concept (Silverstone and Hirsch 1992; Silverstone and Mansell

1996). As Haddon explains, domestication does not reduce the phenomenon of use to a single moment of adoption or rejection by individuals. Rather, adoption is better understood as an ongoing process involving multiple members in a family, where symbolic and family tensions are played out in and around functional uses in the physical domain of the home.

Other researchers have focused on particular elements of mobile use within Western families. For example, Chesley (2005) found that over time, mobile use tended to blur the internal-external boundaries at home, eroding family satisfaction. Ling (1998; 1999a) has explored gender differences in mobile use within families, observing how "fathers own, and mothers loan" mobiles. Others have looked at how mobile use has forced renegotiations of the relationships between children and parents. These negotiations can result in new family rules and norms dictating appropriate mobile use (Ling and Yttri 2005), new strains in the discussions around managing money and finances (Haddon and Vincent 2005), or altered strategies for intrafamily communication in terms of what is said, via what channel, and at what time (Ling 2006).

Similar questions are being asked about family mobile use among non-Western households. For example, Bell (2003) identifies ways in which the relative importance of strong familial bonds over individual autonomy in many Asian contexts might increase the importance of the home as a place where ICT use is determined. Dobashi (2005) has explored "housewives with *keitai*" in Japan, observing how use of mobiles in an environment of multiple tasks and spatially absent fathers has aligned with existing Japanese expectations of gender roles, particularly those of housewife and mother. In de Gournay and Smoreda's (2005) examination of *keitai* use by Japanese women, they find that Japanese women's extensive use of the mobile was not directly connected to mobility nor to a need to move within the city. The women in both the Japanese studies possessed their own mobiles as opposed to sharing their husbands' mobiles, as observed in India (David 2005).

A Changing Middle Class

It is against this backdrop of increased interest in ICT use by non-Western families that we turn to a brief discussion of the Indian middle class. The notion of class itself is symbolically charged and locally contested. In India, class competes with caste and with a simpler rich-or-poor dichotomy for salience (Dickey 2000), but in the past decade, as India has liberalized and become more prosperous, the middle class has become more seductive. Fernandes (2000) explains:

Advertising and media images have contributed to the creation of an image of a "new" Indian middle class, one that has left behind its dependence on austerity and state protection and has embraced an open India that is at ease with broader processes of globalization. In this image, the newness of the middle class rests on its embrace of the social practices of taste and commodity consumption that market a new cultural standard that is specifically associated with liberalization and the opening of the Indian market to the global economy. Images of mobility associated with newly available commodities such as cell phones and automobiles, for instance, serve to create a standard, which the urban middle classes can and should aspire to. (p. 90)

Combined, the idealized "new" middle class (of IT professionals and multinationals) and the struggling "old" middle class (of state banks and bureaucracies) include between 55 million and 250 million people, depending on how broad a definition of middle class one wants to make (Sridharan 2004). Families in the Indian middle class draw their incomes from nonmanual labor, a definition particularly significant in a society that, despite all its rapid economic growth to date, remains numerically rural, agrarian, and poor (Sridharan 2004). Thus, to Dickey (2000), the Indian middle class cannot be distilled to vocation or income. Instead, it is a symbolic, dynamic construct in which consumption markers play an important role.

Even this brief discussion of family communication and the middle class illustrates how the mobile phone now plays a doubly important symbolic role. Echoing Fernandes, others have noted how for users in the developing world the mobile is a powerful symbol of macrolevel trends in globalization and consumer consumption (Donner 2004; Özcan and Koçak 2003; Varbanov 2002; Wei and Kolko 2005). At the same time, it is a symbol of autonomy at the micro-level, afforded to kids, teens, husbands, and wives as they "express themselves." Indeed, Katz and Aakhus (2002) ask us to look beyond the symbolism of the devices, arguing instead that they embody an *Apparatgeist*, a universal quality that enables and compels users to express their autonomy.

In the following studies, we seek to balance the focus between what the mobile changes and what it amplifies and represents. The symbolic power and appeal of the flashy handset is undeniable. However, following domestication, our studies illustrate the manner in which families use mobiles in ways that are more familiar than new. In this sense, the chapter echoes Harper's (2003; 2005) argument that mobiles are not themselves a driving force of large-scale social change and answers Dholakia and Zwick's (2004) call for mobile research that moves beyond the needs and behaviors of Western "road warriors." Nevertheless, the patterns of use we observe, both actual and symbolic, reflect some of the tensions Indian middle-class families confront as they react to a rapidly changing social environment.

Three Cases: Family Finances, Courtship, and Domestic Space

This chapter draws on three projects at Microsoft Research India that examine how mediated communication reflects and shapes Indian middle-class family dynamics during this period of socioeconomic change. The family finances case is drawn from an ethnographic study of the domestication and management of a "communication repertoire" (Haddon and Vincent 2005) among fifty-six lower-middle-income households in Mumbai, Delhi, and Chennai. The courtship case is based on a study of how mobile phones are used to support romantic relationships among young professionals in Bangalore. This study involved twenty members of the "new" middle class, who are working in the IT and business sectors that are driving the economic growth in the region. The domestic space study, also in Bangalore, examines spatial modalities of individual and shared mobile phone use. It is based on twenty-two interviews with "old" middle-class families—specifically young adults ages twenty to twenty-four and housewives—and seventeen interviews with lower-class families. In all three cases, respondents were recruited using a combination of personal invitations, announcements, and snowball (referral) approaches. These studies were based on in-depth interviews, participant observation, and other qualitative methods.

Vignette: Mobile Phones in the Singh Household

Before discussing the three cases, we present a vignette from Rangaswamy's fieldwork that reflects the range of tensions that exist in middle-class families surrounding mobile phones:

Paritosh Singh, a civil servant, lives in a typical government-housing colony in South Delhi and has a daughter and son who have adopted unconventional careers and aspirations. His wife, Parvati, works part-time in a nearby school. She says this money "helps with the extra jam, over and above the bread and butter." They have spent a lot of money, taking loans much beyond their means, to educate their children. Their daughter is a design graduate and is now married and living in Mumbai. She was the first in the family to get a mobile, as soon as she got her first work assignment. Their son, Navneet, lives at home. He has a management master's degree with a bachelor's degree in information technology.

Parvati persuaded her husband to get a mobile:

He will go out to places. He will fill up forms on the Web sites where they will ask for mobile numbers. And it won't look nice that he does not have a number to fill in so we decided he could have one.

Paritosh says,

I finally got one. There is a landline in the house—that is for my wife. Mine is a prepaid connection, so I will receive incoming calls. Even my wife can receive calls on this if it is from her natal family. If I need to make calls, I can [fill up the account]. If I need to call someone from home, I can use the landline. The good thing is that I can receive a call or a message even if I am out of the house. I like a simple phone but I chose this one with a color screen. I thought this looked good.... In the long run, the only big expenditure you incur is that of the instrument. After that there is nothing. The landline will turn out to be more expensive when calling to mobiles.

Navneet has a "poor man's cell phone," a basic Nokia costing Rs3,550 (about USD 77), which was a gift from his sister. To him, mobiles are for receiving calls and sending messages and need no extra features:

My sister had a fancy phone. [But] I am careful with money. . . . I call a cousin who sells SIM cards, and he gives me a card I recharge my phone with. He comes once in 2–3 months. Then I ask my dad to pay. It is logical thinking.

His mother feels it is their duty to fill his top-up. But she contradicts what her son had to say about his frugality with mobiles:

Now he has started misusing it a lot! He fills one card of Rs570, finishes it off in ten days and then sits quietly! Not that he wants to refill it. He manages like that for the next twenty days. When his sister was around, he would happily use her phone, sometimes talk for a long time. Well, she would never allow others [outside the family] to borrow the phone. But she thinks it is natural to share with family.

Paritosh wants his wife to get a mobile, too:

There will be some scheme for government servants—we can make use of that. Then we will not need the landline at all at home. If all three of us put Rs300 card each, i.e., Rs900 total, then what is the need for the landline?

Mobile Phones and Family Financial Decisions

The Singh family illustrates many of the complexities surrounding use of mobiles in the home. One theme is family finances: who pays for telecom, and how much use is appropriate? At a functional level, the trend is toward complexity: affordable personal communication devices augment and sometimes displace traditional landlines, thus allowing a mixed media environment in homes. Hence, our study follows Haddon and Vincent (2005), exploring how families manage their communications repertoire.

The appeal of the flashy handset is undeniable. Many families we spoke to had redistributed financial priorities in order to purchase more expensive, feature-rich handsets. So too is the desire to just keep on talking: despite having "a poor man's" cell phone, Navneet let his spending on calls race ahead of his means. And yet, desire is balanced by patience. Another girl explains why she is content without a camera phone: "It's a big thing that my father got me this handset. I don't have those big aspirations. . . . He is a single earning member of the family. And he has managed so much. . . . If I were earning, yes [I would have shifted to a camera phone]."

Virtually all the families we spoke with paid careful attention to selecting and using the new range of telecommunication options in the most cost-effective way. Clearly, Paritosh does not intend to spend money on too many outgoing mobile calls. Nor are other family members free to spend as they wish, as each family member evaluates the spending habits of the others. Navneet's case is illustrative: he earns his own money and owns a handset, but in his family—as in others in our sample—the responsibility for minimizing the household's overall expenditures is shared even when there was no formal budget; individual family members keep their overall spending in check. Often, this translates into sharing devices, even if the bills and handsets are owned by individuals within the family. As Navneet's mother says, "It is normal to share within a family."

These patterns of owning and sharing are windows into broader family dynamics. In another family we spoke with, the only mobile was owned by the daughter attending college and living at home. While the father had given up his mobile for budgetary reasons, he gave his daughter a phone as a gift for scoring high marks on her high school exams. And according to the mother, they will soon give the younger son a handset of his own "to maintain equality between siblings." Until then, he takes messages on his sister's handset when she is home in the afternoon. We see the age-old parenting challenges of equity versus reward played out around the highly coveted mobile handset.

Though these patterns of allocation, sharing, and mutual responsibility reflect existing family dynamics, mobile and personal communication devices have significantly expanded the opportunity for children to act as technological innovators for the home. One of our respondents, a mother of two teenagers, mentioned that her children are so smart that they manage to stay within budget by chatting more when they receive calls: "They don't waste talk time to chat aimlessly with their friends—they make the friends call them up!" In another instance, the father refused to lend his mobile—the only phone in the household—to his children, disapproving of their wanton chatting. When his children's friends called, he alerted his children, and they would go to a pay booth and converse. In addition, we see more adolescents actively and strategically engaged with family purchases, instructing parents on choices of plans and features, or, in the case of the Singhs, bringing the first handset into the household.

Mobiles for Maintaining Contact in Romantic Relationships

Besides provoking household discussions about finances, the mobile phone also ties together families and couples who are separated from one another. The study reported here is of young professionals, most of whom work at Microsoft's Global Technical Support Center, an exemplar of the IT sector in Bangalore. Specifically, this case considers the maintenance of romantic relationships over distances imposed by the global economy and by cultural gaps between children and parents. Romantic relationships are situated within a social network and may reveal disjunctures that exist in a family especially if the relationship moves toward marriage.

In this study, *family* refers to the individual and his or her parents even if they live in separate households. Most of the participants in this study had come to Bangalore specifically for career opportunities, leaving behind their "native places" to live on their own or with roommates. Yet, ICT such as mobile phones can tether the families together even if they are a world away: participants often communicate with their parents daily by phone, a facsimile of living with them. One woman "starts her day" by walking into the office and making a regular phone call to home at 7:30 p.m. before she starts the night shift. She comes from a "close-knit" family, and she worries that her parents are overly attached to her. The phone thus allows her parents to learn to cope without her, while at the same time keeping in touch.

Young professionals relocating to Bangalore sometimes must support, at least temporarily, long-distance relationships with significant others. The gaps experienced by couples can be more problematic than just space. One steady couple is colocated in Bangalore, but they do not see much of each other because the man's parents live with him. His parents oppose marriage for the two because they are of different castes and communities. This couple stays connected chiefly through phone calls and SMS during occasional breaks at work. Whether we call it absent presence (Gergen 2002), connected presence (Licoppe 2004), or "perpetual contact" (Katz and Aakhus 2002), these attachments over a distance are central to our understanding of mobile communication around the world. For young people suddenly on their own in urban India, these bonds are important indeed.

On one hand, mobile phones are a boon because they facilitate communication and nurture relationships that might otherwise have been impossible, as illustrated by a newlywed couple's arranged marriage. The man agreed to marry the woman because they had many close family ties: their fathers were longtime friends from the army, and they knew each other from schooldays. But he did not know anything about her as an adult, and they lived in different cities, so he gave her a mobile, and they talked on the phone five hours a day. This multiplex use of phones to support both familial and romantic love recalls the various types of love and romance for which mobile phones have become famous in countries such as the Philippines (Ellwood-Clayton 2006).

On the other hand, the mobile phone may also reinforce problems by acting as a bandage for the serious family issues that affect a long-distance relationship. The close, intimate, communication afforded by the mobile phone may strengthen these romantic relationships without addressing problems such as disagreements with parents. Couples might literally "stay together" longer than their parents would like. In the colocated relationship where the man's parents disapproved of the girlfriend, the man was committed to winning over his parents. He said that it would not be possible to marry his girlfriend if the parents were against them because he would always have to be the intermediary between his parents and wife, an uncomfortable situation. For such a relationship to succeed, much negotiation will need to occur to achieve a happy resolution for all.

Mobiles and Domestic Space

Family harmony concerning children's romantic relationships is more broadly reflected in its perception of household space and boundaries. The study informing this section examined the spatial modalities of mobile phone use in urban Bangalore. It examines categories of "inside" versus "outside," which in urban India "form an enduring and gendered spatial polarity," the contrast between which "holds urban residents' concepts of self and other and affects their movements through space" (Dickey 2000, p. 470). For the middle and upper classes, the innermost space of the home encompasses the dichotomies of inside/outside, private/public, like/different, and safe/unsafe (Dickey 2000, p. 470). Establishing boundaries, the home is "the tool for reinstating difference—a difference that must continuously be maintained because it is assailable" (Dickey 2000, p. 482).

The previous section illustrates how mobiles can breach the home boundary by facilitating unapproved romantic relationships, yet mobile use also extends this innermost boundary, allowing family members to carry a piece of home with them when they leave the house. This is not only about emergencies and security (Ling and Yttri 2002), as lower-class interview subjects cited, but about "keeping in touch," as nearly every middle-class interview subject stated. People needed to reach their parents, husbands, and wives to check in with each other. Parents needed to reach children not only for coordination, but for domestic reasons such as a housewife calling a messy daughter to ask where her notebook goes. The middle-class interview subjects moved quickly toward an expressive, hyper-coordinating use of the mobile (Ling and Yttri 2002). This expressive communication was not just with friends but also with family members: when participants spoke of keeping in touch, they specifically meant their families. Moreover, families made exceptions for expressive communication with relatives, even when they enforced rules around appropriate mobile use in contact with the family. For example, Ashita and Nimisha's parents supported their calling a cousin shortly after midnight on his twenty-sixth birthday but forbade Ashita to talk to her unapproved boyfriend too late in the evening.

Rejection of the mobile phone, except when sharing with a husband or child, indicates another instance of domestic boundary and traditional gender role maintenance. Dickey (2000, p. 468) writes, "Modesty and chastity form the dominant cultural ideal for women of all religions, castes, and classes. Because avoiding public display of the self is a key sign of modesty, ideally women should not go outside the home more than necessary." This modesty, for some housewives we interviewed, meant not owning a mobile phone. Ling (1999b) describes mobile "rejecters" as a part of a broader model of mobile acquisition and gender. Here, mobile rejection is about maintaining the domestic fabric.

One traditionally dressed housewife (Lata, age 44) pointed out that she had two landlines and did not need another mobile phone, though her daughters were "forcing" her to use technology (the computer), a word choice used by other male and female mobile rejecters across class lines. Another traditional housewife, whose husband got a mobile in the last month said, "I'll never get my own mobile" (Jyotisana, age 43). For both, it is as though affiliation with landlines is equivalent to representing the domestic boundary, not showing one's face outside. This preference for the place-to-place nature of the landline represents a spatial reinforcement of the traditionally domestic. Yet neither woman even considered it an act of sharing when they used their husbands' mobile phones.

Discussion

In these three case studies, we have seen how mobiles spark new discussions but reflect existing concerns in middle-class Indian families. The finances case illustrates how families are aggressively managing costs and behaviors against an increasingly diverse (and sometimes confusing) set of telecommunication options. They do this in an "Indian" way, where mobiles indicate upward mobility for the family unit, evident from multiple handset ownership as well as the attraction of a "fancy" phone over a basic handset for both parents and children. Mobiles are lent and loaned and bills are collectively paid, reiterating the focus on the family and not the individual. Hence, mobile phone use has entered the balancing act of the family relational dynamic around finances.

The courtship case illustrates how mobile communication has entered into complex, long-standing family discussions and negotiations surrounding children's marriages. Parents expect and exert considerable control over their children, even when they live far away. The mobile can tie together the family when they are geographically scattered, but it can also undermine parental influence by permitting romantic relationships to be conducted under the radar. In this case, the mobile represents and enables both freedom and continued sense of familial obligations that children may experience away from their parents.

The household spaces case shows how mobile communication represents a challenge to the notions of inside and outside and echoes the domestic space. This case illustrates how mobiles challenge, extend, and protect the traditional Indian domestic boundary. It extends the reach of the domestic space for middle-class family members by allowing a continual, mundane keeping-in-touch. In families following more strict customs, a housewife's rejection of mobile ownership reflects the traditional gender directive for modest women to stay close to the home.

In each of these cases, the mobile plays instrumental and symbolic roles in the microenvironment of the Indian middle-class family. Yet, at the same time, the studies illustrate ways in which the broader context is changing as well. The finances case points to the prosperity and consumer choice enjoyed by the idealized "new middle class" but also to the real pressure on Indian families to consume and adapt. Child autonomy in family decision-making and purchasing, symbolized by ownership of and aspirations surrounding mediated communication technologies, is afforded by the new work opportunities generated by the IT-driven economic boom in India. The information-based work is not only entrenching ICTs but also lifestyles that reflect global trends. Unconventional career choices and the desire to acquire personal technologies among young persons are reflective of these trends; they involve and intervene in the process of family decision-making regarding these choices.

The courtship case points to a shifting demographic environment where children are not uniformly living at home until marriage, in part because of participation in offshoring or business process outsourcing (BPO) where Indian offices support overseas partners and customers. This work is typically around the clock and is clustered in a few cities such as New Delhi, Mumbai, and Bangalore. Although young people now can make good salaries working for prestigious international companies, they are under pressure to work nontraditional hours and to relocate. A recurring theme in publications about BPO work is the havoc that overnight hours can play on workers' health and personal relationships (Taylor and Bain 2005). The mobility emblematic of and supported by the new ICTs both necessitates and supports romantic relationships carried out over distances.

The domestic spaces case suggests that the mobile phone will help redefine and extend the boundaries of the home. It will be a locus of gender role definition and young people's building of identity. It will continue to serve as a point of individuation, with families adding one or multiple mobiles or losing their landlines altogether. Yet unlike Japan, where the mobile is an individual object (Thomas 2005), it frequently will be a collective, shared object in India. In the middle class, informal mobile sharing with family and close friends does not serve only an explicitly financial need. This welcomed, collective use of the mobile will extend the way the household operates, keeping the family in touch in a variety of expressive ways.

Conclusion

"Express yourself" and "stay together" are slogans from two Indian mobile providers' advertising campaigns. The phrases represent the complex, sometimes contradictory, meanings that the mobile may hold for Indian families. A mobile handset simultaneously signifies individuality and autonomy as well as family security and cohesiveness. This symbolic tension surrounding mobile phone use within middle-class Indian families can contribute to a broader understanding of the role of mediated communication devices in society. At the same time, the mobile is a powerful tool that can offer a measure of autonomy to children as well as link geographically dispersed families. Indian families are using mobiles in nuanced and sophisticated ways in order to go about their business of daily life while staying connected to one another.

Thus mobile use among middle-class Indian families is theoretically challenging. It is not clear that mobile use is fragmenting the middle-class Indian family into autonomous individuals. In fact, the domestication theory has allowed us—like Bell (2003)—to see how mobile decisions may be made as a family rather than as individuals, and how families are able to adopt and adapt the mobile to coordinate themselves even when dispersed across distances. The mobile can support the goals of the family even if individuals may be simultaneously using it in ways that might undermine those family goals. The bridegroom talking on the mobile five hours a day to get to know his fiancée before their arranged marriage is an example of adopting the mobile phone in ways consistent with traditional constructs of the family's role in romantic relationships. We feel that domestication research should be pursued further to better understand the family as a locus of global cultural changes.

The mobile's arrival in India roughly coincides with the broader effects of economic liberalization that have been felt by middle-class Indian families: increased prosperity (and diminishing job security), increased choice in lifestyles, and increased social fragmentation. This change is not entirely caused by the mobile, but its use certainly is an enabling and complementary factor. Nevertheless, as these brief dispatches from the field suggest, Indian middle-class families will elect to use mobiles in unique, culturally appropriate ways. Seen from the Indian living room, it seems certain that new middle-class users will take advantage of mobiles in ways that may at times be at variance with their families. However, there is also sure to be sharing of the mobile and "staying together." The flexible and mobile around-the-clock lifestyle afforded by ICTs will continue to create new opportunities for users while simultaneously providing mechanisms for coping with these shifts. The new middle-class behaviors enabled by the device will continue to look distinctively Indian.

References

Bell, G. 2003. Other homes: Alternate visions of culturally situated technologies for the home. Paper presented at the Conference on Computer-Human Interaction (CHI). Ft. Lauderdale, Fla.

Betteridge, J. 1997. Answering back: The telephone, modernity, and everyday life. *Media, Culture, and Society* 19: 585–603.

Chesley, N. 2005. Blurring boundaries? Linking technology use, spillover, individual distress, and family satisfaction. *Journal of Marriage and Family* 67: 1237–1248.

David, K. 2005. Mobiles in India: Tool of tradition or change. Paper presented at the Preconference on Mobile Communication at the Annual Conference of the International Communication Association. New York.

de Gournay, C., and Z. Smoreda. 2005. Space bind: The social shaping of communication in five urban areas. In *A Sense of Place: The Global and the Local in Mobile Communication*, edited by K. Nyíri. Vienna: Passagen Verlag.

Dholakia, N., and D. Zwick. 2004. Cultural contradictions of the anytime, anywhere economy: Reframing communication technology. *Telematics and Informatics* 21(2): 123–141.

Dickey, S. 2000. Permeable homes: Domestic service, household space, and the vulnerability of class boundaries in urban India. *American Ethnologist* 27(2): 462–489.

Dobashi, S. 2005. The gendered use of *keitai* in domestic contexts. In *Personal, Portable, Pedestrian*, edited by M. Ito, D. Okabe, and M. Matsuda. Cambridge, Mass.: The MIT Press.

Donner, J. 2004. Microentrepreneurs and mobiles: An exploration of the uses of mobile phones by small business owners in Rwanda. *Information Technologies for International Development* 2(1): 1–21.

Ellwood-Clayton, B. 2006. All we need is love—and a mobile phone: Texting in the Philippines. Paper presented at the International Conference on Cultural Space and the Public Sphere in Asia. Seoul, Korea.

Fernandes, L. 2000. Restructuring the new middle class in liberalizing India. *Comparative Studies of South Asia* 20(1 & 2): 89–112.

Gergen, K. J. 2002. The challenge of absent presence. In *Perpetual Contact: Mobile Communication, Private Talk, Public Performance*, edited by J. E. Katz and M. A. Aakhus. Cambridge: Cambridge University Press.

Haddon, L. 2003. Domestication and mobile telephony. In *Machines that Become Us: The Social Context of Personal Communication Technology*, edited by J. E. Katz. New Brunswick, N.J.: Transaction Publishers.

Haddon, L., and J. Vincent. 2005. Making the most of the communications repertoire: Choosing between the mobile and fixed-line. In *A Sense of Place: The Global and the Local in Mobile Communication*, edited by K. Nyíri. Vienna: Passagen Verlag.

Harper, R. 2003. Are mobiles good or bad for society? In *Mobile Democracy: Essays on Society, Self and Politics*, edited by K. Nyíri. Budapest, Hungary: Passagen Verlag.

Harper, R. 2005. From teenage life to Victorian morals and back: The technological change and teenage life. In *Thumb Culture: The Meaning of Mobile Phones for Society*, edited by P. Glotz, S. Bertschi, and C. Locke. Bielefeld, Germany: Transcript Verlag.

Katz, J. E., and M. Aakhus. 2002. Conclusion: Making meaning of mobiles—a theory of Apparatgeist. In *Perpetual Contact: Mobile Communication, Private Talk, Public Performance*, edited by J. E. Katz and M. Aakhus. Cambridge, UK: Cambridge University Press.

Licoppe, C. 2004. "Connected" presence: The emergence of a new repertoire for managing social relationships in a changing communication technoscape. *Environment and Planning D: Society and Space* 22(1): 135–156.

Lindlof, T. R. 1992. Computing tales: Parents' discourse about technology and family. *Social Science Computer Review* 10(3): 291–309.

Ling, R. 1998. "She calls, [but] it's for both of us you know": The use of traditional fixed and mobile telephony for social networking among Norwegian parents (No. R&D Report 33/98). Kjeller, Norway: Telenor.

Ling, R. 1999a. I am happiest by having the best: The adoption and rejection of mobile telephony (No. R&D Report 15/99.). Kjeller, Norway: Telenor.

Ling, R. 1999b. "We release them little by little": Maturation and gender identity as seen in the use of mobile telephone. Paper presented at the International Symposium on Technology and Society (ISTAS'99) Women and Technology: Historical, Societal and Professional Perspectives. Rutgers University, New Brunswick, N.J.

Ling, R. 2006. "I have a free telephone so I don't bother to send SMS, I call": The gendered use of SMS among adults in intact and divorced families. In *Mobile Communication in Everyday Life: Ethnographic Views, Observations, and Reflections*, edited by J. R. Höflich and M. Hartmann. Berlin: Frank & Timme.

Ling, R., and B. Yttri. 2002. Hyper-coordination via mobile phones in Norway. In *Perpetual Contact: Mobile Communication, Private Talk, Public Performance*, edited by J. E. Katz and M. A. Aakhus. Cambridge: Cambridge University Press.

Ling, R., and B. Yttri. 2005. Control, emancipation and status: The mobile telephone in the teen's parental and peer group control relationships. In *Information Technology at Home*, edited by R. Kraut. Oxford: Oxford.

Özcan, Y. Z., and A. Koçak. 2003. A need or a status symbol? Use of cellular telephones in turkey. *European Journal of Communication* 18(2): 241–254.

Silverstone, R., and E. Hirsch, eds. 1992. *Consuming Technologies*. London: Routledge.

Silverstone, R., and R. Mansell, eds. 1996. *Communication by Design: The Politics of Information and Communication Technologies*. Oxford: Oxford University Press.

Simpson, P. 1987. *Parents Talking Television: Television in the Home*. London: Comedia Pub. Group.

Sridharan, E. 2004. The growth and sectoral composition of India's middle class: Its impact on the politics of economic liberalization. *India Review* 3(4): 405–428.

Taylor, P., and P. Bain. 2005. "India calling to the far away towns": The call centre labour process and globalization. *Work, Employment & Society* 19(2): 261–282.

Telecom Regulatory Authority of India. 2006. The telecom services performance indicators for financial year ending 31st of March 2006. New Delhi: TRAI.

Varbanov, V. 2002. Bulgaria: Mobile phones as post-communist cultural icons. In *Perpetual Contact: Mobile Communication, Private Talk, Public Performance*, edited by J. E. Katz and M. A. Aakhus. Cambridge, UK: Cambridge University Press.

Wei, C., and B. Kolko. 2005. Studying mobile phone use in context: Cultural, political, and economic dimensions of mobile phone use. Paper presented at the International Professional Communication Conference. Limerick, Ireland.

Thomas Molony

Voucha Ya Shilingi 5000, or 50 Cent?

Since the mid-1990s, development agencies have explored how mobile phones and the Internet could help combat poverty and stimulate development in the Third World, especially in Africa (e.g., DOI 2001; Gerster and Zimmermann 2003). According to the UNDP's "Making Technologies Work for Human Development" report, information and communication technology (ICT) "is a pervasive input to almost all human activities: it has possibilities for use in an almost endless range of locations and purposes" (UNDP 2001, p. 35).

The recent uptake of ICTs—and mobile phones in particular—has been remarkable, with an average subscriber growth close to 60 percent between 2000 and 2005 (ITU 2005). Yet, evidence that these technologies are actually contributing to an increase in productivity is scarce (UNCTAD 2004). Even when a case for the macro-impact is made, such as a 2005 Vodafone-sponsored study of the impact of mobile phones in Africa (Waverman, Meschi, and Fuss 2005), authors may not have sufficiently considered changes in economic and infrastructure activities (Eggleston, Jensen, and Zeckhauser 2002; Matambalya and Wolf 2001).

Certainly there are many accounts of people in developing countries putting these technologies to uses that allow them to access resources such as health information and market prices. Unfortunately, though, many of these accounts are anecdotal or are based on unsustainable donor projects. Such evidence is insufficient if we hope to make substantial conclusions regarding the impact of ICT at any general level. In the East African country of Tanzania, the "almost endless range of locations and purposes" (UNDP 2001) that, apparently, many poor people apply ICT to would surely disappoint those seeking examples of exciting ICT applications for replication in other poor countries. Go to any Internet café and Curtis "50 Cent" Jackson's latest single or the goal tally of Chelsea's soccer team's most recent signing is more typical of the information that people are seeking (see, for example, Mercer 2006; Mwesige 2004; Nielinger 2003). Tanzanians who are parting with their hard-earned money to buy *voucha ya shilingi 5000* (USD 4.20 airtime vouchers) to top-up their mobile phone balance so that they can daily seek the wholesale price of maize, though often cited in pro-ICT development reports, are remarkably elusive (see, for example, Molony 2005).

In this chapter I argue that although there are times when individuals use ICT in ways that aid personal or collective development, in much of Africa mobile phones are more commonly put to a nondevelopmental use. I do recognize that while instances of the developmental application of mobile phones may be infrequent (such as selling seasonal farm produce), these uses may well be more "important" to individuals than the nondevelopmental uses that mobile phones are often put to. However, the research that this chapter is based on did not seek to uncover valuative opinions of the developmental "importance" of mobile phones relative to their other possible uses. My thoughts are influenced by various personal encounters; at times this piece does not claim to offer any more than the anecdote itself. It is informed by the views of a wide range of Tanzanians, including multiple interviews with seven government ICT policymakers; fifteen tomato and potato farmers, four intermediary traders, and five Dar es Salaam–based auctioneers involved in domestic perishable foodstuffs marketing; nineteen leaders of groups of workers (and one laborer) in the urban informal construction industry; and eleven carvers and/or workers in carving groups, nine retail suppliers, and seventeen Tanzanian-based retailers involved in the domestic trading and export of African blackwood carvings. I collected their opinions during a series of semistructured interviews conducted over fifteen months of doctoral fieldwork in Tanzania during 2002 and 2003 (Molony 2005). These discussions, complemented by subsequent informal exchanges I have had with numerous Tanzanians from many other walks of life, suggest that those who are interested in the use of ICT in developing countries are often unduly blinkered by literature portraying the developmental potential of ICT. The everyday reality can be quite different. My research *does* reveal in some detail a good number of entrepreneurs who were using ICT in various ways to enhance their business prospects. However, in this chapter I offer an infrequently heard view concerning the ordinary, everyday adoption of mobile phones across Africa, and in particular in Tanzania.

The Uptake of Mobile Phones and the Internet in the "Least Wired Region in the World"

I focus here on mobile phones, and give much less time to the Internet, simply because the former is used far more than the latter in developing countries. It is clear from my research that while the term *ICT* has often been used to refer to the new information and communication technologies of mobile phones and the Internet, the difference in the uptake of the two is so great that, in the African context at least, it is no longer useful to refer to them together as *ICT*. While mobile phone coverage in developing countries is still far from that of industrialized countries, in many areas where coverage is available mobile phones are the only new ICT available. Even in areas where the infrastructure exists to make the use of both technologies possible, for the vast majority of people in developing countries the constraints toward accessing the Internet are so much greater than those of accessing mobile phones (Molony 2006) that placing the two together for analysis as a potential developmental tool is not reasonable. Of the three subsectors from Tanzania that I examined, only the elite retailers of African blackwood carvings in Dar es Salaam use the Internet, and that is largely for e-mail communication with foreign contacts. Where ICT is employed in business in developing countries far more people use mobile phones. This fact is supported quantitatively by Souter et al. (2005), whose vast sample of about 2,300 interviewees in rural areas of India, Tanzania, and Uganda shows only a negligible amount of Internet users, against approximately 75 percent of adults who make use of "telephones" where they are available.

With the spread in mobile wireless networks, telephony is now available for the first time to hundreds of millions of people in settlements across the developing world, either as private subscribers or as users of public access points. The African continent, dubbed the "least wired region in the world," now has the highest ratio of mobile to total telephone subscribers on the planet (ITU 2005). In East Africa, mobile phones are now so much more accessible than fixed lines that in everyday Swahili these new technologies are referred to simply as *simu* (phones), and less as *simu ya mkononi* (handheld [i.e., mobile] phones). The fixed-line phone (*simu ya kawaida*) is now largely disregarded. The frequent explanation from interviewees was that while in many areas mobile phone coverage is still far from ideal, the accessibility, reliability, and price (particularly when texting) of mobile phone use is so superior to the fixed-line service provided by the incumbent Tanzania Telecommunications Company Limited (TTCL) that in both urban and rural areas fixed lines are increasingly ignored as a form of telecommunication. Frustrations with both the fixed-line and the phone kiosk service (*huduma ya simu*) are explained by Maiko Bakari, a middleman who sends carvings from Tanzania's Mtwara region:

Before I bought my mobile phone I used to use TTCL or *huduma ya simu*. But without a mobile phone I couldn't be contacted easily by customers. For example, a customer wanted to give me an order and would phone my brother [on his fixed line] but sometimes he was not there. So I would use the *huduma ya simu*, but that was not reliable either because I didn't have a good relationship with those people.... They can take messages from Dar es Salaam but I wouldn't always get the messages and this would annoy the customers and I would lose out on getting orders. (Bakari, UVWIMA Carving Group workshop)

Despite the inconvenience of having to travel to a phone kiosk, the *huduma ya simu* service is popular among those who wish to make a call but cannot afford their own handset, or for subscribers who are unable to obtain credit vouchers to recharge their

Figure 25.1
Kiosk-shack on stilts to improve reception. Source: Author.

phones. The service is available in almost every location where the mobile phone operators provide a service—primarily in urban areas, along main roads, and in mining and
tourism hotspots. Reception has also been "appropriated" by *huduma ya simu* operators
and individuals in locations that are an unintended consequence of more blanket coverage from transmission towers ("base stations") designed primarily to serve areas such
as main roads. See figure 25.1, *Huduma ya simu*, Kidamali village, southern Tanzania.
The kiosk, one of four in the village, is on stilts to obtain reception from a nearby base
station that was installed to provide coverage along the main Malawi/Zambia—Dar es
Salaam road.

  This ability to connect to the network in rural areas is but one instance of mobile
phones in a developing country being a more "democratic technology" than the Internet. Other reasons would include ease of use—close to a third of Tanzanians can neither read nor write (United Republic of Tanzania 2002)—and affordability—relative to
the Internet. Along with the distinct benefit that mobile phones save costs on travel,
these advantages, however, can be offset by increased social pressure to call relatives
more often (Gamos Ltd. 2005, p. 1). And despite the introduction by mobile phone
operators of payment options such as per-second billing, a free "Call Me" service, and
a credit transfer facility to send airtime from one account to another, mobile phones

are still expensive to use. An average postpaid cross-network call lasting five minutes costs USD 1.85 (United Republic of Tanzania 2005), almost one third of the mean average weekly per capita income of a little over USD 6.00 (World Bank 2005). Nevertheless, the uptake of mobile phones in the world's poorest region has been extraordinary and largely unanticipated.

Tournaments of Value: "Super Glue" Battle Commanders versus the Wabenzi

In Tanzania, mobile phones are so popular because they are cool. The handset itself can be a mark of status for the owner and confers sophistication, particularly in the rural areas where they are less common and suggest connection with the city. Following Ferguson's (1999) description of "cosmopolitans" and "locals" on the Zambian copper belt, Mercer (2006, p. 12) suggests that the consumption of the Internet in Tanzania can be interpreted as a cosmopolitan act. It enables consumers to "reach out . . . and signify affinity with an outside world beyond the local" (Ferguson 1999, p. 212) and to publicly demonstrate their developed selves. In Tanzania, in contrast to the personal computer (which at its most public access point can be visible only to other customers in the Internet café), the mobile phone is a more desirable ICT for Tanzanian cosmopolitans—if not to *use*, then to *own*. The computer may be a prestige object of power to the elite and an instrument for their exercise over the management of social, political, and commercial information and contacts (van Binsbergen 2004, p. 134), but for the nonelite the mobile phone is a more displayable and—by its very definition—a more mobile gadget, which can be shown off to a greater number of people (or targeted individuals) in still more public arenas such as the bar or on the street. The mobile phone is an embodiment of the experience with modernity, so much so that some people will walk with a dysfunctional handset on display in an attempt to participate, if only by simulation, in the increasing practice of digital consumption. In Ghana these "dummy" phones are often sourced from North America where, in a bid to create a market, service providers have offered handsets for free or almost free. These are sent to relatives who demand mobile phones back home, but are not GSM-compatible so cannot be used in Ghana (Alhassan 2004, p. 206). Whether functioning or not, a visible handset is status on display (Donner 2004, p. 4; Varbanov 2002), although, as in many countries, the smaller (and therefore less visible phones) are the most sought after in Africa since they can imply the owner is wealthy, sophisticated, and with-the-times. Again, as in the North, in Dar es Salaam a bulky handset is called a *tofari* (brick), or is given the local name of *mshindi* after a popular brand of household soap sold in twelve-inch blocks. In Accra the owner of the bulky handset is known as a "battle commander," while in Lagos the person is called "super glue" because their old phone cannot be prized out of their hand to purchase a more up-to-date version.

It is difficult to tell whether concern over having a state-of-the-art handset is generated by the fashion-conscious in Dar es Salaam who judge themselves what is cool and what is not, or whether demand is influenced by the marketing staff of mobile phone operators in Tanzania or by foreign handset manufacturers. Most likely it is a combination of the two, although handset marketing outside Tanzania must also play a part in generating demand for what are often perceived as newer and "better" European products. As with cars and beer, handsets produced by European-owned companies such as Ericsson and Nokia are deemed more desirable and better made—and often use the same level of raw sexual imagery in commercials (Townsend 2000, p. 6). As with handsets, mobile phone services in Tanzania are marketed as something cool, in a similar style as cigarette advertising with visual images in newspapers and on TV and billboards are of beautiful, successful urbanites. The themes on the advertising hoardings and front-page banners promote an operator's new area of coverage, advertise new services, or simply promote calling as a means to help users keep in touch with friends and family, one of the main uses of the mobile phone in Africa (Souter et al. 2005). Another parallel with cigarettes is that, as in Europe and North America, in Africa mobile phones can be seen as "cigarettes for the twenty-first century" (Stewart 2004), being highly social in many situations (loudly taking a call when with friends in a bar seems to be acceptable), but highly antisocial in other situations. At the main door of many of Dar es Salaam's mosques, for example, there are signs in Arabic, English, and Swahili telling worshipers to switch off their handsets.

The mix of what influences potential customers to use a particular brand is complex and space does not allow for such an analysis here. Taken with the perceived and real uses of mobile telephony, however, it does seem that there is little doubt that image is all-important to many users. As the owner of a small restaurant in Dar es Salaam explained in a comment that could equally have been said by a parent anywhere in the world, he thinks that image is especially important for Tanzania's youth:

...for the youth these days it's important for them to show that they have money, so they buy these expensive things like mobile phones, wasting their money, so their friends and girls will think they've got a lot of money. Not like us [of my generation], who save the money we earn. They're wasting their money on mobile phones and these expensive things like the Timberland boots my son wears. (Owner, Farmer's View restaurant, Kijitonyama)

It is not only to the urban youth that mobile phones have become weapons in what Appadurai (1986, p. 21) and MacGaffey and Bazenguissa-Ganga (2000, p. 137) later describe as "tournaments of value" in the construction of reputation and status, symbols of the *tamaa* (lust for things) that creates the context of competition and hierarchy in Dar es Salaam (Lewinson 1999, p. 108). I interviewed Moses, a successful middle-aged architect, who once owned a mobile phone that was so expensive its theft was featured in a newspaper article (Keregero 2002), and who always makes a point of upgrading to the latest handset on his visits abroad, "so that I can organize myself well and I don't look like an *mshamba* [peasant] when I'm doing my business" (Moses, Mekon Arch Consult). He purchased his first mobile phone in 1999 but, relative to the early pioneer customers for whom the costs of running a mobile phone in Dar es Salaam were tremendous and the status of owning one was equally high, it is increasingly felt (e.g., Mbogora 2002) that in Dar es Salaam mobile phones are no longer a sign of such wealth. These days they are easy to get a hold of, and are more affordable to use. Yet this has deterred neither Moses nor another of my wealthy interviewees, Abdi. Both men attempt to better their opponents by regularly purchasing the most expensive phones—a practice resembling the consumer habits of Zimbabwean elites who sought status in expenditures on automobiles and the consumption of "European" lager beers during the postwar economic expansion (Burke 1996, p. 183). Both Abdi and Moses fit into this top income quintile, being members of the wealthy, Mercedes-Benz–owning elite, locally known as the *wabenzi* (Shivji 2005). Moses, for example, owns two of the top-of-the-range cars and only drinks Heineken, the most expensive beer on the Tanzanian market. On his next overseas trip he plans to purchase the Nokia N90, currently retailing at more than USD 400.

Mobile phone operators are increasingly reaching into people's disposable incomes, which hitherto alcoholic and nonalcoholic beverages have taken a large slice of. While it is difficult to obtain evidence to prove this, multinational drinks companies are concerned that they are losing profits to the mobile phone industry. One challenge has come in the form of successful bids for the sponsorship of popular events imitating European versions, such as the Vodacom Premier (soccer) League and beauty contests. Throughout the country Vodacom has now taken the place of Coca-Cola in providing many branded road signs.

Branded parasols for street vendors such as this shoe-shine boy in the center of Dar es Salaam (figure 25.2) are one of many advertising methods mobile phone operators are using across Africa. For instance, both Celtel and Vodacom are replacing bus shelters installed originally by beer companies. Mobile phone companies also run social projects, such as providing teaching aids to impoverished schools. Vodacom has even managed to convince COSTECH—the country's main research body into ICT—to allow it to paint the government building in its distinctive brand color scheme (minus their operating name). Interestingly, the facing building is the headquarters of Celtel, Vodacom's main rival.

Mobitel Women on "Compensated Dates"

Where in other parts of Africa product branding has been such that in Zimbabwe toothpaste has threatened to become "Colgate" (Burke 1996, p. 213) in the same way that in Britain the vacuum cleaner has become a "Hoover," no mobile phone operator has yet managed to achieve such intense brand identification that their name is used

Figure 25.2
Public prominence of mobile communication images. Branded parasols for street vendors such as this shoe-shine boy in the center of Dar es Salaam are one of many advertising methods mobile phone operators are using across Africa to promote their services. Source: Author.

as the generic identifier. The closest example of another good or service in Tanzania is the petrol station, which is always *sheli* (Shell) even if it is owned by Caltex or Oilcom. The mobile phone has been used only in less flattering parlance. In Dar es Salaam slang, a physically small and attractive women who can be readily "displayed" in public is known as *portable*, *mobile*, or, after the first GSM operator, *mobitel* (now rebranded as Tigo). Their counterparts in Abidjan, Côte d'Ivoire are similarly known as *cellulaire*, while the Dakarois favor computer parlance: Senegalese girls who sport "fashionable, tight-fitting worn-out jeans, short skirts, blouses and with conspicuously displayed mobile phones" are termed *disquettes* (Nyamnjoh 2005, p. 299).

The sexuality of these mobitel women—or, for that matter, many other African women—is often their best and only resource to bargain with (Haram 2004; Wight 2005). In Tanzania, as across much of the continent, transactional sex is common and by no means always comparable to or associated with prostitution (Luke 2006; Silberschmidt and Rasch 2001), so sex is often given by girlfriends to boyfriends or "sugar daddies" in return for gifts (Dinan 1983) that can be listed as "the Cs": use of a car, a chicken, or (the latest popular item to be exchanged for sex) a cell phone. Cash, of course, is also on the list, although credit for mobile phone airtime is an increasingly acceptable form of payment. In Zambia young girls are willing to have unprotected sex with much older men in exchange for the latest mobile phone and a steady flow of top-up airtime cards. As one young girl puts it, "Many of the girls at my school want those small cell phones, not the big ones, and the only way to get them is to have sex, often without a condom, with these older men" (Dale 2003). (That poverty is not always the motivation, however, can be seen in a very similar case from Japan where girls from middle-class backgrounds involved in *enjo kosai*—compensated dates, prostitution involving high school students and middle-aged men—usually spend their discretionary income on expensive designer clothes and to pay their mobile phone bills (Kingston 2004, pp. 26–27).) Across Africa the mobile phone obtained through these methods is now forcing change within families where parents do not have access to the technology. In Madagascar the mobile phone symbolically enacts a reversal of who holds power and authority within the household. At the same time it is facilitating further relational mobility by acting as "an important form of social capital, marking out a girl as modern and increasing the chances that she finds either one of the rare jobs available or a wealthier man, or both" (Cole 2004, p. 582).

The Economy of Mobile Phone Acquisition and Sale

The handsets that are given as gifts need not necessarily be new. Recipients are often grateful for any means of communicating that allow them to network with others who may act as future sources of gifts. Indeed, the fact that Tanzanians living in rural areas are willing to pay on average 50,000 shillings (USD 42) for a handset (Scanagri 2005, p. 7) when a new mobile cannot easily be purchased for this price suggests that secondhand phones must be acceptable to many users. As users upgrade to new models, old mobile phones are also commonly recycled through donations from friends and relatives in the metropolis and abroad. In Accra this supplies the avant-garde "digital *flâneur*" with the means to "live up to the dream [of modernity] thanks to the power of connection of digital communication technology and the market" (Alhassan 2004, p. 208). Purchased handsets, on the other hand, are frequently acquired from the flourishing "not-new" (i.e., stolen and/or reconditioned, at home or abroad) and "not-original" (i.e., counterfeit) mobile phone market that operates in many of the continent's cities.

As with much of the informal economy, it is impossible to know the scale of this trade since it is difficult to know the legitimacy of a handset that is not purchased from an accredited dealer. It is widely known across Dar es Salaam that many of the "not-new" handsets that are sold in the city are stolen. Some are stolen in Britain and exported by gangs to Africa (BBC 2003), a trade that one mobile phone repairman alluded to when he explained that "secondhand" mobiles he obtains from a Zanzibari "refugee" in the southern English town of Reading (where a large Tanzanian population resides) are shipped to Dar es Salaam via Zanzibar and concealed inside refrigerators stored in containers (Anonymous, Mchafukoge, Dar es Salaam). To meet the huge demand for the mobile phones that police believe fuel crime in Dar es Salaam (Hangi 2002), other handsets are stolen in armed raids that specifically target shops selling mobile phones (*Guardian* (Dar es Salaam) 2004). Pickpockets also target customers at commuter bus stops (Kamwaga 2002). During a spate of druggings in the exclusive Oyster Bay residential area, a bar owner even colluded with thieves waiting outside her establishment.

Conclusion

If ICT are being used to fuel any financial system, it appears to be as much an informal economy of mobile phone handset acquisition and sale as it is part of the wider, formal economy for many in Africa's cities. Thanks to this thriving trade, the challenge with mobile phones is no longer one of persuading people to acquire the necessary hardware to use the technology. Rather it is one of making interconnection affordable for more users as coverage further expands into rural areas. Only then can the majority of Africans regularly employ mobile phones and attempt the even greater challenge of applying the Internet for the "developmental" uses that their northern donors so wish they would embrace.

References

Alhassan, A. 2004. Development Communication Policy and Economic Fundamentalism in Ghana. Unpublished PhD, University of Tampere, Tampere.

Appadurai, A., ed. 1986. *The Social Life of Things: Commodities in Cultural Perspective*. Cambridge: Cambridge University Press.

BBC. 2003. Mobile phone crime blitz launched. http://news.bbc.co.uk/1/hi/uk/3326171.stm.

Burke, T. 1996. *Lifebuoy Men, Lux Women: Commodification, Consumption, and Cleanliness in Modern Zimbabwe*. London: Leicester University Press.

Cole, J. 2004. Fresh contact in Tamatave, Madagascar: Sex, money, and intergenerational transformation. *American Ethnologist* 31(4): 573–588.

Dale, P. 2003. AIDS threat for Zambian girls. http://news.bbc.co.uk/2/hi/africa/3275107.stm.

Dinan, C. 1983. Sugar-Daddies and Gold-Diggers: the white collar single woman in Accra. In *Female and Male in West Africa*, edited by C. Oppong. London: George Allen and Unwin.

DOI. 2001. *Creating a Development Dynamic: Final Report of the Digital Opportunity Initiative*. New York: UNDP.

Donner, J. 2004. Microentrepreneurs and mobiles: An exploration of the uses of mobile phones by small business owners in Rwanda. *Information Technologies and International Development* 2(1): 1–22.

Eggleston, K., R. Jensen, and R. Zeckhauser. 2002. Information and Communication Technologies, Markets, and Economic Development. In *The Global Information Technologies Report 2001–2002: Readiness for the Networked World*, edited by G. Kirkman, P. Cornelius, J. Sachs, and K. Schwab. Oxford: Oxford University Press.

Ferguson, J. 1999. *Expectations of Modernity: Myths and Meanings of Urban Life on the Zambian Copperbelt*. London: University of California Press.

Gamos Ltd. 2005. *The Economic Impact of Telecommunications in Rural Livelihoods and Poverty Reduction*. Reading: Gamos Ltd.

Gerster, R., and S. Zimmermann. 2003. *Information and Communication Technologies (ICTs) and Poverty Reduction in Sub Saharan Africa: A Learning Study*. Richterswil: Gerster Consulting.

Guardian (Dar es Salaam). 2004. Gunmen unleash terror on busy Dar street. http://ipp.co.tz/ipp/guardian/2004/06/12/13096.html.

Hangi, P. 2002. *Mobile Phones Fuel Crime, Police Admit. Dar es Salaam*.

Haram, L. 2004. "Prostitutes" or modern women? Negotiating respectability in northern Tanzania. In *Re-thinking Sexualities in Africa*, edited by S. Arnfred. Uppsala: Nordiska Afrikainstitutet.

ITU. 2005. International Telecommunications Union, Africa regional ICT profile 2004. http://www.itu.int/ITU-D/ict/statistics/ict/index.html.

Kamwaga, E. 2002. Mtandao wa wizi wa simu za mikononi wagundulika Dar es Salaam (Gang of mobile phone thieves revealed in Dar es Salaam). *Mtanzania*, May 12, p. 4.

Keregero, K. 2002. Sekretari abambwa "live" akiuza mtandao wa bosi (Secretary caught red-handed with boss' phone). *Alasiri*, Aug. 19, p. 4.

Kingston, J. 2004. *Japan's Quiet Transformation: Social Change and Civil Society in the Twenty-first Century*. London: Routledge Curzon.

Lewinson, A. 1999. Going with the times: Transforming visions of urbanism and modernity among professionals in Dar es Salaam, Tanzania. Unpublished doctoral dissertation, University of Wisconsin, Madison.

Luke, N. 2006. Exchange and condom use in informal sexual relationships in urban Kenya. *Economic Development and Cultural Change* 54(2): 319–348.

MacGaffey, J., and R. Bazenguissa-Ganga. 2000. *Congo-Paris: Transnational Traders on the Margins of the Law*. Oxford: James Currey.

Matambalya, F., and S. Wolf. 2001. Performance of SMEs in East Africa: Case studies from Kenya and Tanzania (draft). Paper presented at the DESG-IESG Annual Conference 2001. University of Nottingham.

Mbogora, A. 2002. Ushindani wa makampuni ya simu za mkononi wapunguza gharama (Competition between mobile phone companies reduces costs). *Mtanzania*, March 27, p. 9.

Mercer, C. 2006. Telecentres and transformations: Modernizing Tanzania through the Internet. *African Affairs* 105(419): 243–264.

Molony, T. S. J. 2005. Food, Carvings and Shelter: The adoption and appropriation of information and communication technologies in Tanzanian micro and small enterprises. Unpublished doctoral dissertation. University of Edinburgh, Edinburgh.

Molony, T. S. J. 2006. ICT in Developing Countries (No. 261). Westminster, London: Parliamentary Office of Science and Technology.

Mwesige, P. G. 2004. Cyber elites: a survey of Internet Cafe users in Uganda. *Telematics and Informatics* 21(1): 83–101.

Nielinger, O. 2003. Rural ICT Utilisation in Tanzania: Empirical findings from Kasulu, Magu, and Sengerema (Draft version 2.1). Hamburg: Institute of African Affairs.

Nyamnjoh, F. B. 2005. Fishing in troubled waters: "Disquettes" and "thiofs" in Dakar. *Africa* 75(3): 295–324.

Scanagri. 2005. Socio-economic feasibility study in ICT connectivity and services in three selected areas in Tanzania. Paper presented at the Rural ICT Conference. Golden Tulip Hotel, Dar es Salaam.

Shivji, I. G. 2005. Tanzania: What kind of country are we building? http://www.pambazuka.org/index.php?issue=200.

Silberschmidt, M., and V. Rasch. 2001. Adolescent girls, illegal abortions and "sugar-daddies" in Dar es Salaam: Vulnerable victims and active social agents. *Social Science and Medicine* 52: 1815–1826.

Souter, D., N. Scott, C. Garforth, R. Jain, O. Mascarenhas, and K. McKemey. 2005. The economic impact of telecommunications on rural livelihoods and poverty reduction: A study of rural communities in India (Gujarat), Mozambique and Tanzania. Commonwealth Telecommunications Organisation report for UK Department for International Development.

Stewart, J. 2004. Mobiles phones: Cigarettes for the 21st Century. http://homepages.ed.ac.uk/jkstew/work/phonesandfags.html.

Townsend, A. M. 2000. Life in the real-time city: Mobile telephones and urban metabolism. *Journal of Urban Technology* 7(2): 85–104.

UNCTAD. 2004. E-Commerce and Development Report 2004. Geneva: United Nations.

UNDP. 2001. *Human Development Report: making new technologies work for human development*. New York: Oxford University Press.

United Republic of Tanzania. 2002. *Budget Household Survey*. Dar es Salaam: Bureau of Statistics.

United Republic of Tanzania. 2005. Telephone tariffs for Tanzania market (Tanzania Telecommunications Regulatory Authority). http://www.tcra.go.tz/Market%20info/mobile%20tariffs.htm.

van Binsbergen, W. 2004. Can ICT belong in Africa, or is ICT owned by the North Atlantic region? In *Situating Globality: African Agency in the Appropriation of Global Culture*, edited by W. van Binsbergen and R. van Dijk. Leiden and Boston: Brill.

Varbanov, V. 2002. Bulgaria: Mobile phones as post-communist cultural icons. In *Perpetual Contact: Mobile Communication, Private Talk, Public Performance*, edited by J. E. Katz and M. Aakhus. Cambridge, UK: Cambridge University Press.

Waverman, L., M. Meschi, and M. Fuss. 2005. The impact of telecoms on economic growth in developing countries. In *Africa: The Impact of Mobile Phones*, edited by D. Coyle. Newbury: Vodafone.

Wight, D. 2005. Sexual norms and behaviour in rural Tanzania. Paper presented at the University of Edinburgh Swahili Club seminar series. University of Edinburgh.

World Bank. 2005. Tanzania Country Brief. http://web.worldbank.org/WBSITE/EXTERNAL/COUNTRIES/AFRICAEXT/TANZANIAEXTN/O,,menuPK:287345~pagePK:141132~piPK:141107~theSitePK:258799,00.html.

Gerard Goggin

Despite the widespread take-up and use of mobile communications, their cultural dimensions have not received the sustained critical attention they merit. While much work is ahead in understanding mobiles from a social dimension, it is fair to say that the humanities and social sciences have now gone some far to rectifying their lack of attention to the social study of mobile telephony (something lamented by Ithiel de Sola Pool in 1977 about wireline telephony, a situation only slightly improved in the intervening thirty years). However, we lack equivalent studies devoted to the cultural dimensions of mobiles—whereas there are no lack of such cultural accounts for other information, communication, and media technologies.

One reason for this is that culture, even more so than society, was neglected when it came to the telephone (except studies such as Katz 1999, and Martin 1991). With the advent of the mobile, wireless, portable, and personal telephony has become inescapably and visibly part of the cultural realm, and so presses more insistently upon the researcher. This is why there is much interesting and pertinent material about the cultural significance of the mobile dispersed among the fast growing literature. There have been several collections with a broad sweep (Katz 2003; Ling 2004; Castells et al. 2006). There are a number of collections, for instance, that collocate both the social and cultural in their titles (for example, Glotz, Bertschi, and Locke 2005; Kim 2005). This is not surprising given the domain of culture is intimately connected to the constitution of the social sphere (du Gay et al. 1997). However, these links need to be analytically distinguished and traced before the two concepts can be systematically brought together.

In this chapter, I provide an overview of cultural studies of mobile communication and discuss the issues studies in the area have raised. I also suggest what is pressing in terms of new research on cultural aspects of mobiles. Before I proceed, a word about culture.

Culture is a notoriously elusive concept, and freighted with historical and political baggage. In my view, culture is the set of shared (though often contested and contradictory) ideas, symbols, practices, and artifacts that both allow people to make sense of everyday life but also position them in their society. We all enter into culture, in a general sense, as infants and very young children, but then we also learn how to inhabit various other cultures. There are large cultures, associated with and shaped by social relations of nations, classes, races, genders, sexualities, impairment and disability, and other forms of embodiment and subjectivity. Subtended by, intersecting with, and crossing over are various "little" or "micro" cultures, such as those associated with subcultural formations, forms of media cultures, and technologies (for instance, du Gay et al.'s 1997 discussion of the Sony Walkman).

Divergent notions concerning culture, and the very real battles and stakes it calls up, can be observed in the reception of mobile phones and mobile communications. In many societies, mobiles have been both celebrated in a technologically determinist register, or even from the standpoint of the technological sublime, but have also been the subject of scorn, humor, or condescension because they represent the nadir of "Culture" today. The ubiquity, prevalence, and apparently constant resource to mobile communication is often counterposed against other apparently self-evident sources of cultural value and heritage—whether "high" literature, the virtues of "proper" conversation and letter writing, reasoned discussion of an ideal public sphere, or quality movies or photography.

Little Cultures of Mobile Phones

An important strand of the cultural studies of mobile phone literature is concerned with exploring, defining, and interpreting what characterizes the "little" cultures of this technology.

Early studies were preoccupied especially with inventorying and noting the new mobile communicative practices, and how they were creating new cultures of use, of portable and personal voice telephony (for example, the collaborative European research reported in Haddon 1998, and elsewhere; Plant 2002; Haddon and Vincent 2005), and of associated communication and information technologies, such as voice mail, phone books, caller identification, alerts (Jensen, Thrane, and Nilsen 2005)—some of which were being introduced on the new, fixed, intelligent telecommunications networks during the 1980s and 1990s. Especially important here were the cultural implications of the new constructions of place, and new forms of both individuation but also new relations to social collectivities and identities that mobile technologies entailed (Fortunati 2001b and 2002; Katz and Aakhus 2002). Later studies sought to compare mobiles with other contemporaneous media technologies such as the Internet (for example, Fortunati and Contarello 2002; Miyata, Wellman, and Boase 2005) or WiFi (Sawhney 2005).

With the passage of time, there has commenced some historical reconstruction and reconsideration of the development of these fledgling mobile cultures, and how they drew from, and were articulated with, the cultures of precursor technologies; not only the telephone, but also various technologies of radio such as radiotelegraphy, citizens' band radio, pagers, transistor radios, and the Walkman, as well as conceptual inventions prefiguring new forms of mobile communication such as the Dick Tracy video wristwatch. Such cultural histories of mobiles need to be carefully delineated as they are often written from different standpoints (nascent history writing includes Arceneaux 2005; Agar 2003; contributions to Goggin and Thomas 2006 and Hamill and Lasen 2005).

There are a number of studies that look at how the mobile phone was ushered in and its cultures produced, through how such technology and its uses were imagined and narrated through various discourses, not least advertising, branding, symbols, images, language, metaphors, and rhetoric (for instance, contributions to Brown, Green, and Harper 2002; Wang 2005; Kavoori and Arceneaux 2006; cf. the social representations approach of Fortunati, Contarello, and Sarrica 2007 or the symbolic analyses of Campbell and Russo 2003 and Campbell, chapter 12 in this volume). Others have considered the role of media and cultural discourses more generally in the production of the mobile (Burgess 2004; Goggin 2006a; Yung 2005).

There has been a growing focus on the figure of the user as actively consuming, if not cocreating, and producing mobile technology. One of the attributes of the mobiles in its mass distribution phase coinciding with second-generation mobile technology was its potential for user customization through changing faces, downloading wallpaper, choosing ringtones, or adorning the handset. Such customization and the importance of the user in the sense-making and domestication of mobile telephone has not escaped scholars (Hjorth 2005; Hjorth and Kim 2005) or thoughtful industry practitioners.

Exploration of the intricate and often locally inflected cultural practices that have developed with different types of mobile communication technologies have emerged as an important facet of understanding design (a motif in Lindholm, Keinonen, and Kiljander 2003, for example). There is a burgeoning, fertile, and eclectic body of work and practice around mobile design, with ethnographic inquiry featuring in the attempt to scrutinize mobile culture to recursively improve design of new technologies (for instance, various chapters in Harper, Palen, and Taylor 2005; or Yue and Tng 2003).

One of the most striking and fully developed areas of cultural inquiry is text messaging, where there have been many fascinating studies. The most comprehensive account, and indeed one of the best cultural studies of mobiles we have, is Kasesniemi's 2003 *Mobile Messages*, which not only captures mobile messaging culture fully blown in its Finnish incubation, but is a highly illuminating account of second-generation GSM (the Global Standard for Mobile communications, developed in Europe) mobile culture.

Perhaps it is because text messaging developed seemingly out of nowhere, unenvisaged as a major feature of digital mobiles by those who devised text messaging—yet avidly used, with great significance, by users. The reasons for the "success" of text messaging, as compared to say the damp squib of Wireless Access Protocol (an early form of mobile Internet), has been discussed at length. Here various commentators have had recourse to theories that articulate the relationship between the social and the technological, such as in the traditions of social studies of science and technology (Fortunati 2005), but also theories in which culture itself is central, such as actor-network theory (e.g., Taylor and Vincent 2005). Text messaging has become something of a synecdoche for mobile culture itself. Texting is celebrated for its ability to form collectivities, or even shape national identities and polities, as in notably the coup d'text that putatively brought down the Philippines's President Estrada (Rafael 2003; cf. Pertierra et al. 2002), Howard Rheingold's smart mobs (2002), others' flash mobs, or even race riots (Goggin 2006b).

Mobile Phones In and Across Cultures

Another important if not preponderant strand of research is interested in how mobile phones have developed in different cultural formations. Most salient are the national studies of mobiles, which were crucial to laying a foundation for mobile research. Katz and Aakhus's pioneering collection not only invaluably curates a number of these studies but also encourages comparative consideration (to mention only a few: Fortunati 2002a; Mante 2002; Robbins and Turner 2002). An exemplary study of national mobile culture is Ito, Okabe, and Matsuda's detailed *Personal, Portable, Pedestrian: Mobile Phones in Japanese Life* (2005). What the studies collected in this book do so well is to bring together various levels of discussion of mobile culture in a nuanced, historically and culturally contextualized treatment of the *keitai* (Japanese for the Internet-enabled mobile). As mobile cultures have taken firmer hold and have grown in complexity, there are now some important comparative, cross-national cultural studies of mobiles (Katz et al. 2003; Leonardi, Leonardi, and Hudson 2006; or various contributions to Kim 2005).

While many countries still await their detailed accounting in the literature on mobiles, the national as a category has certainly loomed large. What has been less well covered, if at all in some cases, are other social forms and their cultural correlates. There have been some studies of minority cultures, for instance, explorations of African American or Latino mobile use in the United States (respectively, Heckman 2006 and Leonardi 2003), but there is much scope for the recognition of multicultural mobility, or how mobiles are deeply implicated in contemporary cultural diversity and hybridity. Culture identities and practices of sexual minorities (and indeed majorities) were the subject of a fine, groundbreaking collection (Berry, Martin, and Yue 2003), but work since then has been difficult to find—despite the function of sex and intimacy in popular conceptions of, and many anecdotes devoted to, mobiles (see Elwood-Clayton 2005).

Similarly, there have been surprisingly few fine-grained, or theorized, studies about mobiles' role in connecting diasporic communities (something that has been discussed elsewhere in fine media studies of TV, video, and diasporas, as well as newer work that considers the Internet). There are a few studies on transnational cultures (for instance, Uy-Tioco 2007, which discusses the role of the mobile in globally distributed parenthood), but not as much as might be expected given the centrality to the modern world of foreign workers, international travel (in its widely varying forms), migration, refugee flight, and movements of capital and humans.

There are other studies into cultural aspects of mobiles scattered across the literature, though they indicate the importance of taking such ventures further. Youth culture has been compulsively studied (Caronia and Caron 2004; Ito, Okabe, and Matsuda 2005; Kasesniemi 2003; Kasesniemi and Rautiainen 2002; Lorente 2002), while the cultures of older people have been largely overlooked. While there has been little explicitly written on race and mobiles, for instance (even compared to the small literatures on this topic in, say, Internet studies, Rice and Katz 2003), we do have some important studies that take into account gender (Green 2002; Lacohée, Wakeford, and Pearson 2003; Hjorth and Kim 2005; Lee 2005; Shade 2007). Such studies underscore the importance of gender in the development of mobiles, but also the subtlety of how these processes unfold to shape the technology, to construct users and their uses in everyday life. There is also now the beginnings of critical discussion on disability (Goggin and Newell 2006; Power and Power 2004) that finds the mobile intimately involved in the constitution of disability as a cultural constitution—but also points to normative notions of disability and ability as influential in the shaping of mobiles as cultural technology. Cultural studies of mobiles and spirituality, faith, and religion is another important yet relatively neglected topic, and show the enduring centrality of faith communities, and their transformations, to how mobiles are imagined (though thought-provokingly canvassed in Campbell 2007; Elwood-Clayton 2003; Katz 2006).

Mobile Media Cultures

As mobile phones have not only infiltrated themselves into communication, but also media, entertainment, and information, so we already have important cultural soundings of these developments. In the wake of the excitement surrounding text messaging, there have been important studies of multimedia messaging (MMS) (such as Ling and Julsrud 2005; Oksman 2005; Scifo 2005). Also concerned with visual cultures of mobiles have been a growing stock of studies of camera phones and mobile photography (see contributions to Ito, Okabe, and Matsuda 2005; Lee 2005; Rivière 2005). Another area of considerable interest has been the articulation of mobiles, Internet, and online technologies and cultures with new forms of digital photo and image representation, sharing, and exchange—for instance, exploration of mobile blogging, or moblogs (Döring and Gundolf 2005). Sound, music, and auditory cultures have also been explored (for example, Bull 2005), though only to a minor extent so far. Less well represented are cultural studies of mobile film and video, and also the new developments in mobile television (on the latter, see Goggin 2006a). Third-generation mobiles have often been discussed, but more often than not through the rose-colored lens of utopian discourses—critical work is badly needed here. There is much work and theorizing in artistic circles on locative media and mobile art, but still too little in the academic literature (De Souza e Silva 2006).

Cultural Futures for Mobiles: A Research Agenda

In this brief chapter, I hope to have given the interested reader a snapshot of the state of cultural studies of mobiles. Clearly, my treatment here can only be indicative and selective, rather than comprehensive. It is also provisional and partial, not least given that it is largely based on the Anglophone literatures. This said, I shall finish by briskly proposing a research agenda in this area.

Firstly, there is much work to do in documenting and theorizing the distinctive features of mobile phone culture, and the various particularities and features of the "little" cultures it comprises. The best exemplar for this work so far is perhaps the intense interest and debate centered on text messaging. We have uneven bodies of work on other aspects of mobile phone culture, such as camera phones, customization, and ringtones, but here further research, synthesis, and theoretical inquiry is required.

With mobile phone culture becoming mobile *media* cultures (Nilsson, Nuldén, and Olsson 2001), there will now be even larger domain requiring investigation—and also a pressing need to engage with concepts of and research traditions pertaining to media. With the novelty of, and hype ushering in, developments such as mobile television, mobile video, or mobile gaming, we need not only historically and culturally situate mobiles in relation to their borrowings from and relations with TV, radio, or print, but also newer media. Here then, mobile studies can profitably engage with the fields of film, TV, journalism, games, and Internet studies. Given the blurring of mobiles, wireless technologies, and pervasive computing, it will be fruitful to continue and intensify the cross-fertilization of conversations across these literatures and undertakings.

Secondly, there is a need to broaden and intensify the cross-cultural work on mobiles—to understand the social and cultural construction, appropriation, and domestication of mobiles, and how our use of these devices have modified our notions of communication. While there are important studies available we still require 1) further, extended, systematic and comprehensive studies of the insertion and shaping of mobiles in national cultural contexts (as counterparts to the work available on Finland, Japan, and the Philippines); 2) work that grapples and gauges the implication of mobiles in the dynamics of cultural diversity and hybridity, especially in multicultural

societies—focusing, for instance, on cultural minority, migrant, and refugee populations; on the cultures of class and race; and on cultures associated with various forms
of identity (such as sexuality, gender, race, and disability).

Thirdly, once the peculiar features of mobile culture is grasped, and how mobiles
have been taken up and made sense of in various society is understood, then mobiles
can be placed in the larger arc and field of culture. How do the cultures of mobile technologies relate to culture more generally? Here there is a need for the work on mobiles
to join other innovative cultural research in discerning the role of mobile communication in contemporary social and cultural transformation, whether social conflict (riots)
or dissent (protests), or social production and reproduction (work, households). An important venture here is exploring mobiles as part of popular culture, and through this
also, mobiles' implication in the changing nature of the relationships among, indeed
the viability of sustaining distinctions across, high or elite culture, middlebrow culture,
and "low" working-class or marginalized cultures. Imperative here is attention to the
implication of mobiles in contests of values, and, in a general sense, in the moral
clashes over ordained and accepted public uses.

Another underdeveloped problematic is mobiles and cultural citizenship (Lillie
2005), as might be thought through the notion of "mobile commons." Also important
is understanding the part of mobiles in large-scale and small-scale communicative
architectures that underpin cultures and national or global or local conversations that
used to be associated with newspapers (with the rise of nationalism), or TV and radio
(especially in the twentieth century), or the Internet at the beginning of the twenty-
first century.

Finally, there has been only a limited range of voices, themes, and cultural locations
represented in mobile communication study thus far. Unsurprisingly, mobile scholarship and commentary, as well as the diffusion of mobile devices themselves, have been
concentrated in wealthy countries. At least there has been a modicum of attention
accorded to mobile communication in many countries in Europe, North America,
Asia, as well as a number of other countries, which for various reasons have been the
source of fascination for their mobile uses (such as Japan, Korea, or the Philippines, and
China being the latest case in point here). However, there are many other countries,
especially in the developing world, where mobiles are important cultural technologies
despite issues of income, cost, and affordability (Donner 2005), yet these are invisible,
or perhaps illegible, even in the vibrant, interdisciplinary, and cosmopolitan world of
the mobiles scholarly community.

Of course, it is not simply a question of mobiles researchers from the better-
resourced countries in the West and East studying, writing, and discussing mobile cultures of other societies and places, though, if the fraught power and other relations
here are recognized, such work could assume great importance. There are difficult politics of knowledge at stake, and imposing issues of voice and representation. There are

questions, too, of acknowledging the cultural specificity of our theories, histories, concepts, and methods, as has been articulated in the new wave of research on Asian mobiles and modernities (see, for instance, Bell 2005; Law, Fortunati, and Yang 2006; Lin 2005; McLelland 2007; Pertierra and Koskinen 2007). When we reflect upon the condition and politics of culture played out and debated elsewhere, this seriously stands to problematize cultural studies of mobile communications—but also to greatly recast them and spur them on.

References

Agar, J. 2003. *Constant Touch: A Global History of the Mobile Phone*. Cambridge: Icon Books.

Arceneaux, N. 2005. The world is a phone booth: The American response to mobile phones, 1981–2000. *Convergence* 11(2): 23–31.

Bell, G. 2005. The age of the thumb: A cultural reading of mobile technologies from Asia. In *Thumb Culture: The Meaning of Mobile Phones for Society*, edited by P. Glotz, S. Bertschi, and C. Locke. Bielefeld: Transcript Verlag.

Berry, C., F. Martin, and A. Yue, eds. 2003. *Mobile Cultures: New Media in Queer Asia*. Durham, N.C.: Duke University Press.

Brown, B., N. Green, and R. Harper, eds. 2002. *Wireless World: Social, Cultural and Interactional Issues in Mobile Communications and Computing*. London: Springer-Verlag.

Bull, M. 2005. The intimate sounds of urban experience: An auditory epistemology of everyday mobility. In *A Sense of Place: The Global and the Local in Mobile Communication*, edited by K. Nyíri. Vienna: Passagen Verlag.

Burgess, A. 2004. *Cellular Phones, Public Fears, and a Culture of Precaution*. New York: Cambridge University Press.

Campbell, H. 2007. What hath god wrought: Mobile faith and the culturing of the cell phone. In *Mobile Phone Cultures*, edited by G. Goggin. London: Routledge.

Campbell, S. W., and T. C. Russo. 2003. The social construction of mobile telephony: An application of the social influence model to perceptions and uses of mobile phones within personal communication networks. *Communication Monographs* 70(4): 317–334.

Caronia, L., and A. H. Caron. 2004. Constructing a specific culture: Young people's use of the mobile phone as a social performance. *Convergence* 10(2): 28–61.

Castells, M., M. Fernandez-Ardevol, J. L. Qiu, and A. Sey. 2006. *Mobile Communication and Society: A Global Perspective*. Cambridge, Mass.: The MIT Press.

Contarello, A., L. Fortunati, and M. Sarrica. 2007. Social thinking and the mobile phone: A study of social change with the diffusion of mobile phones, using a social representations framework. In *Mobile Phone Cultures*, edited by G. Goggin. London: Routledge.

De Sola Pool, I., ed. 1977. *The Social Impact of the Telephone*. Cambridge, Mass.: The MIT Press.

De Souza e Silva, A. 2006. Interfaces of hybrid spaces. In *The Cell Phone Reader: Essays in Social Transformation*, edited by A. Kavoori and N. Arceneaux. New York: Peter Lang.

Dong, S., ed. 2005. *When Mobile Came: The Cultural and Social Implications of Mobile Communication*. Seoul: Communication Books.

Donner, J. 2005. Research approaches to mobile use in the developing world: a review of the literature. *Proceedings of International Conference on Mobile Communication and Asian Modernities*, edited by A. Lin. City University of Hong Kong, Kowloon.

Döring, N., and A. Gundolf. 2005. Your life in snapshots: Mobile weblogs (moblogs). In *Thumb Culture: The Meaning of Mobile Phones for Society*, edited by P. Glotz, S. Bertschi, and C. Locke. Bielefeld: Transcript Verlag.

du Gay, P., S. Hall, L. Janes, H. Mackay, and K. Negus. 1997. *Doing Cultural Studies: The Story of the Sony Walkman*. Milton Keynes: Open University; Thousand Oaks, Calif.: Sage.

Elwood-Clayton, B. 2003. Texting and God: the Lord is my textmate—folk Catholicism in the cyber Philippines. In *Mobile Democracy: Essays on Society, Self and Politics*, edited by K. Nyíri. Vienna: Passagen Verlag.

Elwood-Clayton, B. 2005. Desire and loathing in the cyber Philippines. In *The Inside Text: Social, Cultural and Design Perspectives on SMS*, edited by R. Harper, L. Palen, and A. Taylor. Dordrecht: Springer.

Fortunati, L., and A. Contarello. 2002. Internet-Mobile convergence: Via similarity or complementarity? *Trends in Communication* 9: 81–98.

Fortunati, F., M. Sarrica, and A. Contarello. 2007. Social thinking and the mobile phone: A study of social change with the diffusion of mobile phones, using a social representations framework. *Continuum: Journal of Media and Cultural Studies* 21: 149–163.

Fortunati, L., J. E. Katz, and R. Riccini, eds. 2003. *Mediating the Human Body: Technology, Communication, and Fashion*. Mahwah, N.J.: Lawrence Erlbaum.

Fortunati, L. 2001. The mobile phone: An identity on the move. *Personal and Ubiquitous Computing* 5: 85–98.

Fortunati, L. 2002a. Italy: Stereotypes, true and false. In *Perpetual Contact: Mobile Communication, Private Talk, Public Performance*, edited by J. E. Katz and M. Aakhus. Cambridge: Cambridge University Press.

Fortunati, L. 2002b. The mobile phone: Towards new categories and social relations. *Information, Communication & Society* 5: 513–528.

Fortunati, L. 2005. The mobile phone as technological artefact. In *Thumb Culture: The Meaning of Mobile Phones for Society*, edited by P. Glotz, S. Bertschi, and C. Locke. Bielefeld: Transcript Verlag.

Glotz, P., S. Bertschi, and C. Locke, eds. 2005. *Thumb Culture: The Meaning of Mobile Phones for Society*. Bielefeld: Transcript Verlag.

Goggin, G. 2006a. *Cell Phone Culture: Mobile Technology in Everyday Life*. London: Routledge.

Goggin, G. 2006b. SMS riot: Transmitting race on a Sydney Beach, December 2005. *M/C Journal* 9(1). http://journal.media-culture.org.au/0603/02-goggin.php.

Goggin, G., ed. 2007. *Mobile Phone Cultures*. London: Routledge.

Goggin, G., and J. Thomas, eds. 2006. Mobile histories. Special issue of *Southern Review: Communication, Politics & Culture* 38(3).

Goggin, G., and C. Newell. 2006. Disabling cell phones. In *The Cell Phone Reader: Essays in Social Transformation*, edited by A. Kavoori and N. Arceneaux. New York: Peter Lang.

Gottlieb, N., and M. McLelland, eds. 2003. *Japanese Cybercultures*. London: Routledge.

Green, N. 2002. Who's watching whom? Surveillance, regulation and accountability in mobile relations. In *Wireless World: Social, Cultural and Interactional Issues in Mobile Communications and Computing*, edited by B. Brown, N. Green, and R. Harper. London: Springer-Verlag.

Haddon, L., ed. 1998. *Communications on the Move: The Experience of Mobile Telephony in the 1990s*. Stockholm: Telia.

Haddon, L., and J. Vincent. 2005. Making the most of the communications repertoire: Choosing between the mobile and fixed-line. In *A Sense of Place: The Global and the Local in Mobile Communication*, edited by K. Nyíri. Vienna: Passagen Verlag.

Haddon, L. 2003. Domestication and mobile telephony. In *Machines that Become Us: The Social Context of Personal Communication Technology*, edited by J. E. Katz. New Brunswick, N.J.: Transaction Publishers.

Haddon, L. 2004. *Information and Communication Technologies in Everyday Life: A Concise Introduction and Research Guide*. Oxford and New York: Berg.

Hamill, L., and A. Lasen, eds. 2005. *Mobile World: Past, Present and Future*. London: Springer.

Harper, R., L. Palen, and A. Taylor, eds. 2005. *The Inside Text: Social, Cultural and Design Perspectives on SMS*. Dordrecht: Springer.

Heckman, D. 2006. "Do you know the importance of a skypager?": Telecommunications, African Americans, and popular culture. In *The Cell Phone Reader: Essays in Social Transformation*, edited by A. Kavoori and N. Arceneaux. New York: Peter Lang.

Hjorth, L., and H. Kim. 2005. Being there and being here: Gendered customising of mobile 3G practices through a case study in Seoul. *Convergence* 11: 49–55.

Hjorth, L. 2005. Odours of mobility: Japanese cute customization in the Asia-Pacific region. *Journal of Intercultural Studies* 26: 39–55.

Ito, M., D. Okabe, and M. Matsuda, eds. 2005. *Personal, Portable, Pedestrian: Mobile Phones in Japanese Life*. Cambridge, Mass. and London, England: The MIT Press.

Ito, M. 2005. Introduction: Personal, portable, pedestrian. In *Personal, Portable, Pedestrian: Mobile Phones in Japanese Life*, edited by M. Ito, D. Okabe, and M. Matsuda. Cambridge, Mass.: The MIT Press.

Jensen, M., K. Thrane, and S. J. Nilsen. 2005. The integration of mobile alerts into everyday life. In *Mobile Communications: Re-negotiation of the Social Sphere*, edited by R. Ling and P. E. Pederson. London: Springer.

Kasesniemi, E. L., and P. Rautiainen. 2002. Mobile culture of children and teenagers in Finland. In *Perpetual Contact: Mobile Communication, Private Talk, Public Performance*, edited by J. E. Katz and M. Aakhus. Cambridge: Cambridge University Press.

Kasesniemi, E. L. 2003. *Mobile Messages: Young People and a New Communication Culture*. Tampere: Tampere University Press.

Katz, J. E. 1999. *Connections: Social and Cultural Studies of American Life*. New Brunswick, N.J.: Transaction.

Katz, J. E. 2006. *Magic in the Air: Mobile Communication and the Transformation of Social Life*. New Brunswick, N.J.: Transaction Publishers.

Katz, J. E., ed. 2003. *Machines that Become Us: The Social Context of Personal Communication Technology*. New Brunswick, N.J.: Transaction Publishers.

Katz, J. E., and M. Aakhus. 2002. Conclusion: Making meaning of mobiles: A theory of Apparatgeist. In *Perpetual Contact: Mobile Communication, Private Talk, Public Performance*, edited by J. E. Katz and M. Aakhus. Cambridge: Cambridge University Press.

Katz, J. E., M. Aakhus, H. D. Kim, and M. Turner. 2003. Cross-cultural comparisons of ICTs. In *Mediating the Human Body: Technology, Communication, and Fashion*, edited by L. Fortunati, J. E. Katz, and R. Riccini. Mahwah, N.J.: Lawrence Erlbaum.

Kavoori, A., and N. Arceneaux, eds. 2006. *The Cell Phone Reader: Essays in Social Transformation*. New York: Peter Lang.

Kim, S. D., ed. 2005. *When Mobile Came: The Cultural and Social Impact of Mobile Communication*. Seoul: Communication Books.

Lacohée, H., N. Wakeford, and I. Pearson. 2003. A social history of the mobile telephone with a view of its future. *BT Technology Journal* 21: 203–211.

Law, P. L., L. Fortunati, and S. Yang, eds. 2006. *New Technologies in Global Societies*. Singapore: World Scientific.

Lee, D. H. 2005. Women's creation of camera phone culture. *Fibreculture Journal* 6. http://journal.fibreculture.org/issue6/issue6_donghoo.html.

Leonardi, P. M. 2003. Problematizing "new media": Culturally based perceptions of cell phones, computers and the Internet among United States' Latinos. *Critical Studies in Media Communication* 20: 160–179.

Leonardi, P. M., M. E. Leonardi, and E. Hudson. 2006. Culture, organization, and contradiction in the social construction of technology: Adoption and use of the cell phone across three cultures. In *The Cell Phone Reader: Essays in Social Transformation*, edited by A. Kavoori and N. Arceneaux. New York: Peter Lang.

Lillie, J. 2005. Cultural access, participation, and citizenship in the emerging consumer-network society. *Convergence* 11(2): 41–48.

Lin, A., ed. 2005. *Proceedings of the International Conference on Mobile Communication and Asian Modernities*. Hong Kong, June 7–8.

Lindholm, C., T. Keinonen, and H. Kiljander, eds. 2003. *Mobile Usability: How Nokia Changed the Face of the Mobile Phone*. New York: McGraw-Hill.

Ling, R., and P. E. Pederson, eds. 2005. *Mobile Communications: Re-negotiation of the Social Sphere*. London: Springer.

Ling, R., and T. E. Julsrud. 2005. The development of grounded genres in Multimedia Messaging Systems (MMS) among mobile professionals. In *A Sense of Place: The Global and the Local in Mobile Communication*, edited by K. Nyíri. Vienna: Passagen Verlag.

Ling, R. 2004. *The Mobile Connection: The Cell Phone's Impact on Society*. San Francisco, Calif.: Morgan Kaufmann.

Lorente, S., ed. 2002. Juventud y teléfonos móviles. Special issue of *Revista de Estudios de Juventud* 57.

Mante, E. 2002. The Netherlands and the US compared. In *Perpetual Contact: Mobile Communication, Private Talk, Public Performance*, edited by J. E. Katz and M. Aakhus. Cambridge: Cambridge University Press.

Martin, M. 1991. *Hello Central?: Gender, Culture, and Technology in the Formation of Telephone Systems*. Montreal: McGill-Queen's University Press.

McLelland, M. 2007. Socio-cultural aspects of mobile communication technologies in "Asia": Examples from the recent literature. In *Mobile Phone Cultures*, edited by G. Goggin. London: Routledge.

Miyata, K., B. Wellman, and J. Boase. 2005. The wired—and wireless—Japanese: webphones, PCs and social networks. In *Mobile Communications: Re-negotiation of the Social Sphere*, edited by R. Ling and P. E. Pederson. London: Springer.

Morley, D. 2003. What's "home" got to do with it? *European Journal of Cultural Studies* 6: 435–458.

Nilsson, A., U. Nuldén, and D. Olsson. 2001. Mobile media: The convergence of media and mobile communications. *Convergence* 7: 34–38.

Okada, T. 2005. Youth culture and the shaping of Japanese mobile media: personalization and the *keitai* Internet as multimedia. In *Personal, Portable, Pedestrian: Mobile Phones in Japanese Life*, edited by M. Ito, D. Okabe, and M. Matsuda. Cambridge, Mass.: The MIT Press.

Oksman, V. 2005. MMS and its "early adopters" in Finland. In *A Sense of Place: The Global and the Local in Mobile Communication*, edited by K. Nyíri. Vienna: Passagen Verlag.

Pertierra, R., and I. Koskinen, eds. 2007. *The Social Construction and Usage of Communication Technologies: European and Asian Experiences*. Singapore: Singapore University Press.

Pertierra, R., E. F. Ugarte, A. Pingol, J. Hernandez, and N. L. Dacanay. 2002. *Txt-ing selves: Cellphones and Philippine Modernity*. Manila: De La Salle University Press. http://www.finlandembassy.ph/texting1.htm.

Plant, S. 2002. *On the Mobile: The Effects of Mobile Telephones on Social and Individual Life*. www.motorola.com/mot/doc/0/234_MotDoc.pdf.

Power, M., and D. Power. 2004. Everyone here speaks TXT: Deaf people using SMS in Australia and the rest of the world. *Journal of Deaf Studies and Deaf Education* 9: 333–343.

Rafael, V. L. 2003. The cell phone and the crowd: Messianic politics in the contemporary Philippines. *Public Culture* 15: 399–425.

Rheingold, H. 2002. *Smart Mobs: The Next Social Revolution*. Cambridge, Mass.: Perseus.

Rice, R. E., and J. E. Katz. 2003. Comparing Internet and mobile phone usage: Digital divides of usage, adoption, and dropouts. *Telecommunications Policy* 27: 597–623.

Rivière, C. 2005. Mobile camera phones: A new form of "being together" in daily interpersonal communication. In *Mobile Communications: Re-negotiation of the Social Sphere*, edited by R. Ling and P. E. Pederson. London: Springer.

Robbins, K. A., and M. Turner. 2002. United States: Popular, pragmatic and problematic. In *Perpetual Contact: Mobile Communication, Private Talk, Public Performance*, edited by J. E. Katz and M. Aakhus. Cambridge: Cambridge University Press.

Sawhney, H. 2005. Wi-Fi networks and the reorganisation of wireline-wireless relationship. In *Mobile Communications: Re-negotiation of the Social Sphere*, edited by R. Ling and P. E. Pederson. London: Springer.

Scifo, B. 2005. The domestication of the camera phone and MMS communications: The experience of young Italians. In *A Sense of Place: The Global and the Local in Mobile Communication*, edited by K. Nyíri. Vienna: Passagen Verlag.

Shade, L. R. 2007. Mobile designs as gendered scripts. In *Mobile Phone Cultures*, edited by G. Goggin. London: Routledge.

Taylor, A. S., and J. Vincent. 2005. A SMS history. In *Mobile World: Past, Present and Future*, edited by L. Hamill and A. Lasen. London: Springer.

Uy-Tioco, C. 2007. Mobile phones and text messaging: Reinventing migrant mothering and transnational families. In *Mobile Phone Cultures*, edited by G. Goggin. London: Routledge.

Wang, J. 2005. Youth culture, music, and cell phone branding in China, *Global Media and Communication* 185–201.

Yu, L., and T. H. Tng. 2003. Culture and design for mobile phones for China. In *Machines that Become Us: The Social Context of Personal Communication Technology*, edited by J. E. Katz. New Brunswick, N.J.: Transaction Publishers.

Yung, V. 2005. The construction of symbolic values of the mobile phone in the Hong Kong Chinese print media. In *Mobile Communications: Re-negotiation of the Social Sphere*, edited by R. Ling and P. E. Pederson. London: Springer.

Mobile Music as Environmental Control and Prosocial Entertainment

James E. Katz, Katie M. Lever, and Yi-Fan Chen

Personal music players, typified currently by iPod and other MP3 technologies, have enjoyed a global surge of popularity. The widespread adoption of the technology has also spawned a variety of normative critiques that characterize users as morally dubious and assert that their behavior is a social corrosive that undermines the pleasures of public life (Rosen 2005; cf. du Gay et al. 1997). These technologies certainly have the potential to isolate individuals from the ambient environment in general and other humans in particular: this very quality seems to be an informing principle of the technology's design. It also is a major driver of its popularity. Moreover, just as with the mobile phone (Fortunati 2005; Ling 2003; Sugiyama 2006), the technology has important status-display functions, especially among young people.

In this chapter, we extend findings concerning mobile phones to another form of personal communication technology, namely digital music players. (The related topics of mobile digital video, TV, and games are treated in chapter 30.) We also aim to transcend this topic and suggest that these tools are not infrequently used as interpersonal-bridging and community-building artifacts. We need to be careful, though, to point out that we are not saying that this is the primary aim of most adopters, at least initially. Rather we assert that many users arrive at this point during the course of their social routines and interactions that involve the music technology. We are also cognizant of the extremely limited nature of the data we are able to present in support of this hypothesis and agree it remains highly tentative.

As to the success of digital music technology, we can cite numerous impressive statistics about worldwide sales, and instead offer a precise empirical demonstration of the technology's popularity. The data are from a Rutgers University Center for Mobile Communication Studies research team that did observations of digital technology on a college campus (Chen and Lever 2006). In their two years of unobtrusive observation ($N = 11,307$), Yi-Fan Chen assisted by Katie Lever found rapidly rising "teledensity" levels between the two years. Although mobile phones were the most common form of mobile technology manifested, mobile music players were also prominent. In fact the percentage of users of MP3 players nearly doubled from 4.8 percent in 2005 to 8.5

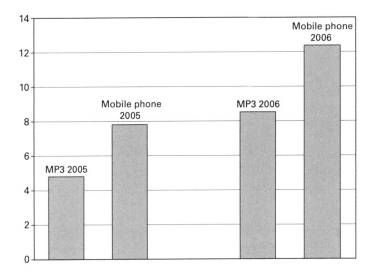

Figure 27.1
"Teledensity" of MP3 players and mobile phone players on campus by year as a percentage of all passersby observed. N = 4,562 in 2005; N = 6,745 in 2006. Note: Twelve individuals appeared to be using both an MP3 player and a mobile phone in the 2005 observation, and twenty-eight in the 2006 observation; they were included here as MP3 users.

percent in 2006, as shown in figure 27.1. That about one in twelve people on campus were "plugged in" to their portable music players suggests the enormous popularity of the technology. Preliminary observations in late 2006 as part of the third wave of the study suggest that the percentage of users continues to climb.

One intriguing finding was that some individuals appear to be using both MP3 players and mobile phones at the same time. While their numbers remain quite small, they nonetheless increase over time. In 2005, twelve individuals were observed using both and, in 2006, twenty-eight people were so observed. Noteworthy too was the relative lack of a gender divide in using the technology. In 2006, 40 percent of MP3 users observed were women, so at least at the highest level of abstraction, there does not seem to be dramatic differences in the distribution of possession of MP3 players. (Still, women were less likely to be MP3 player users.)

Given this statistical backdrop, we can interrogate the meaning of personal mobile music technology. Bull (2004) asserts that MP3 players are "the first cultural icon[s] of the twenty first century" and as such will alter the "way people manage their experience of music in urban space." These technologies provide what Bull calls an ongoing "soundtrack to the world" (Bull 2004). Because of its inherent portability, users can have music (and video) accompany most of their daily activities. However, like many other ICTs of earlier generations, critics question whether the portability of this technology will create a nation of strangers (Katz and Rice 2002; Rosen 2005). However, we found that mobile music users are modifying these tools from a purely personal indulgence and environmental isolation to a means of self-expression. Moreover, as noted later, they are using these potentially isolating tools in ways that are actually community- and social network–building mechanisms.

Theoretical Perspectives

Studies examining the symbolic meaning of mobile phones have flowered (Sugiyama 2006). These studies have diverged from traditional analyses of the functional applications of technologies and have incorporated more of a social-constructivist perspective. Thus, underpinning our study, which focuses on mobile music technology rather than the mobile phone, is the notion that individuals may not only use technology according to original design principles but also may reconstruct them to serve socially embedded and symbolic needs. *Apparatgeist* theory argues that norms regarding technology use are continuously being modified, often quite creatively, by users within social environments to serve expressive interests (Katz and Aakhus 2002). Moreover, it seems that sometimes users of technologies imbue them with special meaning and emotional valences. The increasing integration with the physical body, social meaning, and individual identity is captured by the phrase "machines that become us." Users are extending their physical attributes to include the technologies that have been increasingly connected to one's being. This allows for an additional means of self-expression.

Traditionally, fashion has been considered a communication mechanism that communicates one's sense of self through the use of clothing (Sugiyama 2006). Many scholars have also argued that fashion is a social process, wherein individuals conceptualize what others are wearing, construct the meaning of the attire, and then decide whether or not to adopt the clothing for themselves. Technologies are used as fashion devices to connect the body with the personality and the externally perceived attributes, and thus become part of an integrated package (Sugiyama 2006).

From this stance, we argue that mobile digital music users transcend the intended purpose of these digital technologies and integrate novel practices more consistent with their social control and integration needs but that at the same time may be divergent and even contrary to the technology's original design thesis. Also, we seek to demonstrate that, like other areas of social competitiveness, such as jewelry, home décor, and clothing, users find the technology to be a tool of hierarchy and a manifestation of values. Technologies such as the mobile telephone or MP3 player serve to renegotiate perceptions of personal space versus that related to private space (Katz and Aakhus 2002), and also provide a new way to represent the self to others and create an admixture of selves with those of others.

Hearing from Students

To explore these questions, the Rutgers team interviewed a total of forty-three university students on three different occasions in the spring of 2006. The focus group data were taped with three different tape recorders and was later transcribed informally by three different researchers. We found that the penetration rate within our focus groups was approximately 68 percent when we asked students whether or not they owned an iPod or MP3 player. In fact, in one focus group eleven students were iPod or MP3 player users while two were not, thus demonstrating the prevalence of the digital music technologies.

Conferring Status

We found that MP3 and iPod users were initially largely motivated to purchase these tools by their peers. For example, when we asked respondents to explain how they found out about iPod and MP3 player technologies, a female student answered:

From my roommate freshmen year. I remember moving in and she was listening to it. I thought that it was so cool. I asked for one and I didn't get it. I asked for one every holiday and still didn't get one.

Another student echoed this claim through his indication that he would not have been as apt to purchase an iPod technology if his friends had not promoted it so heavily. Related to this theme of social influence, we asked students what motivated them to purchase an iPod technology as opposed to a non-iPod, MP3 player. A female student replied:

I'll be totally honest—it's pretty. I only got it because it looks better than the other ones.

The discussion leader next asked: "How many bought it because of the looks?" to which one student answered:

I had friends in high school who had like the first generation, like the huge ones that looked archaic and ugly.

Another female student mentioned that her interest was piqued while exercising at the gym, having seen other patrons using the technology. Additionally, an interesting comment came from one student who said that she thought she was one of the last people to not have an iPod. To be sure, individuals' perceptions of what is popular and trendy feed into their decisions to adopt a technology and thus demonstrate that tribalism plays a role in adoption.

Environmental Control

A pervasive theme that emerged during focus group discussions was that college-aged individuals use digital music technology in ways that serve to create a personal environment and control the access of others to themselves. When asked when he uses his iPod, a male participant said:

Just to get through bus rides, because without it, I think that I would go crazy.

Interviewer: What about the bus rides do you usually go crazy about?

You're usually standing, it's packed, and everyone is having their own conversations.

When asked if he was using the iPod to cut himself off, he said, "no, but there are a billion conversations going on."

Said one female student:

For me, it's sort of like a zone . . . [when] I go to class and in between, I put my headphones on. If someone I see tries to stop you it kind of like interrupts my uh, I don't know . . ." When asked if this prevented people from interacting with her, she said, "Not anybody that I'm close with or a good friend.

Gyms are popular sites for using the technology, for both isolating and environmental-conditioning effects. One woman who was asked if she were dependent on the iPod responded, "Like in the gym especially. I can't go to the gym without listening to music."

Multitasking is also common: One female said, "I always put one earphone in, so you can do two things at once."

Social Collaboration

MP3 players and iPods can also be used to facilitate interactions and symbolize closeness. For instance, one young woman said that she shared her iPod earbuds with her boyfriend while on vacation. Another female echoed this sentiment: "I actually bought a splitter so like if I go on vacation with someone, I can plug it in and two headphones go with it." To our question "Have you become dependent upon [your music device?]," another female student responded:

My best friend and I walked out of our house together to go to [another campus] and I forgot my iPod. When we were walking out, I just made her give me one of the ear things and we walked to the bus stop just like that and we sat next to each other just like that—we looked really cool.

To this same question, another young woman added that when her car radio broke, she and her passenger decided to share the earpieces to one iPod.

This sense of sharing and community building was illustrated by several other students and through various sharing activities. For example, when discussing the musical downloading process, one male said that he had text messaged an individual asking about a certain song that he wanted to download and was able to foster a friendly conversation through this inquiry. Another female student said: "I put a new song on my iPod and I asked my friend to listen to it." A male respondent echoed this claim: "Definitely. Same situation, if I get a new song, I share it with my friends." One female student said that she was able to deepen the bonds of her relationship with her "tech-savvy" boyfriend through her discussion of Podcasts, as she thought that this particular discussion would be of interest to him. Another student added that his fraternity house had iTunes music applications linked so that all members of the house could listen to the others' music. He added that this system allows individuals to share with others new music that they find interesting.

Our focus group sessions drove home the heavily social nature of the digital music experience. This included collective sharing of techniques and resources for downloading music and especially for participation in online communities. As but one of many illustrations, members of a fraternity may share music through their local house network. On a more personal level, and of greater import for the theoretical questions we are considering, there is sharing of the more intimate embodiment of digital music. As mentioned, this occurs by sharing earbuds, as in the instance of the girl saying that she frequently shared her earbuds with her boyfriend while on vacation. Other individuals also recalled sharing their iPod headphones with friends while eating at the campus student center, or while waiting at the bus stop. Thus sharing is not only a symbolic and reputational endeavor, but at least for some, occurs in a way that raises unsavory hygienic questions.

Another common theme that emerged from our focus groups echoed Bull's claim that MP3s and iPods can be a "soundtrack to the world" (2004, p. 4). For instance, one student voiced that his iPod served as his "soundtrack to life." Another student indicated that in environments such as that of the airport he would use the iPod during a "boring day," while one of his colleagues added that she enjoyed listening to music when sick. All three of these situations serve as examples of how individuals may be in uncomfortable situations that they remedy through their use of digital music.

Despite many positive remarks and much sharing of digital music technologies, a good number of students also said that they found others' use of iPod and related technologies "obnoxious," especially when the music was being played loud enough to be heard.

But beyond the irritation factor is the question of the consequences for digital music usage for the ambient social environment. In one focus group of seventeen students, thirteen students said they felt that the iPod isolates individuals, while four said that it did not. Of those who did not see the use as problematic, one student said:

The only time I listen is when I'm in the car driving. But all of my friends have it. They listen to it and I'll just read a magazine. It's not a big deal to me because I'm not a big music person.

Those who saw digital music devices as isolating included users, and indeed there was explicit recognition of, and appreciation for, this dimension. One female noted that she perceives that her iPod has the potential to isolate her from others and vice versa, and takes steps to moderate its enthralling effects:

It allows me to become more antisocial. It allows me to tune out reality and I realize that more and more—I'm kind of sinking into my iPod. I have to make more of a conscious effort [so] I usually leave one earbud out.

Further, several students indicated that they did not feel as though they were isolating themselves to the extent that they perceived that they could still "pause" the music so as to be able to interact with someone if they were to be approached. When asked whether MP3 or iPod players can allow a user to ignore someone else, one female responded: "I wouldn't hesitate to come up to someone with an iPod." However, after a moment she went on to say that this dynamic may differ when considering acquaintances. She said in those instances "you are kind of taken out of the equation of having to say hello to them."

Another finite and interesting theme that emerged during our focus groups is that students construct their own usage practices and sentiments about their technologies, thus very much working in accord to Apparatgeist theory. For example, one student indicated that when having a bad day she would frequently use her iPod technology in public arenas, thus allowing her to "escape" from her environment. Our respondents also discussed the attachment that they feel to their digital music technologies and how they feel "when people around [them] are using them":

I was really upset. I went to Florida on winter break and I forgot my iPod...my flight [was] delayed an hour. It was the most painful and uncomfortable thing to just sit there staring for like an hour and a half waiting for the flight to take off and I felt so lost without my iPod on the beach and by the pool. It wasn't fun.

We followed this response by asking whether this individual felt jealous that other people had music technologies. She responded by saying, "Yeah, they had their laptops and everything. I had nothing." Another female echoed this concept by saying that when one is without one's MP3 and iPod player, "you have nothing in your hand—it's awkward."

Discussion and Implications

Social influence occupies a central role in the decision to purchase digital music players. This influence extends not only to fashion but also even to a degree of tribalism. As discussed, respondents also indicated various levels of goal pursuit and achievement. For example, several students indicated that they use MP3 and iPod players during their trips to the gym in an effort to prevent boredom. Clearly, students are able to overcome what some see as an isolating technology so as to share the tool (both physically and verbally), and their music with others.

We also found that our respondents employed many different means to get music for their MP3 and iPod. Regardless of the level of discomfort associated with sharing the music, whether it is through bypassing copyright guidelines, or having to sit next to a friend who is driving a car, students are making extreme efforts to share their music with others.

Numerous studies have sought to explain the varying ways in which individuals adopt and use technology across gender lines. For instance, when the first personal computers arrived on the home front, many argued that women were reluctant to use them because such use would contrast with traditional conceptions of gender identity (Katz and Rice 2002). Scholars of technology have long noted that women tend toward social uses while men often use technology for more utilitarian reasons (Fischer 1992). In our study of mobile music players, both males and females said they shared their players with others. This uniformity of a "sharing" use may be due to the nature of the technology (it is a recreational tool), but our findings may begin to demonstrate that there are no true variations in adoptions and use among men and women. However, additional research is necessary to expand upon this realm of research. The sample of students we had was small, and not at all representational.

As one related piece of evidence, albeit indirect, we note the "silent raves" phenomena, namely preorganized events for people who "flashmob" to a busy public place (such as a railway station). At the appointed time, they begin dancing to their own MP3 player– or mobile phone–provided music, listened to on earphones. Prearranged semisecretly by e-mail or text message, the event is similar to 1960s "happenings." After a certain amount of "play time," the gesticulating but silent dancers will stop and rejoin the crowd. Further, the silent rave underscores the inherently interpersonal and prosocial aspects of a technology that appears to have been designed to achieve precisely the opposite effect. However, for reasons not entirely clear, none of these activities seemed to be part of the repertoire of the students in our focus groups. Neither did they seem to have any interest in participating in such activities. Without trying to make too much of this apathy toward flashmobbing, it does suggest that the phenomena is an experiment in potential and novelty rather than a new social form. At the very least, it is worth a more systematic investigation.

Conclusions

Mobile music players are heavily used as instruments of environmental control. But environmental control is not limited to the sound that one hears and blocks out. It also includes the projection of one's image to others in the ambient environment. In this regard, the MP3 and iPod devices are important aspects of the image one creates and projects. Equally clear, these devices help screen out and protect one not only from unpleasant sounds but unpleasant people as well.

Yet these technologies are not exclusively tools that can be used to protect and isolate. Although subsidiary in their frequency and importance, they can also be used to build bridges and connect people to their friends. And, on a symbolic level, they can not only project prestige, but also represent friendship, involvement, and even love.

In many ways, mobile music players potentially and actually isolate users. However, our focus group data indicate that some students are in fact using digital music technologies as community-building tools. We found that some students are desirous of sharing creative solutions to accessing music and operating file-sharing systems. They are also ingenious in collaborating with fellow users, both online and offline, to share their musical tastes and learn from others about new groups and resources.

The interests are not limited to music and online resources alone. Notably, we also found that MP3 and iPod users are sharing their earbuds with others. This is significant because earbuds are designed to fit only one person, and have a highly personal (and not necessarily hygienic) aspect to them. The point then is that users are not channeled into narrow behaviors but rather are creative reusers and modifiers of their technology. More fundamentally, all these practices suggest that users are reforming both the meanings and uses of their digital music technologies.

The iPod has become a synecdoche for urban cool in the early twenty-first century. "Tuned in, turned on, and spaced out" is the surface commercial image. But, from the user's viewpoint, this image seems to be read more accurately as "socially embedded" and increasingly "personally connected." As wireless technologies for these music devices improve, so too will their capacity for social connectivity. Music, more than ever, will be a bridge, if not for international understanding, then at least for interpersonal communication and community solidarity.

References

Bull, M. 2004. iPod professor tunes into music on the move. [Press Release]. Sussex University: http://www.sussex.ac.uk/press_office/media/media389.shtml.

Chen, Y.-F., and K. M. Lever. 2006. Teledensity: A study of gender differences in the use of mobile communication technology on a college campus. Paper presented at the International Communication Association on the 56th Annual Conference. Dresden, Germany.

du Gay, P., S. Hall, L. Janes, H. Mackay, and K. Negus. 1997. *Doing Cultural Studies: The Story of the Sony Walkman.* Milton Keynes: Open University; Thousand Oaks, Calif.: Sage.

Fischer, C. S. 1992. *America Calling: A Social History of the Telephone to 1940,* Berkeley: University of California Press.

Fortunati, L. 2005. The mobile phone as technological artefact. In *Thumb Culture: The Meaning of Mobile Phones for Society,* edited by P. Glotz, S. Bertschi, and C. Locke. Bielefeld: Transcript Verlag.

Katz, J. E., and R. E. Rice. 2002. *Social Consequences of Internet Use: Access, Involvement, and Interaction.* Cambridge, Mass.: The MIT Press.

Katz, J. E., and M. Aakhus. 2002. Conclusion: Making meaning of mobiles—a theory of Apparat-geist. In *Perpetual Contact: Mobile Communication, Private Talk, Public Performance*, edited by J. E. Katz and M. Aakhus. Cambridge: Cambridge University Press.

Ling, R. S. 2003. Fashion and vulgarity in the adoption of the mobile telephone among teens in Norway. In *Mediating the Human Body: Technology, Communication, and Fashion*, edited by L. Fortunati, J. E. Katz, and R. Riccini. Mahwah, N.J.: Lawrence Erlbaum.

Rosen, C. 2005. Playgrounds of the self. *The New Atlantis* 9 (Summer): 3–27.

Sugiyama, S. 2006. Melding with the self: The social process of fashion and symbolic meanings of the mobile phone as constructed by youth in Japan and the United States. Doctoral dissertation, Rutgers University, New Brunswick, N.J.

28 | Supernatural Mobile Communication in the Philippines and Indonesia

Bart Barendregt and Raul Pertierra

Literature on the new media generally ignores new media's use for spiritualistic pursuits. Yet this aspect of novel technologies reveals much about embedded practices and their cultural construction in particular spatial and temporal contexts. In this chapter, we look at representations of the cell phone in Southeast Asian societies focusing especially on the ways new mobile technologies have been represented in the Philippines and Indonesia. These representations have drawn on a long tradition of communicative exchanges between people and the supernatural. New technologies often elicit unintended and unexpected practices from their users. In many parts of Asia, mobiles have become a new writing tool, their main use being to transmit SMS. That these messages sometimes involve the dead or other supernatural beings is not surprising since these multiple realities happily coexist in many Asian societies, including technologically advanced ones.

In the northern Philippines, as in most parts of Asia, the souls of the recently deceased are believed to hover near their earthly dwelling for a period of days or weeks. During this time, communication between the dead and their living kin are frequent. Messages from the dead are conveyed in myriad ways, from the strange chirping of birds to the presence of an unusual number of fireflies. This is also done by conversations between a medium and the deceased. A favorite relative goes into *naluganan* (trance) and asks the dead person's soul what they need for the journey into the afterlife. Previously, such conversations were easily managed but since many villagers now work overseas, special arrangements have to be made. The mobile phone is a handy technology in such circumstances. The medium, often a relative who may be abroad, having been informed about the situation, goes into trance. In the village, the mobile phone is placed on a favorite item of the deceased. After the trance, the medium conveys the necessary information to their kin in the village.

One example of the almost ordinary experience of supernatural communication in the Philippines was recently provided by Jaime Licauco, a well-known journalist and commentator on spiritualistic matters. In his regular newspaper column he reprinted a bereft man's letter:

Two weeks ago, Carmina, who became my girlfriend through text messaging, died. After her death, she started sending me text messages. During her wake, she continued to text me, telling me how much she loved me. In one of our conversations, she told me she was not dead. After her burial, I thought she would stop sending text messages. I was wrong. She would usually send me text messages around 11 in the evening or at midnight. Her cousin is now using her SIM card, per Carmina's request. I asked her cousin to turn off the phone but Carmina was still able to get through to me. It's funny that she found herself in different places, like her home or a resort.

I want to ask: Should I continue talking to her? Please enlighten me on this strange event. Thank you. (Licauco 2006)

Licauco advised the young man to inform his dead girlfriend that he still loved her but that she should accept her new condition and move on. Similar cases of supernatural communication via mobile phones have been reported for Indonesia and Africa (Barendregt 2006). Many Filipinos have similar stories, even if not as graphic or explicit, about recently deceased kin. These stories usually mention receiving calls at unusual times but with no caller, batteries suddenly running out, or the phone being moved or misplaced. These events, in the appropriate context, are readily interpreted as involving attempts by the dead to communicate with their living kin.

Comparable supernatural contacts abound in stories from Central Java, Indonesia. The strange stories pages of a Javanese weekly newspaper in early 2002 contained a piece on a mysterious phone in one of Yogyakarta's public payphone centers. The writer was asked to dial the number 11378 and to ask for a person named Endang or Dewi, and when she did so, she was told politely that Endang was not in. When trying again now asking for other persons a similar thing occurred, as she was told that the person had just left. More scarily, the booth's display unit [the meter that records the duration of calls] did not record the use of any pulses. The writer of the piece was left wondering where in the city of Yogyakarta would one have a five-digit phone number, but to the street vendor it was clear—the woman must have called the cemetery. How else could one explain the fact that whoever one asks for has always just left? It must be that the phone number connects one to a grave or to the *alam gaib*, the world beyond. Most Asians are usually amused by such stories.

The frequency of such seemingly uncanny experiences, regularly recounted in popular media, indicates that something more important than mere amusement is going on. We can see such tales as shedding light on the conceptions about the new technology that the telephone and especially its mobile variant represent, and on how some Asians encounter modernity.

Technology and Culture

The conceptualization of the relationship between technology and culture always has been problematic. On the one hand, all technology is a product of its contextual culture, but on the other, technology can precipitate cultural change in unpredictable ways. While technology does not itself determine sociocultural change, it opens up new conditions of possibility hitherto unavailable (Katz and Aakhus 2002). In a parallel but opposite direction, modern technology such as mobile phones brings about changes in the inner-world of their users (Pertierra et al. 2002) that have significant social and cultural consequences. Paradoxically, mobile phones encourage a more privatized and personalized orientation to the world when used in public, and the opposite orientation when used in private. Yet even under both conditions of public and private use, they enable a discursive intimacy hitherto difficult if not impossible in traditional societies such as Indonesia and the Philippines. Moreover, private orientations may quickly coalesce into collective actions through the rapid transmission of information. These seemingly easily mobilized collectivities or smart mobs are capable of the microcoordination of their hitherto unconnected participants, with sometimes significant political ramifications, as discussed in the chapter by Rheingold. The mobile phone seemed to have reached its apogee as a symbol of the new open Indonesian society when Indonesia's President Susilo Bambang Yudhoyono declared that citizens could now directly contact him by texting. He made public his private phone number and promised that from now onward the text messages and phone calls he would receive from the people would serve as "his eyes and ears." A similar texting service was available in the Philippines in 2002 shortly after Gloria Arroyo took over the presidency following President Estrada's downfall. Known as "text Ate Glo," it encouraged citizens to communicate in intimate (kinship) terms with President Gloria Macapagal Arroyo.

The global condition exacerbates this tension between society and technology since technology can now rapidly spread to cultures far removed from its origins. While this chapter deals with the seemingly idiosyncratic supernatural uses of mobile phones in Indonesia and the Philippines, there are also numerous commonalities of their uses with those found throughout the world. While technology if adopted can fit into its cultural environment, its uses may also reflect transcultural and universal features. Technologies may express the zeitgeist and hence may be described as *Apparatgeist* (Katz and Aakhus 2002). Such claims may be made of mobile phones not only because of their rapid spread globally but also because of their common enabling effects, a phenomenon which has been widely reported, including in other chapters in this volume.

Similar to its introduction in the West, the first steam locomotive initially unsettled the tranquility of the Asian countryside yet quickly established itself as integral to commerce and everyday life. Around 1870 the construction of the first railways in Java was started, and in 1890 the first steamships plied the trade routes. In contrast to the evolutionary introduction experienced in the West, the sudden appearance of steam technology in Java was interpreted in a religious way. Javanese peasants not only worshipped but also feared these first steam engines and made offerings to their machine's supposed spirits (Pemberton 2003). Rudolf Mrázek (2002, p. xv–1) explains how the introduction of technologies had the capacity to defamiliarize and transform the routine into the extraordinary: in the Dutch East Indies "as people handled, or were handled by, the new technologies, their time, space, culture, identity, and nation came to feel awry." A kind of floating modern space developed detached from the rest of society in which colonial engineers used technologies to draw a line between themselves and the non-Dutch population. Similar stories of an emerging modern space but also the resultant alienation among segments of the society abound in the early twentieth century with the introduction of the camera and "talking machine" into various Asian societies (compare with Weidman 2003). The belief soon emerged that such machines could capture reality "as it is" (cf. Gunning 1995). Along a similar vein, Rafael (2003, p. 400) discusses how the telephone in the colonial Philippines was supposed to finally uncover what had been hidden, enabling the people to directly file complaints with the leaders of the free nations in the West. The telegraph, the motorcar, and other inventions were soon introduced in the colonies, reflecting as well as exacerbating the rapid changes of late modernity. In such shifting conditions, people seek ways to better anchor themselves, and we would argue that the mobile phone responds to this need.

Following such global changes, it is no surprise that worldviews and other orientations of the *habitus* have been significantly affected. But most of these changes arose out of earlier structures and kept their traditional form if not their precise substance. The new technologies affected traditional religions, causing them to evolve into new and exotic forms. Cargo cults in Oceania are not the only examples of these puzzling modifications of belief systems. The spread of millenarian cults worldwide, of bizarre and futuristic groups such as Jonestown and Scientology, as well as the use of media like TV, the Internet, and SMS (short message service) to spread the faith, have become routine aspects of contemporary life, not only in the Western world but also in most Asian societies.

While many of these new religious movements are a response to the material changes brought about by modernity, they also express new ways of relating to the mundane world. Heidegger (1977) has argued that technology not only affects the world outside our existence but also enters into our being-in-the-world in new ways. We are thus "in the world" differently, opening up new possibilities of being and becoming. New technologies allow us to relate to ourselves and to others in novel ways. Technology is not only mechanical materiality or a body of techniques that stands in an exterior relationship to human subjectivity. Technology is also *techne*, the application of knowledge that connects us intersubjectively to the material and the supernatural worlds. It enables new ways of being in the world (including the afterworld), thereby revealing to us our human possibilities. As a recent conference on mobile phones concluded, "the machine becomes us" (Katz 2003).

Historically, the spread of organized religion has always been closely linked to the growth and proliferation of new technologies. The Gutenberg press and more recently, radio and television have become important channels for experiencing as well as for spreading the faith. Mobile media and associated services have been no exception to this. Other chapters in this volume and elsewhere (Katz 2006) discuss this relationship between technology and religion more fully.

While the contribution of technological innovations to economic change is widely known and accepted, how new technologies impact on the experience of the sacred and the supernatural in Asia is poorly understood. The relationship linking society, technology, and experiences of the supernatural in Asia involve complex structures whose details cannot be tackled here. But an examination of the mobile phone will shed useful light on aspects of this complexity.

Mobile Phone Revolutions

Mobiles are an intimate technology with a truly global reach. It is a technology not only close at hand and often close at heart but also largely perceived as closed to uninvited others. However, the fear of uninvited callers also marks the vulnerability of mobiles for dangerous intrusions. As a consequence of this technology, the boundary between self and other is precariously balanced and possibly porous.

In Indonesia, crooks use a text variant of a scam known as "SMS terror," harassing calls or messages that come at any time of the day, often containing obscene contents. Since the tsunami disaster of late 2004, moreover, numerous SMS rumors have on occasion sparked panics in parts of Indonesia, prompting thousands to flee their homes for higher ground. In 2004, SMS messages rumoring that three cars filled with explosives had been brought into Jakarta spread through the capital. This turned out to be a hoax (ADP 2004, 2005). Similar rumors, threats, and scams are also regularly reported in the Philippines (Pertierra et al. 2002). Most recently during the Lebanon crisis, text messages circulated around Manila informing Filipino Muslims that Israel intended to bomb Mecca.

For many Indonesian and Filipino users who have emigrated from the countryside, mobiles connect the rural world to its urban counterpart and even beyond, effectively erasing their differences. A librarian in a Javanese university acts as a middleman between students and a shaman in his home village. Students ask him to contact the shaman about predictions for the next day's (illegal) lottery. Coming from a village in the hinterland of Java gives this librarian's SMS messages an almost magical aura; the students using the librarian's cell phone not only to directly contact the Javanese shaman but also the world of the supernatural.

Overseas work is now almost routine for many of our informants and mobiles connect them to meaningful others, allowing domestic workers in Hong Kong or seamen in distant ports to maintain an absent presence in their home communities, including participation in traditional rituals. Synchronic communication allows subjects to share a simultaneous present, (Pertierra 2006) anywhere anytime with anyone, even with the recently deceased.

In Indonesia and the Philippines one cannot overstress the representational value of mobile phones. Owning a cell phone is essential for a modern and hip lifestyle, especially among members of the young urban middle class. Talking about the decline of the generation of 1998, when Suharto was forced to resign after ongoing protests, Indonesian student activists in one newspaper recently complained that disappointingly this generation has quickly fallen for market capitalism. The mobile is no longer a weapon of the weak but the ultimate symbol of hedonism and consumerism (*Jawa Pos* 27 December 2004). According to these student activists, the mobile revolution has turned out to be mainly a consumer's revolution. TV quizzes with call-in possibilities have led Islamic hardliners to a call for a fatwa on bets placed through SMS.

Southeast Asians presently associate new and mobile media with certain social problems. Many moral anxieties are projected on the new media, which in Indonesia is often associated with three *P*s: pornography (on screensavers and exchanged by Bluetooth), piracy (video, CD, and DVD), and political subversion (communist and *jihad* videos). There are similar though less heightened fears about the new technologies in the Philippines. These and other moral fears remind us how modernity is perceived by many Asians and how this is reflected in the use of new mobile technologies.

Asian Counter-Modernities

Most Indonesians and Filipinos eagerly accept technology. Moreover, for urban Southeast Asians, modernity has become equivalent to mobility. But some people are also suspicious of the new technologies, seeing them as a form of enslavement as much as liberation. Modernity has produced its own resistances, particularly in former colonies with experiences of domination and exploitation. Some Asians associate Western technology with neocolonial domination even if they realize the centrality of technology for contemporary life. For them, modernity is often incomprehensible.

Some Indonesians refer to themselves as *anti-ponsel* (anti cell phone) or *anti-HP* (*Manusia anti-ponsel*, "the anti cell phone man," *Matra* magazine, April 2005). We interviewed people who referred to the "robotization of mankind." These narratives temper much of the enthusiasm that usually accompanies the new communications media. The fear of technology is a common phenomenon and not only tells us much about how such novel technologies are embedded in existing practices but also about how such a novel technology comes to be constructed in a particular context, time, and place. The mobile phone stands as an icon of the more sinister aspects of modernity just as much as it represents modernity's advantages.

New technologies often give rise to a fascination for the novel but simultaneously arouse suspicions, which, as many accounts show, are often linked to the world of spirits or supernatural phenomena. Jeffrey Sconce (2000, p. 209) presents an overview of what he calls "fantastic folktales generated by the world's most uncanny innovation: electronic media," which not surprisingly starts in 1844 with Samuel Morse and the birth of telecommunications such as the electromagnetic telegraph. Sconce illustrates how a parallel discourse soon developed, namely that of the spiritual telegraph used by spirit mediums that were said to be in contact with the dead (Sconce 2000, p. 23). Spiritualists and the devotees of new technologies often had in common the hope that the new apparatus could shed light on the spirit world, uncovering what the human eye could not see. As Sconce argues, enduring well beyond a fleeting moment of naïve superstition at the dawn of the information age, the historical interrelationship of these competing visions of telegraphic "channeling" continues to inform many speculative accounts of media and consciousness today.

Technologies serve either as uncanny electronic agents or as gateways to another dimension. There are many parallels between late-nineteenth-century and late-twentieth-century stories of these uncanny technologies. At the dawn of the twenty-first century, people in many societies, including those in Southeast Asia, seem to associate the mobile phone with yet another possibility of an electronic elsewhere: the media's occult power to give form to sovereign electronic worlds. Once again "electricity facilitates transmutations between mind and media, allowing the inanimate to become sentient and the sentient to become a ghost in the machine" (Sconce 2000, p. 205). Virtual reality and postcorporeal identities are simply newer versions of an alternative world made possible by the new media.

A Ghost in My Mailbox!

In 2002 the Indonesian newspaper *Bernas* (6 March 2002) included a story on a *mailbox hantu* (voice mail ghost) that had appeared in the Klaten area of Central Java. Numerous people had received a mysterious text message that suggested that if they called the cell phone number 0812838xxxx they would hear a ghost talking to them. After calling the number, people indeed reported hearing a terrifying female voice alternately crying and laughing. The voice sounded like a *kuntilanak*, the ghost of a woman who died in childbirth and a creature that is often mentioned in popular supernatural reality shows. This number became so popular that even young children asked their parents for money to call the ghost in the phone. Owners of public phone centers responded to the demand by placing the number on their booths saying, "If you want to hear a kuntilanak, dial 0812838xxxx."

Similar stories were reported in other parts of Indonesia during the period of our fieldwork (2004). An SMS message is received urging the recipient to dial a number.

Having done so one hears strange sounds, awful voices, dogs barking, a door squeaking, and a lot of hysterical shouting. Some people have reported this Ghost in the Phone to the local authorities. In one case, on the island of Madura, the police contacted the local telecom company to find out who the owner of the ghost number was (*Jawa Pos/Radar Madura* 17 April 2004). The owner, however, denied involvement with otherworldly practices. The police asked the telecom company to disconnect his number, but strangely enough, the paper suggests, when one calls the number, the baleful voices can still be heard.

Many of these fears are, of course, encountered elsewhere. There have been reports in Nigeria that people died after receiving certain mysterious calls. Jane-Francis Agbu (2004) recounts the case of a woman who claimed to have received a call that almost led to her death. The phone flashed the name of a relative but without the number. A similar incident was reported in Lagos that same year. A young man received a call from a number 0172021127. Suddenly he shouted "Blood of Jesus, Blood of Jesus" before he collapsed. Fortunately bystanders were able to revive him. A company in Lagos warned its employers about these killer numbers and posted them in its notice board: "Please beware of these strange GSM numbers: 0801113999, 08033123999, 08032113999 and 08025111999. In short any number that ends with 333, 666, 999. They are killing! This is nothing but reality, you are warned" (Agbu 2004). The power of numbers or words to kill is familiar throughout the traditional world. They exercise a deadly illocutionary force.

All throughout Asia a similar belief exists in unlucky or lucky numbers. Phone numbers that include the digit 4 are considered to bring bad luck by some Javanese, as the Javanese word for *four* sounds similar to the word for *coffin* and is therefore considered to cause early death. Beautiful numbers (*nomor cantik* in Indonesian) are primarily numbers that are easy to remember, with, for example, the last four digits referring to the date of one's birth or wedding. There is, however, also a special category of alleged lucky numbers. Although the popularity of such numbers in Indonesia is sometimes explained with reference to older Javanese traditions of so-called *primbon*, numerological systems, these latter numbers are most popular among people of East Asian descent all throughout Southeast Asia but also in mainland China. They are often willing to pay a lot of money for such phone numbers, which they base on *Hong Shui* or *Feng Shui*. These numbers are not distributed through the usual channels, but are often sold to the highest bidder by providers. In the early 2000s, newspapers and specialized Web portals regularly advertised such numbers, and today shopping malls specializing in mobile phones and accessories, such as Jakarta's Roxymas mall or Bangkok's MBK shopping center, still sell such phone numbers at often incredible prices.

We suggest that the fears associated with the new technologies such as the cell phone and the PC are due to their rapid rate of change. Cell phones use a not-yet-crystallized technology that needs to be updated every two or three years, with consumers often hardly aware which generation of mobile technology they are using. The very novelty of the technology such as answering or fax machines and computers adds to its mystification. In addition, people are puzzled by complex instruction manuals in English, which they get with their phone, particularly when buying cheap illegally imported mobiles. These cheap mobiles are bought by buyers with poor English-language skills and who are already insecure in using the new technology. These fears and uncertainties are reflected and at times exaggerated in popular culture.

Popular Ghost Culture

Many of these mentioned phone ghost stories are often portrayed in film, radio shows, or popular lore. In our view these are predominantly tales of changing times and places, where everything is in flux. Interestingly, many of such stories and anecdotes are not located in traditional or rural spaces as are ghost stories of the West. In Asia, phone ghost stories are usually staged in large anonymous cities or in newly con-structed suburbs without yet a history of themselves, satisfying what Marc Augé (1995) has called "non-places." These Asian stories refer to unnamed sites or unknown times. Within these unnamed sites the mobility of the phone stands for a portable memory place, a mobile hybrid in which modern futuristic tools are portrayed as merg-ing past beings pointing at unresolved businesses. It is this mobile hybrid that helps people come to terms with the uncertain present.

This need is also addressed by shops selling the telephone as ghost-capturing devices. In the Indonesian city of Yogyakarta, one store selling cell phones is called Dunia Akhirat, the world of beyond. The shop specializes in cellular technology but also sells electronic ghost-busting devices to assure a peaceful and comfortable household. Indo-nesian advertisements likewise praise so-called *dukun seluler*, cellular shaman, who take care of your mobile phone and necessary accessories, while other shamans directly link cell phone technology with certain forms of magic: after all, both use a certain fre-quency (*Metro BalikPapan* 22 April 2005). Such stories are widespread in Asia and pop-ular culture seems quick to cash in on such ideas. In early 2005, a new mobile game Real: Another Edition was launched in Japan. It is a cellular phone game in which play-ers can earn ghosts by simply capturing them at designated locations by using their cell phone cameras. Every time you catch a ghost a text message is automatically sent to your phone mailbox telling you where to find the next ghost or simply making fun of you and challenging the mobile phone user to look further.

Miller (2001, p. 120), although referring to Victorian haunted houses, points out that certain objects can easily possess their owner: "the objects around us can embody an agency that makes them oppressive and alienating and may in turn be projected in a personified form as the ghosts that haunt us." Ghosts, he argues, become representa-tives of superseded eras and modes of thought that still need a place to dwell. By endowing the cell phone with its own agency this yet incomprehensible technology becomes a possible link between past and future. The mobile phone seems to be the best anchor in a society that is constantly in flux and increasingly mobile in character. Its portability roots a mobile identity.

Conclusion

Phone ghost stories deal with society's oldest fears, that of the unknown. For many Asians, the phone is indeed still a mystery that taps into the popular imagination. The capacity to transmit messages across the ether has always evoked supernatural notions. In societies where communication with the dead is an almost routine affair, the pervasiveness of mobile phones simply draws on these common beliefs. These practices are restricted neither to rural areas nor to lower segments of society. Rather, they are popularly used in shows on national television and in contemporary films, as well as among the rich, famous, and powerful.

Bliss Cua Lim (2001, p. 287) argues that ghost stories call our calendars into question: "The temporality of haunting, through which events and people return from the limits of time and mortality, differs sharply from the modern concept of a linear, progressive, universal time." Not coincidentally, the theme of many supernatural phone stories deal with lingering ghosts or the deceased communicating with the living, the phone often becoming a modern tool invoking the unresolved past to avenge itself or at least to make itself present in the present. Different concepts of time govern such electronic elsewheres, and the ghost in the phone genre, like other ghost stories, blurs time: the past as a ghost appears in the present while the future seems to become increasingly uncertain. Time itself is a dwelling place and hence a common site (e.g., an old house) may contain several temporalities.

In learning to appreciate the new, people seek help from something from the past, here a ghost in the phone, to help deal with these new times, new technologies, and new practices. Ghost stories, like other much-discussed tactics of nostalgia (cf. Chow 1993; Appadurai 1996, p. 76) are yet another strategy of coping with progress. Technophobes may fear technology, but phone ghost stories stand not so much for a fear of modern technology as for a broader anxiety about a rapidly changing environment. These Asian phone ghost stories could be considered a metacomment on modernity itself and the mobile phone becomes one of its main icons.

In this chapter we discuss aspects of new technology that are often neglected in the literature, which itself is largely based on the Western experience of technological change. Asia presents another reality, one where technology coexists with deep beliefs of the supernatural. Despite Asia's varied technological development, from Japan and Korea to Indonesia and the Philippines, a common thread is found. New technologies can coexist and even assist old beliefs. It is inevitable that these two parallel worlds—

the technological and the supernatural—intersect and dialogically engage. While these narratives may be seen as part of a discourse of counter-modernity, they may more accurately be seen as a continuing dialogue of modernity, using elements that have been contingently displaced in its Western variant.

References

ADP. 2004. Polisi Persempit Ruang Gerak Pelaku Peledakan. *Kompas*, September 2004, Metropolitan Section.

ADP. 2005. Terjadi Lagi, Ancaman Bom Via SMS. *Kompas*, January 24, Metropolitan Section.

Agbu, J. F. 2004. From "Koro" to GSM: "Killer Calls" scare in Nigeria: A psychological view. *CODESRIA Bulletin, 2004* (Nos. 3 and 4): 16–19.

Appadurai, A. 1996. *Modernity at Large: Cultural Dimensions of Globalization.* London: University of Minnesota Press.

Augé, M. 1995. *Non-places: Introduction to an Anthropology of Supermodernity.* London: Verso.

Barendregt, B. 2006. Mobile modernities in contemporary Indonesia: Stories from the other side of the digital divide. In *Indonesian Transitions*, edited by H. S. Nordholt and I. Hoogenboom. Yogyakarta: Pustaka Pelajar.

Castells, M., M. Fernandez-Ardevol, J. L. Qiu, and A. Sey. 2004. *The Mobile Communication Society: A Cross-cultural Analysis of Available Evidence on the Social Uses of Wireless Communication Technology.* Research Report Annenberg School for Communication.

Chow, R. 1993. A souvenir of love. *Modern Chinese Literature* 7(2): 59–78.

Cua Lim, B. 2001. Spectral times: the ghost film as historical allegory. *Positions* 9(2): 287–329.

Eisenstein, E. 1979. *The Printing Press as an Agent of Change.* London: Cambridge University Press.

Gergen, K. J. 2002. The challenge of absent presence. In *Perpetual Contact: Mobile Communication, Private Talk, Public Performance*, edited by J. E. Katz and M. Aakhus. Cambridge: Cambridge University Press.

Giddens, A. 1990. *Consequences of Modernity.* Palo Alto: Stanford University Press.

Gunning, T. 1995. Phantom images and modern manifestations: Spirit photography, magic, theater, trick films and photography's uncanny. In *Fugitive Images*, edited by P. Petro. Bloomington: Indiana University Press.

Heidegger, M. 1977. *Time and Being.* J. Macquarie and E. Robinson, trans. Washington: SCM Press.

Katz, J. E., and M. Aakhus, eds. 2002. *Perpetual Contact: Mobile Communication, Private Talk, Public Performance.* London: Cambridge University Press.

Katz, J. E. 2003. Do *Machines Become Us?* In *Machines That Become Us: The Social Context of Personal Communication Technology*, edited by J. E. Katz. New Brunswick, N.J.: Transaction Publishers.

Katz, J. E. 2006. *Magic in the Air*. Piscataway, N.J.: Transaction Publishers.

Kopomaa, T. 2000. *The City in Your Pocket: Birth of the Mobile Information Society*. Helsinki: Gaudeamus.

Licauco, J. 2006. *Philippine Daily Inquirer*, Jan. 3, C4.

Markus, G. 1997. Antinomien der kultur, Lettre Internationale. *Herbst*: 13–20.

Miller, D. 2001. Possessions. In *Home Possessions: Material Culture Behind Closed Doors*, edited by D. Miller. Oxford: Berg.

Mrázek, R. 2002. *Engineers of Happy Land: Technology and Nationalism in a Colony*. Princeton: Princeton University Press.

Pemberton, J. 2003. The specter of coincidence. In *Southeast Asia Over Three Generations: Essays Presented to Benedict R. O'Gorman Anderson*, edited by A. Kahin and J. Siegel. Ithaca: Cornell University Press.

Pertierra, R. 1988. *Religion, Politics and Rationality in a Philippine Community*. Honolulu: Hawaii University Press.

Pertierra, R., E. F. Ugarte, A. Pingol, J. Hernandez, and N. L. Dacanay. 2002. *Txt-ing Selves: Cellphones and Philippine Modernity*. Manila: De La Salle University Press.

Pertierra, R. 2005. Mobile phones, identity and discursive intimacy. *Human Technology* 1(1): 23–44.

Pertierra, R. 2006. *Transforming Technologies: Altered Selves—Mobile Phone and Internet Use in the Philippines*. Manila: De La Salle University Press.

Rafael, V. 2003. The cell phone and the crowd: Messianic politics in the contemporary Philippines. *Public Culture* 15(3): 399–425.

Rheingold, H. 2002. *Smart Mobs*. Cambridge, Mass.: Basic Books.

Sconce, J. 2000. *Haunted Media: Electronic Presence from Telegraphy to Television*. Durham: Duke University Press.

Weidman, A. 2003. Guru and gramophone: Fantasies of fidelity and modern technologies of the real. *Public Culture* 15(3): 453–476.

29

Boom in India: Mobile Media and Social Consequences

Madanmohan Rao and Mira Desai

India is the world's fastest growing mobile market. India's growth has been so strong that in 2005 it became the world's sixth-largest market, and by 2010 should be second only to China. After liberalizing regulations, and against a backdrop of increasing personal wealth, mobile subscriberships are growing rapidly, far outstripping fixed lines. India's mobile phone subscriber base in 2005 grew at an astounding 47 percent to include more than seventy-five million in 2006, and by spring of 2006 had reached eighty-two million. By 2010, the mobile phone subscriber base in India should be more than a quarter billion, with a 24 percent cellular penetration rate. This would be up considerably from the current penetration rate of less than 8 percent of its billion-plus citizens (Reuters 2006).

Research funding for wireless technologies is pouring into Indian R&D labs, numerous wireless startups have emerged, and outsourcing of enterprise wireless application development is playing to India's strengths as an international software resource. With a rich potential content base, a wide range of premium SMS services have been launched. Significantly, several initiatives also have been launched to bridge the digital divide via wireless access, especially in rural areas, but in terms of handset accessibility as well.

In its early years in the mid-1990s, the mobile phone was largely used by the privileged sector of India, due to high device costs and call tariffs. All this changed at the turn of the twenty-first century when India witnessed entry of new players who began sharply competing with pricing and branding strategies underscored by aggressive advertising campaigns. The mobile phone has become not just a white collar tool but increasingly a blue collar and even "collarless" business tool. Unsurprisingly, they have been adopted by affluent youth and managers. More surprisingly for a formerly elite tool, they are now possessed by people in marginally compensated trades such as carpentry, peddlers, pedicab drivers, and domestic servants.

A major factor in the mobile market is the youth segment, for whom the mobile phone has become variously a status symbol, security blanket, and fashion statement. Accordingly, the cellular operators have launched different kinds of branding campaigns targeted at these various groups. For instance, in Karnataka (one of the four southern states of India) its oldest wireless service provider, Spice Telecom, linked up with the state's Department of Posts to market its "Uth" prepaid card and recharge coupons for the youth segment. Other carriers have linked up with television programs and other ways to give a sense of brand to their services.

All this mobile communication is having a tremendous impact on the quality of life in India as well as how Indians participate in various aspects of their society. Here we describe some aspects of how proliferating mobile communication is affecting life in the domains of civil society, media involvement, and some sociological aspects of users and usage.

Civic and Political Impact

India is embracing the idea of electronic government (e-government) and its corollary, mobile government (m-government). The fact that about 70 percent of India's population lives in small towns and villages requires that novel recombinations be explored of traditional and new technology to maximize their joint potential for advanced communication services.

An intriguing example of this recombination is occurring in Karnataka, where the government has computerized all land records. While this is a boon for those seeking economic development and real estate transfers, the data remain stored in servers at district headquarters. These are remote and virtually inaccessible for those in villages who have neither phone lines nor cell towers. To enable local access, a store-and-forward wireless broadband network has been created using a mobile access point (MAP). This is placed aboard standard passenger buses that pursue their regular routes. With MAPs, it is possible to transmit information between district headquarters and a village. In practice, then, the system works as thus:

A villager can request information about their land records (or other services) through a PC in a WiFi-enabled village kiosk [WiFi stands for "wireless fidelity": a radio-based protocol for transmitting information]. The request will be stored in the computer until a bus with a MAP passes and collects the information wirelessly. The information will then be transferred to the district headquarters when the bus is within range of the WiFi-enabled systems based at headquarters. The villager gets their response when the bus "delivers" the information back to the PC in the village kiosk. This can include delivery of land record and related service transactions. (Lallana 2004)

Yet despite their initial utility, this and other imaginative but cumbersome systems will likely be phased out as India's mobile communication infrastructures continue to grow. Carriers expect by 2007 to deploy cellular networks above all of India's 5,200 towns and at least half its 600,000 villages. According to plans, this roll-out should provide coverage for 75 percent of the Indian population. Government ministers, aware of the economic potential of mobile communication, are urging providers to put more phones in the hands of the poor. One recent proposal by India's minister of telecommunications, Dayanidhi Maran, challenged mobile phone companies to begin offering a USD 22 handset (*Cellular-News* 2006). These and other steps are addressing India's digital divide. But even without such initiatives, there is enormous enthusiasm for mobile communication among all segments of India's society, as is discussed below, even though affordability remains a challenge despite ever-dropping rates.

Turning from infrastructure to services, the m-government initiative is allowing much more rapid flow of Internet access. In a growing number of cities, updates about problems concerning public safety or welfare can be delivered via SMS to citizens. Public concern over the ability to notify residents of problems was greatly elevated after the devastating 2004 Indian Ocean tsunami; many believed that thousands of lives could have been saved had their been an SMS warning system in place.

Of course beyond the mega-disaster, there is on-going frustration due to traffic jams as well as electricity and water service disruptions that plague much of India's urban and rural infrastructure. Notices are sent, for instance, in case of water shortages and unscheduled power cuts. SMS alerts can be originated by municipal officials in Indian cities such as Pune. This was put into practice in Bangalore, where the police department sent a city-wide alert via SMS urging public calm when a famous film star was kidnapped by a notorious bandit. And many government schools now disseminate their examination results by SMS as well as by e-mail and online. Hence, for the highly competitive examinations ranging from the Karnataka state's Secondary School Certificate (SSC) board exams to the nationally administered Central Board of Secondary Education (CBSE) board, tens of thousands of anxious student test-takers (and their parents) can receive their exam results via SMS.

An array of other time-saving, convenient services is being rolled out. For instance, India has one of the world's largest railway networks, and many Indian families make at least one trip by train each year. The Indian Railways offers ticket confirmation via SMS for users who sign up for the notification service at the Web site. The Northern Railway has devised an SMS-based system for passenger inquiries. For a tiny fee of Rs 2 (about two U.S. cents) they can get information concerning the arrival and departure of any train in the region. Updates and delay alerts are also part of the system.

Valuable information can also be obtained from public sources. As an example, the New Delhi police have introduced an SMS-based inquiry service for potential used cars buyers. The police offer a helpline to check on the legal status of the vehicle in question. Communication via SMS is becoming two-way: New Delhi's government-run water board has opened a complaint hotline through which users can submit their concerns via SMS.

Until recently it took years to get a phone line in India. Government services would creak along slowly and sloppily, if at all. Against this backdrop, there is high enthusiasm for what mobile communication can offer. This includes greater efficiency in democratic processes. As but one illustration, SMS has been used actively by political parties to campaign in India's recent national election.

In terms of civic culture, and the reproduction of cultural forms, the mobile phone's role seems to be growing. As elsewhere, social relationships are increasingly moderated and expanded (and sometimes contracted) through the use of mobile communication. In India, there is great concern about caste and class, perhaps to a degree unparalleled elsewhere. Parents are often deeply involved in supervising the behavior of their children as well as in making marital arrangements for their children (though with the coming of modernity and economic development, reflected in the mobile phone as icon, this role seems to be shrinking). Hence, there has been a tremendous proliferation of mobile dating sites. As well, mobiles have been used for flirting and circumventing parental authority. Mobile divorce, especially among the Muslim community, has raised problems precisely similar to those reported in Malaysia and Singapore (see chapter on Singapore, 21 in this volume, for details).

Although we do not have direct survey data concerning mobile use, a December 2005 online survey of 6,365 Indians is telling nonetheless in terms of both mobile usage for news updates and more generally for the rich nature of spiritual pursuits of Indians. In the first instance, 17 percent of the respondents use their mobile phones to read news. This is further evidence indicating that there is significant cross-over between media platforms. In the second instance, namely highlighting the superstitious and religious interests of Indians, the survey found that 26 percent look online for astrological predictions, 18 percent for religious or spiritual information, and 14 percent for matrimonial offerings (IAMAI 2006).

The inclusion of mobile phones in areas of Indian life is not proceeding without some trepidation. Issues of fraud, disruption, and appropriate conduct afflict Indians just as much as they do people in other countries, if not more so. While India has a freer environment for political expression than some of its Asian counterparts, an unusual case arose in December 1994 with the arrest of a New Delhi schoolboy who was responsible for filming and distributing an MMS clip featuring oral sex. While this case received international attention, it was by no means the last. New cases of highly inappropriate behavior being captured by mobile technology continue to proliferate (Rizwanullah 2006). Yet such highly visible cases allow a public reaffirmation of values, and allow public chastisement of the offenders.

There can be offenses of a lesser sort, unofficial but that nonetheless offend social arrangements and public sensibilities. As illustration, a piquant combination of observations is given in figure 29.1 concerning mobile communication in India's parliament highlights the informal and intrusive nature of mobiles. Particularly noteworthy in this context, though is that India's parliament, like Germany's, has technical means that are supposed to prevent the use of mobile communication. Obviously, it does

Figure 29.1
M-etiquette.

NEW DELHI: A cell phone ring during business hours in the Lok Sabha on Monday irked Speaker Somnath Chatterjee, who then launched an unsuccessful search for its owner.

As the phone rang, Chatterjee asked whose phone it was and directed the owner to hand it over to the marshals. Although the marshals were walking up and down, no MP was seen handing over a phone to them.

Although there is no rule barring MPs from bringing their mobile phones inside, Chatterjee had requested them to switch them off. The jammers inside the parliament complex make them inactive automatically.

But MPs are quite often seen fiddling with their phones inside the house, more so the young MPs who remain busy playing games on their most modern equipment.

"I never switch off my phone. But it rings only when jammers do not work," said an MP. *India Times* (2006)

---

not always work, and even when it does, there seems to be plenty of distraction offered even by noncommunicating mobile devices.

In a final note on the civic and social integrative aspects of mobile communication, it is clear that mobile devices have already become well integrated into religious practices. Indian festivals like Diwali, Dugra Puja, Eid, and Navratri are convenient occasions around which to host SMS gaming, contests, greetings, and downloads. Reliance IndiaMobile reported ten million Navratri-specific downloads in 2003, including *garba* event listings and recipes. And more than fifty thousand users responded to an instant poll on the question "Up to what time should playing loud music be allowed during Navratri?" Temples and various religious orders have begun integrating mobile communication into their institutional processes including announcements, reminders, and the receipt and delivery of prayers.

Media Impacts: SMS Polling and Beyond

Mobile communication processes of course operate alongside many different media, but as the technology become widely available it is increasingly being intermixed with them. To some extent, it is even becoming integrated with them. TV viewing is generally a strongly family-oriented activity in India. It has become paired with SMS, giving television programs more of an interactive feel that seems to impress India's media-obsessed viewers.

Popular TV shows have assimilated SMS into their programs by offering polls, games, and quizzes. These seem to provoke a highly positive response. Hence, a news program that ran a viewer poll about whether a particular minister should be allowed to stay in office drew quickly 130,000 SMS votes. The wildly successful TV quiz show *Kaun Bange Crorepati? (Who wants to be a millionaire?)* launched a play-along service via SMS. In the first few seconds on the air, it was deluged with more than 500,000 messages, bringing down the SMS processing center.

SMS of course is increasingly incorporated into the lifestyle of Indians. One indicator, which also reflects the branching out into mobile content from "old media" sources, is the venerable print newspaper, the *Times of India*. In 2002, its portal, *IndiaTimes*, launched an infotainment service with the shortcode 8888 (in Asia often considered a lucky number). Available to GSM as well as CDMA networks, it receives an estimated 1.8 million requests every day for SMS horoscopes, dating, cricket information, jokes, and other branded content. T. N. Prabhu, *IndiaTimes* managing director for SMS content and downloads, says he classifies his service users into four personality types: community activist, info-seeker, fun-monger, and buyer.

Sports are quite important to Indians, as for most nations. Cricket, though, stands head and shoulders above other sports. In fact, it contributes more than 60 percent of the SMS traffic in India, outside of the person-to-person messaging category, according to Chennai-based Badri Seshadri, CTO of the Web and wireless cricket information company CricInfo (www.cricinfo.com). Services currently offered by CricInfo include live text content for SMS, live text content for IVR operation through text-to-speech conversion, audio content, SMS alerts, and WAP content (Rao 2003). Also, CricInfo provides real-time delivery via SMS of international scores to numerous operators in India. To support this intensive activity, it has developed relationships with mobile operators and intermediaries in the UK, Australia, Pakistan, and Sri Lanka.

Tapping into India's cricket craze from another angle, namely mobile gaming, operators like Hutch, one of India's largest networks, offer Java-based mobile games including an exclusive Rahul Dravid Game. (Rahul Dravid is a famous Indian cricketer and member of the Indian cricket team since 1996.) In early 2004, this enormously successful game was clocking an average of 120,000 mobile games downloaded every month by Hutch users. This represented a fiftyfold increase from one year earlier. Games can be downloaded for a price of Rs 50 to Rs 99 or about USD 1.10 to 2.25. Neither is the interest in cricket restricted to SMS. MMS is getting into the act. Hutch offers live MMS video replays of sports and entertainment events, and has found especially popular the highly followed India-Pakistan cricket series. In fact Hutch offers more than 400 mobile gaming products (*Indiantelevision* 2004).

Hutch is by no means alone in generating cash flow from mobile services, and providers are exploring many potential targets. Some have launched a mobile version of the *tambola* (lottery) game. SMS-based lottery services have much potential in India, as the offline lottery business is estimated to be worth more than Rs 50,000 a year

(USD 10 billion) annually. The idea is that mobile phones can tap into segments unreached by the PC-based Internet or VSAT terminals, let alone physical locations.

As TV channels claim they have evolved from being mere content-providing media to interactive media through tools such as the Internet and SMS, industry observers say SMS has become a major source of revenue for them. In fact, we have been told that the offering TV channels get 30 to 50 percent of the revenue from each SMS received. When multiplied with the total number of SMSs received, this is a great deal of money. Despite the lucrative nature of SMS participation, however, TV channels regard SMS less as a revenue generation model and more as a marketing tool to increase viewer numbers and involvement. According to one network executive,

SMS as a source of revenue generation is secondary. It is more importantly a marketing tool to get viewers hooked on to the channel. In fact, we use it as a medium to interact with our viewers. It enables us not only to send information to the viewers, but also get valuable suggestions from them.

SMS is certainly an effective marketing tool as it is a sure-shot touch point with the consumer, where one can directly reach into the consumer's inbox and get due exposure. The medium allows for a lot of flexibility and innovation. We have been using SMS as a marketing tool for all our initiatives ranging from interaction to information, connecting with consumers and getting valuable insights. (Shashidhar 2005)

Rediff Newshound provides a service on WAP-enabled mobile phones. Claimed to be the first service of its kind in India, Newshound provides a capsule of news headlines across categories such as business, entertainment, cricket, and health updated every five minutes. These headlines are aggregated and categorized from more than a thousand news sources. The Rediff Newshound service can be accessed free of charge using WAP on any GPRS-enabled phone, the company said (Bhattacharyya 2006).

Viren Popli, senior vice president (Interactive Services), Star India, says the channel is planning to look at GPRS in a big way in the future:

As we go ahead, we will try to enable viewers to download news clips, recap of fiction and a variety of other information on their mobile phones. During the Mumbai floods, we had used a lot of footage sent by people from their mobile cameras. For all you know we may soon have mobile reporters! (Shashidhar 2005)

The use of value-added services may still be low in India, but when it comes to sending photos via MMS, Indian males are the most avid users in Asia. And those on post-paid connection are more likely to use MMS for sending photos than their prepaid counterparts, reports Rashmi Pratap in Mumbai (Shashidhar 2005).

"You are more likely to MMS a photo in Asia if you are an Indian, male, aged 25 to 36 and on a post-paid contract," according to an international survey on mobile user habits. In India, about 37 percent subscribers on postpaid contracts are likely to MMS photos, while only 27 percent on prepaid connections will do the same, according to the survey by SmartTrust, a mobile device management firm. Prepaid users form a vast majority of consumers in India and most of them are younger and have limited spending power.

"This explains the larger number of postpaid users opting for MMS," says the author of the survey, Tim De Luca-Smith, communications manager at SmartTrust. As against India's 37 percent, only 27 percent of those on postpaid connections in China will use the MMS for sending photos and 15 percent in Hong Kong.

But nearly 50 percent of mobile-using Indians store photographs on their phone, while 20 percent store it on their computers. A total of 30 percent send it via MMS, while around 17 percent use the computer to distribute photos. Indian users are also most likely to print their photos than any other Asian market covered in the survey (*IndiaTimes* 2005).

Other forms of mobile services and devices are proliferating. Take one example: the mobile game. Industry estimates are that the mobile gaming market in India will grow from USD 26 million in 2005 to USD 336 million by 2009—a growth of over a tenfold in a few years. As evidence of the enthusiasm, the inaugural N-Gage QD Championship, which Nokia organized in 2004 to promote its N-Gage QD game decks, garnered participation from 26,000 gamers from 40 cities across India. Unlike in the West where the PC platform is a preferred mode for gaming, the Indian market is increasingly mobile-platform-centric. One of the reasons for this is the cultural nature of the Indian family where using media is a collective activity, and mobiles give individualistic freedom of operation.

In sum, the Indian situation reflects increasing integration of mobile technology across a spectrum of activities. Significantly, it is integrating into older forms of media consumption rather than displacing them. Next we examine some detailed survey results concerning the mobile phone user in India.

Social Impacts: Family and Youth Dynamics

There are some intriguing cultural differences between and within various countries in terms of the reception and use of mobile communication. In this section, we briefly highlight an industry-sponsored study and review findings from one of our own.

In 2004, Siemens surveyed mobile lifestyles in nine countries—China, Malaysia, Thailand, Indonesia, Philippines, Taiwan, Australia, Hong Kong, and India. More so than in the other countries, Indians consider their phone to be an extension of their personality. Fashion consciousness was reflected in the figure that 68 percent of Indians said that chasing after the newest models is an essential way of keeping up with the latest trends. As to the appropriate behavior dimension, 70 percent said that they pick up their phone wherever they are. Dependency was reflected by the statistics that

63 percent responded that they have to go back home if they forget their mobile phone. As to so-called cell phone addiction, almost half reported that they constantly check mobile phones if they do not receive an SMS or call for a long time (*India* 2004).

In 2004, a survey was conducted of mobile phone users in Mumbai by chapter coauthor Mira Desai. It aimed at examining socioeconomic and behavioral correlates of mobile phone usage. The reason we chose Mumbai was that it seemed emblematic of rapidly modernizing cities and typifies the new India (Patel and Thorner 1995). Data were collected using purposive sampling. Researchers intercepted mobile phone users on the streets of Mumbai City as well as its suburban areas. They sought to use quota sampling to assure gender balance and cover diverse user populations. In all 320 mobile users interviewed for the study, all of whom identified themselves as owners of the devices. Their age ranged between 14 to 80 years with a mean of 31 and a standard deviation (SD) of 13.78.

As shown in table 29.1, the mobile phone is getting diffused in the lives of Indians irrespective of their age, class, or religion. As table 29.1 indicates, people belonging to various religious, educational, or marital status used mobile phones. ("Others" category in religion includes Sikh, Buddhist, or Jewish.) A large proportion of the users,

Table 29.1

Mobile Users in Mumbai

	Percent
Religion	
Hindus	83
Muslims	6
Christians	4
Jains	2
Others	4
Education	
Illiterate	1
Secondary School Certificate/ Higher Secondary Certificate	18
Undergraduates	24
Graduates	32
Postgraduates	25
Marital status	
Unmarried	55
Married	43
Single/divorced	2

Source: Authors
N=320

Table 29.2

Distribution of Family Monthly Income for Mobile Users

Family monthly income	Percent
Rs.<10,000	15
Rs. 10,000–25,000	21
Rs. 25,000–55,000	37
Rs.>55,000	27

Source: Authors

N=320

Table 29.3

Percentage of Possession of Status Products of Mobile Users

Status product ownership	Percentages (multiple responses allowed)
Home computer	73
Laptop	15
Credit card	64
Car	63
None of the above	14

Source: Authors

N=320

however, are young, unmarried, student or higher-educated citizens. At the same time, while it is impossible to generalize, the high proportion of students, and even three people who reported being illiterate, is noteworthy.)

To understand the economic profile of mobile users, the family monthly income and ownership of status products like computer, laptop, credit card, and car were queried. As evident, though the mobile is owned by users from diverse economic backgrounds, it is still a commodity heavily weighted to the economically more prosperous classes (table 29.2), with a quarter of mobile users having monthly incomes of more than Rs 55,000 (USD 1,250). Still, 15 percent of the respondents had a monthly family income of less than Rs 10,000 (USD 225). There was heavy computer ownership among the respondents, though 14 percent did not own any of the status products; these results are detailed in table 29.3. Some families had up to eight mobile phones among their members, and as expected they belonged to the high income group.

The average months of mobile phone usage was 23.05, indicating that the adoption had taken place mainly in the past two years or so, but the SD was 24.11, suggesting there was a long tail representing early adopters. The users interviewed included two users, a male and a female in their fifties, who had started using a mobile in the last month. From our sample, and as reflected in other surveys, males tend to have had longer experience owning the technology compared to females.

We used the survey to probe attitudes toward mobile phone technology. When asked about the most important use of mobile phones in their life, the majority responded that it was for staying connected (70 percent). Next in frequency was security (14 percent) and job-related (12 percent) and for status (7 percent). The remaining answers included uses such as making a fashion statement, status symbolism, peer pressure, and convenience. (A few respondents gave more than one reason.)

Few users reporting negative influences of mobiles said they have become "careless," it "has made them addicted to talking/SMSing," "it is waste of time," they "can not sleep properly due to mobile," they have "increased expenses," they are "tired of bill," it is an "unwanted tracking device," or it is a "nuisance." "Others" in changes due to mobiles included diverse responses such as they "feel relaxed and anxious because of mobile," "become more talkative," "feel confident," "become responsible and inform home if late," and have less public phone call expenses, and the "mobile is part of my life," they "made life easy," they "have become mobile because of mobile phone." "Other" in uses of mobiles included they "can talk while traveling," it "helps because stay in hostel," "style use of mobile," and it is "time saving."

In terms of behavior, the proportion of users switching off the mobile "whenever not to be disturbed" increases while those "never" keeping the mobile switched off declines as function of increasing number of the years of mobile ownership. This could be a function of maturation and aging, or perhaps change in usage norms over time.

There certainly was evidence that the mobile phone is an important part of the users' life. To examine dependence of users on the technology, we asked "can you live without your mobile" in response to which close to half (41.9 percent) of the users answered that they could not. Sixty percent of the users reported that they never leave their home without the mobile phone. All but 5 percent of users reported replying to every incoming call. More than half (55 percent) of the users say they never switch off their mobile, even at night. In terms of adhering to rules, it is illuminating that only about one third switch their mobiles off when they are in places where doing so is required.

There are other less direct ways that the mobile is affecting the lives of users. For 10 percent "meeting friends" had "lost meaning" as they talk on mobile phones and hence do not need to meet in person. About one third accepted that mobile phones had made them casual about things as "they do not give/take details before leaving home," as they feel that they are connected. For half of the users, maintaining a printed or written telephone book is an outdated activity as their mobile phone is doing that task.

Another area of psychological implications of mobiles was anxieties associated with the technology. The users were asked two questions: do you "check your mobile when others' cell phone ring" and do you have "anxieties about unwanted/missed calls on your mobile." A majority (65 percent) of the users check their mobile phones when the phones of others ring and many (59 percent) face anxieties about unwanted/ missed calls. But when asked if they feel the mobile phone as a nuisance in their life only 36 percent agreed.

Sociologically speaking, a third of users saw mobile phones affecting their family relationships. They offered diverse evaluations of the nature of the impact. Some said it was positive, such as getting to speak to the family members more often due to mobile access; feeling bonded, secure, reassured, closer; remaining in touch with family members who are oceans apart; family knowing one's location; speaking to the family at any time. Negative expressions indicated ones mostly by youngsters, such as "my parents keep check on me" and "I have to lie to them," "my mum keeps calling," "I spend all time on phone instead of people at home," or "spend less time with my family as I speak more on phone." One sixteen-year-old reported that he has to keep his father happy so that the father will pay the boy's mobile bill on time! A thirty-seven-year-old working mother stated that she gets to speak to her son more on the mobile phone than in person.

Respondents were asked how mobile phones had influenced their life. Two thirds said its effect had been positive. Effects offered included always being connected, having status among friends, maintaining privacy, and being able to talk whenever to whomsoever as desired. Also mentioned were remaining in touch with friends, family, and boss, and increased business. The negative influences were reported as lack of privacy, having to attend to unwanted calls, being forced to keep with the trend, being asked by parents to keep it so they could remain in touch, and being forced to attend all calls anywhere, anytime, especially due to job requirements.

Conclusion

On the civic and political fronts, SMS and other forms of mobile communication are increasingly used to inform and persuade citizens as well as to provide services to them. These technologies are also offering new feedback channels so that officials can identify and address problems. Emergency notification and contact are important if infrequent uses; convenience and social cohesive uses would seem to promise much. On the media front, SMS and mobile communication is helping service providers, including those who make television programs, offer more interactive services for India's media-fixated consumers. Mobile communication is also being used in a variety of social capacities including courtships and managing on-going relationships. Many uses that have sprung from the public have been directed toward both creative and disruptive purposes.

As to the cultural tapestry, our first-hand research of mobile phone users in Mumbai suggests that the most important use of the mobile phone in their life for the majority of users was connectedness, followed by security and as a job requirement. Several measures indicated that a high level of social and psychological dependency on the mobile has developed. Yet most users have a positive attitude toward the technology's effect. They feel it has eased their ability to speak to family members and feel more bonded to friends. Among the negative effects, we can report (with an admittedly wry smile) that many negative views of the mobile phone offered by youngsters focus on the mobile phone giving their parents excessive control.

Given its fast growth rate and flexibility, mobile communication is in a unique position to help a developing country such as India. Also with its IT-powerhouse status, India's mobile ecosystem is an ideal testbed for media and communications researchers. The diversity in India's socioeconomic fabric makes it ripe for research and analysis on mobile impacts on urban and rural users, higher and lower economic groups, and educated and illiterate users. The lessons learned can have applicability to both more-developed and less-developed nations, and to the unfolding processes of increasingly jumbled cultures.

References

Bhattacharyya, P. 2006. Rediff India launches "Newshound" for mobile phones. http://www.dmeurope.com/default.asp?ArticleID=14033.

Cellular-News. 2006. Explosive growth continues for India's mobile-phone market. http://www.cellular-news.com/story/16396.php.

India. 2004. Cell phones; indispensable to modern life: study. http://www.webindia123.com/news/m_details.asp.

Indiantelevision. 2004. Hutch expands mobile gaming market with 400 products. http://www.indiantelevision.com/mam/headlines/y2k4/apr/aprmam18.htm.

IndiaTimes. 2005. Indians top in sending photo via MMS. IndiaTimes Infotech. http://infotech.indiatimes.com/articleshow/msid-1312819,prtpage-1.cms.

IndiaTimes. 2006. Mobile phone ring irks LS Speaker. http://timesofindia.indiatimes.com/articleshow/1448036.cms.

Internet and Mobile Association of India (IAMAI). 2006. Report on online banking. http://www.iamai.in/IAMAI_Report_on_Online_Banking_2006.pdf.

Lallana, E. 2004. E-government for development. http://www.e-devexchange.org/eGov/mgovapplic.htm.

Patel, S., and A. Thorner, eds. 1995. Bombay—Mosaic of Modern Culture. Delhi: Oxford University Press.

Rao, M. 2003. *The Asia-Pacific Internet Handbook, Episode IV: Emerging Powerhouses*. Delhi: Sage.

Reuters. 2006. India adds 1.11mn CDMA users in Feb. http://www.ciol.com/content/news/2006/106030502.asp.

Rizwanullah, S. 2006. Aurangabad students held for oral sex MMS. *The Times of India*, March 8, p. 12.

Shashidhar, A. 2005. The medium and the message. *The Hindu Business Line, Internet edition*, Aug. 25. http://www.thehindubusinessline.com/catalyst/2005/08/25/stories/2005082500010100.htm.

30

Mobile Games and Entertainment

James E. Katz and Sophia Krzys Acord

Studies of mobile gaming can yield insight into the effects of proliferating mobile technology at the levels of individual psychology, social organization, and the public sphere. Often dismissed as mere entertainment, mobile games, like the mobile phone itself, have continued to blur boundaries between work, leisure, family life, and public space. This chapter examines the phenomenon of mobile gaming (generally though not exclusively games played on mobile phones), from individual, stand-alone games to mass, location-based gaming. We also briefly inspect the burgeoning field of mobile video entertainment, including TV and "mobisodes." In interrogating both the demographic and qualitative aspects of mobile entertainment, this chapter proposes conceptual dimensions for understanding mobile gaming, including its consequences for social life.

Games are naturally occurring learning environments, which "provide a representational trace of both individual and collective activity and how it changes over time, enabling the researcher to unpack the bidirectional influence of self and society" (Steinkuehler and Williams 2006, p. 98). While play creates social order, it simultaneously is "a free activity standing quite consciously outside 'ordinary' life as being 'not serious,' but at the same time absorbing the player intensely and utterly" (Huizinga 1950 [1944], p. 13). Indeed, from some vantage points, play is even considered morally suspect: a pure waste of time, skill, and resources (Caillois 1961 [1958], pp. 5–6). The pure essence of play, as freedom, spontaneity, and physiological and psychological intensity, provides a unique means of unpacking rituals of mobilization in the twenty-first century.

In early human history, and still today in hunter-gatherer societies, gaming involved necessarily mobile players and simulated hunting, fighting and survival skills. As societies became sedentary, so too did many gaming practices. However, in the twentieth century, Western society dramatically remobilized (via personal transportation and communication technologies), vastly expanding physical mobility, with concomitant sociological effects (Urry 2000, p. 53). While *mobilization* is the process by which mobile technologies are folding themselves into the fabric of our economies, social lives, and communities, the resulting individual and collective displacements are themselves new opportunities for creativity. In fact, the individualistic essence of the mobile phone has already altered social interactions to be more gamelike, according to Kopomaa (2000), implicating the essential play elements of real-time action and surprise.

Mobile Gaming: Global Trends

In 1997, Nokia embedded the game Snake in their mobile phones (Castells et al. 2004), thus launching the mobile gaming era. Geographically embedded multiplayer games, such as Botfighters, were popular in Europe and Asia by 2001, but the mobile gaming industry did not gain a foothold in the United States until Sprint launched the mobile gaming community Game Lobby in late 2003. Despite initial enthusiasm from experts, including predictions that 80 percent of all Western wireless users would play mobile games (Kleijen, de Ruyter, and Wetzels 2003), the actual uptake has been far more modest.

The Asia Pacific region (primarily Japan and South Korea) has led the world in mobile gaming adoption, followed by Europe (particularly Scandinavia, the UK, and Italy) (Chau 2006). Only 11 to 12 percent of U.S. wireless subscribers play games (Rainie and Keeter 2006), versus 40 percent of those in Korea and even higher numbers in Japan (M:Metrics 2006). Messaging services only account for 30 percent of mobile revenue in Japan and Korea, while they generally account for 70 to 80 percent of revenue in Europe and 80 percent in North America (Sharma 2006). Although the cultural use of the mobile phone as a communicative versus leisure device may be partially explained by the availability or cost of technology, it is also strongly related to issues of lifestyle. Moreau, Sanchez, and Niu (2004) attribute this phenomenon to the Japanese enthusiasm for arcade and console games, the Korean attachment to PC games, the successful launch of mobile data networks in northern Asia, ample phone subsidization, and the relatively unfragmented handset market in those countries. The reason gaming has taken off in Asian countries, mGAIN (2003) believes, is that BREW, HTML, and Java games enable a richer gaming experience compared to SMS and WAP games that work in the GSM-network in Europe and the United States. This of course implies that Asian gamers are looking for a mobile gaming experience that is more immersive and similar to console gaming.

Cultural Variations in Gaming Lifestyles

Gender

Contrary to stereotypes and trends in console gaming, survey data from 2003 to the present confirms that women are as likely to play mobile games as men (McAteer 2005; I-play 2005; M:Metrics 2006). However, these same sources note that women are significantly less likely than men to download mobile games (nearly two to one), and males eighteen to thirty-four are the most active game downloaders (M:Metrics 2006). These data are explained in several ways. Women gamers may favor one game consistently, while men constantly look for something new (Ruff 2004). Or perhaps women prefer card, puzzle, and arcade games, which tend to be embedded on the mobile device, while men prefer more action-oriented games, which must be downloaded (McAteer 2005). However, we offer a third (nonorthogonal) possibility: as has happened with many other technology-use practices, there is simply an initial gender gap that will diminish over time.

Age
Although mobile gaming is primarily a youth-driven activity, it is slowly gaining widespread popularity. In a detailed examination, the 2006 Pew Internet and American Life survey found that mobile game use is directly related to age: of individuals with game-enabled mobiles, 47 percent of those under 30 play games, 21 percent of those 30–49, 17 percent of those 50–65, and only 8 percent of those over 65 (Rainie and Keeter 2006). However, when individuals without game-enabled mobiles were asked if they would like the capability, the results were inversed—only 9 percent of those under 30 replied favorably, versus 18 percent of those over 65, indicating that at least one quarter of the mobile gaming age discrepancy is explained by the fact that younger people are simply more likely to have game-enabled mobiles. In addition, the average age of the mobile gamer is older in Asia, ranging from 19 to 35, confirming mGAIN's (2003) analysis that game playing has been stigmatized as child's play in Europe and North America, while electronic games are part of the overarching media culture involving all ages in Asia.

Race and Ethnicity
When compared with typical U.S. wireless subscribers, mobile gamers are more likely to be Hispanic, black, or Asian (NPD Group 2005). The 2006 Pew survey found that 20 percent of white individuals with game-enabled mobiles play games, versus 29 percent of blacks and 40 percent of English-speaking Hispanics (Rainie and Keeter 2006). As with other leisure practices, these figures may be significantly related to patterns of urban life and pre-existing leisure activities.

Gaming Preferences
Casual arcade, puzzle, and card games such as Tetris and Bejeweled consistently top all demographic brackets in the United States, while action games, particularly multiplayer games, are more popular in Europe. However, the Swedish game producer It's Alive encountered great resistance when it launched location-based Botfighters in Germany and Denmark, both countries with strong traditions of personal privacy that

were reluctant to embrace a game based on identifying the geographic locations of other physical players (de Souza e Silva 2006). Mobile users in Japan and South Korea are also the most active game downloaders—over twice as active as North Americans and Europeans (Menon et al. 2005; M:Metrics 2006). In addition, it must be noted that less than half of U.S. mobile subscribers have devices capable of playing complex games, and likely belong to the same group of young males who already invest in electronic gaming.

Gaming Habits

Mobile gaming has multiple foundations: prior gaming practices, mobile lifestyles, social attitudes toward gaming, and cultural perspectives on the mobile device. Consumers with prior gaming practice on other devices are twice as likely to play mobile games, notably handheld gamers and console gamers, and PC and Internet gamers to a lesser extent (NPD Group 2005). Mobile games seem to have initially followed the model of Internet or PC games, drawing large crowds of women and teens, and requiring an overall low investment in gaming equipment and time resources.

Mobile gamers are more likely to be those with more mobile lifestyles. They universally play games to kill brief periods of time or alleviate temporary boredom (Moore and Rutter 2004), with average gaming sessions around eleven minutes (NPD Group 2005). However, while Americans tend to make calls to fill in these free moments (Rainie and Keeter 2006), mobile gaming in Europe and Asia is linked to the greater use of hands-free public transportation (Geser 2004; mGAIN 2003). It is of little surprise, then, that individuals who rely heavily on their cell phones are twice as likely to play games as those with a landline as well (Rainie and Keeter 2006; NPD Group 2005).

Despite lesser overall frequency, Americans are more competitive mobile gamers than Europeans; they are twice as likely to play to beat previous high scores, and tend to play more often and for longer periods (I-play 2005). While a growing segment of American gamers is likely driving up these usage statistics, Japanese and Korean gamers of all ages place high, personal values in mobile gaming, viewing it as enjoyment and social activity (Fife et al. 2006; Kang, chapter 31 in this volume). In Asia, gaming may be viewed as a general, social lifestyle, whereas in Europe or the United States it is characteristic of a hardcore gaming sector, the very young, or people looking for distraction during occasional free time.

Finally, cultural views about the mobile device are also integral to understanding gaming practices. The majority of North American mobile users view their mobile as a device for work, communication, and voice telephony (Menon et al. 2005, p. 20). In contrast, the Japanese generally perceive mobiles as singular devices simultaneously enabling communication, entertainment, and leisure (Fife et al. 2006). Hence cultural, environmental explanations trump other interpretive frameworks.

In summary, usage data demonstrate that game playing is a widespread source of pleasure across many demographic categories, not the restricted domain of boys and young men as is typically represented in the media. Rather than being an extension of PC and console games, mobile gaming has its own ethos and attractions. Although Asian and European mobile gamers are more interested in multiplayer action games and gaming as leisure activity, Americans overwhelmingly see mobile gaming as a casual (though serious!) way to pass time.

Mobile Gaming Typologies: New Social Modalities?

There is evidence that mobile gaming is becoming a more social technology as the industry evolves. Although originally thought of as a personal technology (Kleijnen, de Ruyter, and Wetzels 2003), mobile gaming's popular developments revolved around simpler games that let gamers "chat" with each other, engage in tournaments, and post scores on community boards. Three main social modalities emerged through which people engage in mobile gaming.

Hardcore Gamers

Hardcore gamers are more likely to be young and male, have a console gaming background, see gaming as a leisure activity or hobby, prefer complex action and adventure games, download games, and be swayed by games with advanced graphical and design capabilities (Anderson 2002; Ruff 2004). They are isolate-achievers, playing games to challenge themselves and win.

Casual Gamers

Casual gamers, who have always composed the majority of the market, are more likely to be distributed along the age dimension and between the genders. They see mobile gaming as a way to pass time, and prefer easy-to-use, fun games (Anderson 2002; Chau 2006). For these users, games fall into the pattern of digital technologies that function to drive out the ennui of daily existence (Turkle 1995). However, this solitary play is only productive of culture in a limited way (Huizinga 1950 [1944], p. 47), and the key point of the mobility paradigm is that mobile people will have the opportunity to create new, unique experiences. Thus, it is unsurprising that social gaming is growing—fueled by converts from casual mobile gaming, console gaming, and PC gaming, who are all interested in the new communication opportunities afforded by mobile technology.

Social Gamers

This third and rather innovative group, the social gamers, composes roughly 40 percent of the current market (Cifaldi 2006). Social gamers are more likely to be women and to be interested in making connections, communicating, and being part of a community, both online and in physical space (Ruff 2004). Perhaps the strongest evidence of the collective use of mobile technologies and the construction of social networks is the development of multiplayer, location-based games in the mold of Botfighters. While there have been some experiments in this genre of spontaneously collaborative, always-on, and massive multiplayer games, smaller situated gaming experiences are the norm for experimental pervasive game design (McGonigal 2005). These games often have a one-off deployment, and are designed for specific local communities. There are some successful geopositional amusements, which are an offshoot of treasure hunting. These often combine GPS (geopositioning location technology) with the search for trinkets. One leading game site, geocaching.com, claims as of October 2006 to have more than thirty thousand active members and more than three hundred thousand "geo-caches" (stashed trinkets) around the globe. These activities can sometimes cause public bemusement or even serious alarm (Vogel 2005). While a fun hobby, there seems to be only limited adoption of massive social-location gaming despite continuing, often-high visibility experiments (see McGonigal 2005 for a description of some of these).

To summarize, games, and now mobile games, allow people to pursue three main types of relationships with competitive activities: hardcore, casual, and social. Mobility allows users ascribing to any of the three gaming modalities to play games in new physical and emotional environments, and in the case of social gamers, introduces an abundance of new human material. As a result, through mobility, broad new vistas of opportunity for physical and social gaming interaction are created, and new, multiple nuances are possible as players interact with their environment in very different ways via games.

While populations of casual and hardcore gamers will never disappear, our subsequent interest lies with the expanding social gaming modality and the prospect of mobile gaming as a uniquely novel technology. The key attribute of mobile gaming is that it combines gaming engagements with the mobile interfaces dedicated to communication between players (such as SMS, text messaging, and mobile chat), as well as locating them in the intimate mobile device (Licoppe and Inada 2006). Multiplayer public activities may remain less popular in the United States than in Europe and Asia presumably due to time, technology, and lifestyle constraints. Despite this variation, ethnographic data from Europe and Asia are vital in understanding the potential consequences of these games for public space, individual psychology, and social life.

Sociological Trends through Mobile Gaming

Baber and Westmancott (2004) demonstrate that merely adding the variable of mobility to a multiplayer card game changes the nature of play and alters the social aspects of gaming. Mobile gaming activities, therefore, have real implications for the mobile phone's impact on individual and social life. Perhaps, as people have different, pseudo-virtual, and often asynchronic competition with other players, more of the pleasure of being a public person is draining out of the public sphere. Or alternatively, public space itself is being transformed through innovative, proactive social practices. To assess these hypotheses, we present some of the ethnographic data on social, location-based mobile gaming, and then propose some sociological interpretations.

Case Studies of Mobile Gaming

Moji in Japan Licoppe and Inada (2006) found that Moji players often took longer detours, or aboveground transportation, to avoid pausing their game during work commutes or running errands. Additionally, couples played together, inventing new modes of gaming that let them specifically interact as a couple (such as spamming each other with SMS to distract each other from picking up virtual objects). Most importantly, mobile gaming transformed encounters with strangers into important, interesting events. Players actively acknowledged others in close proximity with text messages (versus the normal face-to-face polite ignoring, as often occurs between affiliated, but not friendly, individuals in public). Moji players also collaboratively communicated in this way to avoid face-to-face meetings, which presented opportunities for social power dynamics to dramatically alter the gaming relationship (such as an encounter between a young female player who avoids meeting an older male player in a subway car).

Jindeo in Tokyo While playing Jindeo, Licoppe and Guillot (2006) deduced that the screens of the mobile devices became a public space in which connected users were sometimes mutually aware of each other's movements. As for Licoppe and Inada (2006), texting played a vital role in the success of mobile gaming, as players commented directly on their positions and displacements, so as to construe reflexively their mobility as accountable to other players. In particular, there was an enormous interest in mediated encounters, which stimulated brief but intense text messaging. Again, mobility was demonstrated as inherently social in nature, requiring extensive interactional work by the gamers.

Can You See Me Now, by Blast Theory, in Tokyo and Cardiff One player reported that there were moments when the players can hear each other in real space, thus composing sensitive nuances in the virtual-physical relationship with other players. As one player described, "I had a definite heart-stopping moment when my concerns suddenly switched from desperately trying to escape, to desperately hoping that the runner [other player] chasing me had not been run over by a reversing truck (that's what it sounded like had happened)" (Blast Theory 2006). As with Moji and Jindeo, the awareness of and concern for other players was inseparable from any goal-directed play.

Botfighters, by It's Alive, in Sweden For Sotamaa (2002), the gaming experience altered one's surroundings as much as one's surroundings configure the gaming experience. While mobile gaming overwrote new meanings onto public spaces, the emotional territories of a city, such as the "ghetto" or "rich area," impact the gaming experience located there. In addition, there has been one violent experience cited with the use of Botfighters. In 2001, a player vacationing in Sweden located and eliminated five nearby players, who then joined together and physically beat him in retaliation. Psychological identification is so intense in the games that when one Botfighter "kills" another, the latter player can resume play after "recharging" their robot's batteries online (Sotamaa 2002).

Uncle Roy All Around You, by Blast Theory and the Mixed Reality Lab, in London As this game paired online and mobile players together to move around a city and rescue strangers, Rowland et al. (2004) found that the mobile gamers placed great trust in the online gamers, and although the latter were free to hinder the mobile gamers, few did. Rowland et al. (2004) also found a crucial blurring of the distinction between game and nongame—the distorted view of the city superimposed by the gaming interface made the mobile players question their everyday relationship to it, prompting them to cross the usual boundaries of public behavior by getting into strangers' cars and other gaming actions.

The bulk of literature on mobile phones, gaming, and public space (including Ling, chapter 13 in this volume; Koskinen and Repo 2006) takes an interactionist approach to the phenomenon, drawing from Goffman's (1963, and other) notions of individual performance on the public stage, to state that focusing on personal business in a public space sends a message of disrespect for the social gathering. Yet, such highly normative approaches are insufficient for describing new, innovative media use. As an alternative, we offer an ethnomethodological approach, examining mobile gaming as exactly that: the commonsense practice of playing a game, or a way to kill time in public. For Garfinkel (1967), "common culture" is the "the socially sanctioned grounds of inference and action that people use in their everyday affairs and which they assume that others use in the same way" (p. 76). But this common culture is continually altered by every successive action, which contributes to determining this attitude of daily life. In other words, while mobile gamers may play on socially understood codes of nonbehavior (such as using earphones or disobeying pedestrian signs) to signal nonparticipation in public space, in doing so, they are creating new paths of meaningful activity that will come to alter public, social interactions. Rather than be conditioned by available modes of interaction, our individual and collective reality constitutes the world (Mehan and Wood 1983 [1975], p. 189). Instead of reacting negatively to what we perceive as the mobile game player's disregard for established social, public conventions, we need to understand mobile gamers as coproducers of the space that they move through, by examining a number of relevant sociological variables.

Embodiment in Mobile Gaming

Whereas video game players have always demonstrated physical reactions to the virtual events on their screens, mobile gaming takes this one step further: it is a fully embodied activity. The player's emotions are a vital component of the mobile gaming experience, and his or her body is an "input device" into the game (Grüter, Mielke, and Oks 2005). Despite being split between two roles, embodied passerby and equipped player, mobile gaming action is hybridized between the gaming frame and the urban context, which compose a single mediated reality in which the gamer moves and lives (Licoppe and Guillot 2006). Because the mobile device is an incredibly intimate technology, far from being mind-numbing, social mobile gaming engages psychical awareness. As Hall (1966) points out, people's feelings about being properly oriented in space run deep, and thus the mobile gaming experience constructs new relations between the players and their physical environments precisely because the gamers are experiencing these surroundings as embodied individuals.

Gameplay and Intimate Life

Mobile phones were originally hailed as a means to re-establish locality, or shield oneself from the alienation of urban life by escaping into the narrower realm of familiar relationships with close kin or friends (Gergen 2002). In contrast, mobile gaming replaces this private, intimate conversation with practices aimed at communicating with strangers and exploring the wider world. By adding a self-indulgent function to an already-intimate technology, mobile gaming further interferes with what Ling (2004) terms one's "vital fields" (work, family, and friends).

Gameplay and Public Space

Embedded mobile games offer a "third space" beyond the workplace and home for informal sociability (Steinkuehler and Williams 2006), which transforms the mobile screen into a new type of public space of mediated co-presence (Lofland 1998). Yet rather than abandoning physical community, as occurred with the TV (Putnam 2000), the mobile gamer is located in the physical community and uses public space for informal social interaction. Although mobile phones tend to help their possessors transcend urban life by inscribing the game and their interactions over it, the paradigm of mobility means that people will nevertheless experience their surroundings in new, poetic, and surprising ways (du Gay et al. 1997). Directed by the game, rather than their personal habits, players may bring about a revaluing of public space.

Gameplay and Community/Civic Life

Video and PC gaming (Tamborini et al. 2004), as well as the mobile phone (Katz 2006), have been accused of spurring antisocial behavior. So what happens if you combine them? Because of the issues of embodiment already discussed, and the ability to physically locate other players, competition in mobile games could have physically violent consequences, as illustrated with the Botfighters example. However, with this one exception, the ethnographic data demonstrates that mobile gamers are aware of the distinctions between relationships in the game and in real life.

In general, mobile gaming continues to be a relational tool of reinforcement for urban communities of interest, built around the unique notion of "play." Opposed to the cold, impersonal picture of urban life painted by Lofland (1998) and Simmel (1971 [1903]), mobile gamers depend on cooperation for goal-attainment. As Huizinga (1950 [1944]) reminds us, play promotes the formation of social groupings that stress their difference from the common world (p. 49), and thus perhaps a more real danger is that mobile games make a breach between players and nonplayers. As Urry (2000) points out, "corporeal mobility is an important part of the process by which members of a country believe they share a common identity bound up with the particular territory the society lays claim to" (p. 49). The conflations of public and private, physical and virtual gaming space have possible impacts for nationalism and community building in terms of giving different people radically different experiences and usage patterns of public space.

We may ask at this point, is mobile gaming a social technology? The mobile, communicational, spontaneously playful, and physically intimate aspects of mobile gaming are central to its success and all imply one thing for the user: the game-device hybrid becomes a means of physical and emotional intimacy. Although social gamers insist that it is the communicative possibilities of the game that motivate their play, largely using SMS to communicate with fellow gamers, Geser (2004) points out that SMS allows for equilibrated "economic exchange," versus phone calls, which are "social exchanges." Furthermore, the user-centered structure of the mobile communication network means that entertainment is simply an optional practice integrated with the owner's social practice as fits his or her lifestyle (Castells et al. 2004). So is there anything truly social about it? Despite opportunities for interaction via gaming and text messages, and even real-life meetings through location-based games, mobile gaming is still largely linked to online community activity. While there are some examples of creating informal connections with others (what Steinkuehler and Williams (2006) call "bridging" in MMOGs), the bulk of actual communicative practices are temporary and superficial.

The true play spirit is spontaneity and carelessness, marked by tension and uncertainty (Huizinga 1950 [1944], p. 197). Play is distinct from ordinary life (it is sacred versus biological), and contamination by the latter runs the risk of corrupting and destroying its very nature (Caillois 1961 [1958], p. 43). Mobile gaming, then, "throws the ball out of bounds" by conflating games and everyday life. Although the ubiquity of mobile gaming may enhance the key variable of spontaneity, the high degree of centralization found in most gaming environments eliminates some of the pure play quality, reducing play to low-maintenance social interaction or rule-following.

Mobile Entertainment and "Mobisodes"

Interest in the mobile phone as leisure device is being further explored through the development of mobile TV. Again, individuals in the Asia Pacific region are more than three times as likely as elsewhere to use such mobile media (LogicaCMG 2005). In the United States, at least, there appears to be very low use of and interest in video entertainment on mobile phones; only 2 percent of those with video-capable mobiles use the feature (14 percent without the feature would like it), a disposition inversely correlated with age (Rainie and Keeter 2006). However, a London British Telecom trial of mobile TV found that 73 percent of users were willing to pay for the service, and that the average usage was sixty-six minutes a week with a regular phone and three hours a week with a purpose-built phone (*IEE Review* 2006). Again, such variables as cultural activity, interest, and technological availability strongly influence mobile TV adoption.

Worldwide trials of mobile TV demonstrate that mobile TV does not perform a new or unique function in social life. In general, users watched TV to avoid boredom or endure waits (Repo et al. 2003). Entertainment analysts said that the *24* "mobisodes" trialed by the Fox network in 2005 were only successful because they were linked to a favorite TV show. Similarly, in Germany soap operas were the most popular programs for mobile TV by far (not much movement occurs, making the small-screen format less objectionable). While some users employed traditional viewing habits, such as younger users consuming full-length movies, others employed more inventive tactics such as positioning the mobiles on their desks at work or discreetly below the dashboard while driving.

It may be that some users become bored with the content of the videos and TV programs, and worry that the video sound may disturb other people in public areas (Repo et al. 2003; Koskinen and Repo 2006). Others are undeterred by such concerns, or even use video as a pacifier. For instance, one woman in the London British Telecom study used the mobile to quiet her child at the supermarket (*IEE Review* 2006). Public, spontaneous karaoke sessions were also noted as popular in a Finnish study (Koskinen and Repo 2006).

Overall, the results from these studies reinforce our earlier conclusions about mobile gaming: consumers remain concerned that entertainment services will interfere with the mobile's primary use as a tool for social interaction. Thus, it hardly appears as if TV episodes on mobile phones will do for home and public space what the TV did for the family and neighborhood, namely stunt them (Putnam 2000).

A Future for Cooperative Mobile Games?

Ultimately, mobile users appear not to take gaming as seriously as they take communication. Mobile gaming, even for social gamers, seems more about locating oneself in a social environment and less about actually forming relationships with other people sharing that environment. While the communicative function of mobile phones understandably leads people to explore social contact during their gaming experiences, ultimately, it is only a way to kill time for most gamers, and thus such communication, while value-laden, is not transformative.

Cooperative mobile gaming will continue to develop in Asia and Europe, although less than 5 percent of Americans are interested in multiplayer games (Cifaldi 2006). Although there are emerging social and hardcore gaming groups, issues of playability (small screen sizes), bandwidth, accessibility, desirable games, and high costs remain an impediment, especially for older users. In sum, it seems that mobile gaming will continue to develop along the lines of simple, local games with social capabilities.

This leads us to three conclusions: First, while perhaps relevant to the early adoption of complicated digital technologies, popular notions of "gendered technologies" are not highly relevant to the everyday consumption of readily usable technologies. Mobile games have more in common with easy-to-use devices like the TV remote control or microwave, on which people from all demographic groups became rapidly dependent, but for which there remain some variations by gender.

Second, surveys point to possible stratifications along ethnic and racial divides, and certainly gender is a significant factor in predicting usage patterns. In the United States at least, it appears that ethnic "minorities" tend to be heavier users of gaming technology, which is also the case for mobile phone usage in general. Little public debate has ensued concerning a digital divide in terms of mobile amusement, which in any case would appear to have a gap in the opposite direction than is generally discussed.

Third, mobile gaming technologies also do not seem to have the isolating effect pessimists had predicted (just as they had in the early days of the Internet). Socially inclined people like and use them.

New technologies are not adopted because they are new but because they make possible new uses and services (Castells et al. 2004; Moore and Rutter 2004). Connect-the-dots, travel bingo, and license-plate poker games have been part of long road trips for years, just as word search, crosswords, and now sudoku books are common on commutes. Similarly, there have been scavenger hunts and role-playing games for centuries. Even mobile video episodes had their precursors in the nineteenth-century portable slide projectors known as Magic Lanterns. Perhaps what is more likely is that by embracing new digital technologies, children and adults alike can act mature while remaining immersed in a world of gameplay. People can engage in more game-playing while on the move, and, as with SMS, people will explore new ways of expressing themselves and interacting with each other, complementing physical communities (Katz and Rice 2002). Thus, despite some of the social and artistic experiments to the contrary, and the pleasure of pushing limits of technology, mobile games and video really do more of the same in terms of their social functionality. They have not yet produced profound social transformations, and are unlikely to do so anytime soon. They are, however, another element in the realm of the moral and normative order, giving opportunities for people to express themselves and attain status and solidarity in a "public realm," even while the main moral economy continues to operate along the lines of families, friends, and colleagues.

References

Andersen Project Team. 2002. Digital content for global mobile services. http://cordis.europa.eu/econtent/studies/stud_mobile.htm.

Baber, C., and O. Westmancott. 2004. Social networks and mobile games: The use of Bluetooth for a multiplayer card game. In *Proceedings from MCI 2004: Mobile Human-Computer Interaction 2004*, edited by S. Brewster and M. Dunlop. Heidelberg: Springer Berlin.

Blast Theory. 2006. Can you see me now? http://www.blasttheory.co.uk/bt/work_cysmn.html.

Caillois, R. 1961(1958). *Man, Play and Games*. New York: The Free Press.

Castells, M., M. Fernandez-Ardevol, J. L. Qiu, and A. Sey. 2004. The mobile communication society: A cross-cultural analysis of available evidence on the social uses of wireless technology. Report from the International Workshop on Wireless Communication Policies, Los Angeles.

Chau, F. 2006. Mobile gaming aims for mass market. *Wireless Asia*, Sept. 19. http://www.telecomasia.net/article.php?id_article=1744.

Cifaldi, F. 2006. The future of mobile games: CES panelists on mobile opportunities. *Gamasutra*, Jan. 27. http://www.gamasutra.com/features/20060127/cifaldi_01.shtml.

du Gay, P., S. Hall, L. Janes, H. Mackay, and K. Negus. 1997. *Doing Cultural Studies: The Story of the Sony Walkman*. London: Sage.

Fife, E., M. Hillebrandt, K. Chung, and F. Pereira. 2006. The diffusion of networked gaming in the United States and Korea. Paper presented at the 5th Annual Global Mobility Roundtable. Helsinki, Finland.

Garfinkel, H. 1967. *Studies in Ethnomethodology*. Englewood Cliffs, N.J.; London: Prentice-Hall.

Gergen, K. J. 2002. The challenge of absent presence. In *Perpetual Contact: Mobile Communication, Private Talk, Public Performance*, edited by J. E. Katz and M. A. Aakhus. Cambridge: Cambridge University Press.

Geser, H. 2004. Towards a sociological theory of the mobile phone. http://socio.ch/mobile/t_geser1.htm.

Goffman, E. 1963. *Behaviour in Public Places: Notes on the Social Organization of Gatherings*. New York: The Free Press/Macmillan.

Grüter, B., A. Mielke, and M. Oks. 2005. *Mobile Gaming—Experience Design*. In Proceedings from Pervasive: The 3rd International Conference on Pervasive Computing. Munich, Germany.

Hall, E. T. 1966. *The Hidden Dimension*. Garden City, N.Y.: Doubleday & Company.

Huizinga, J. 1950 (1944). *Homo Ludens: A Study of the Play-Element in Culture*. Boston: The Beacon Press.

IEE Review. 2006. Mobile TV proves a hit with users. *IEE Review* 52(2): 15.

I-play. 2005. US gamers play to win. http://www.iplay.com/article.do;jsessionid=D6FAA3DCF612EC0F3786D580CEA10633.tomcat1_iplay1?NID=97&NYID=2005.

Katz, J. E. 2006. *Magic in the Air: Mobile Communication and the Transformation of Social Life*. New Brunswick, N.J.: Transaction Publishers.

Katz, J. E., and R. E. Rice. 2002. *Social Consequences of Internet Use*. Cambridge, Mass.: The MIT Press.

Kleijnen, M., K. de Ruyter, and M. G. M. Wetzels. 2003. Factors influencing the adoption of mobile gaming services. In *Mobile Commerce: Technology, Theory and Applications*, edited by B. E. Mennecke and T. J. Strader. Hershey, Pa.: Idea Group Publishing.

Kopomaa, T. 2000. *The City in Your Pocket: Birth of the Mobile Information Society*. Helsinki: Helsinki University Press.

Koskinen, I., and P. Repo. 2006. Personal technology in public places: Face and mobile video. Working paper 94, National Consumer Research Centre. http://www.kuluttajatutkimuskeskus.fi/files/4866/94_2006_workingpapers_personal_technology.pdf.

Licoppe, C., and R. Guillot. 2006. ICTs and the engineering of encounters: A case study of the development of a mobile game based on the geolocation of terminals. In *Mobile Technologies of the City*, edited by M. Sheller and J. Urry. London: Routledge.

Licoppe, C., and Y. Inada. 2006. Emergent uses of a multiplayer location-aware mobile game: The interactional consequences of mediated encounters. *Mobilities* 1(1): 39–61.

Ling, R. 2004. *The Mobile Connection: The Cell Phone's Impact on Society*. San Francisco: Elsevier.

Lofland, L. H. 1998. *The Public Realm. Exploring the City's Quintessential Social Territory*. New York: Aldine de Gruyter.

LogicaCMG. 2005. May/June 2005 Survey of 1000 adults in each territory covering Europe (Germany, Italy and UK), Asia Pacific (Malaysia), South America (Brazil) and North America (USA). http://corporate.sms.ac/industryresources/news_sports_ent.htm.

M:Metrics. 2006. What ails the mobile game industry? http://www.mmetrics.com/press/PressRelease.aspx?article=20060502-ailsgaming.

McAteer, S. 2005. (X marks the spot.) So Y are games targeted at men? *Wireless Review* 22(6): 26.

McGonigal, J. 2005. SuperGaming! Ludic networking & massively collaborative play. http://www.stanford.edu/class/ee380/Abstracts/050223.html.

Mehan, H., and H. Wood. 1983 (1975). *The Reality of Ethnomethodology*. Malabar, Fla.: R. E. Krieger.

Menon, N., M. Page, M. Watt, and S. Bell. 2005. Mobinet 2005. http://www.atkearney.com/shared_res/pdf/Mobinet_2005_Detailed_Results.pdf.

mGAIN. 2003. *Benchmark Literature Review*. Mobile entertainment industry and culture (report IST-2001-38846, D6.2.1).

Moore, K., and J. Rutter. 2004. Understanding consumers' understanding of mobile entertainment. In *Proceedings from Mobile Entertainment: User-Centered Perspectives*, edited by K. Moore and J. Rutter. Manchester, UK.

Moreau, A., J. C. Sanchez, and H. Niu. 2004. Mobile gaming: Excitement on the move. *Alcatel Telecommunications Review* 4(1): 43–48.

NPD Group. 2005. Mobile games: Who's playing? http://www.wgworld.com/site/article/mobile-games-whos-playing.

Putnam, R. D. 2000. *Bowling Alone: The Collapse and Revival of the American Community*. New York: Simon and Schuster.

Rainie, L., and S. Keeter. 2006. Cell phone use. *Pew Internet Project Data Memo*. http://www.pewinternet.org/PPF/r/179/report_display.asp.

Repo, P., K. Hyvönen, M. Pantzar, and P. Timonen. 2003. Watching mobile videos: User experiences from Finland. In Proceedings from the Second International Conference on Mobile Business. Vienna.

Rowland, D., M. Flintham, S. Benford, N. Tandavanitj, A. Drozd, and R. Anastasi. 2004. On the streets with Blast Theory and the MRL: "Can You See Me Now?" and "Uncle Roy—All Around You." In *Proceedings from Mobile Entertainment: User-Centered Perspectives*, edited by K. Moore and J. Rutter. Manchester, UK.

Ruff, C. 2004. Building a mobile gaming community. *Electronic Gaming Business* 2(21).

Sharma, C. 2006. Worldwide wireless data trends—Mid year update 2006. http://www.chetansharma.com/midyearupdateww06.htm.

Simmel, G. 1971 (1903). The metropolis and mental life. In *Georg Simmel on Individuality and Social Forms*, edited by D. Levine. Chicago: University of Chicago Press.

Sotamaa, O. 2002. All the world's a Botfighter stage: Notes on location-based multi-user gaming. In *Computer Games and Digital Cultures Conference Proceedings*, edited by F. Mayra. Tampere: Tampere University Press.

Steinkuehler, C. A., and D. Williams. 2006. Where everybody knows your (screen) name: Online games as "third places." *Journal of Computer-Mediated Communication* 11(4): article 1.

Tamborini, R., M. S. Eastin, P. Skalski, K. Lachlan, T. A. Fediuk, and R. Brady. 2004. Violent virtual video games and hostile thoughts. *Journal of Broadcasting & Electronic Media* 48(3): 335–350.

Turkle, S. 1995. *Life on the Screen: Identity in the Age of the Internet*. New York: Touchstone.

Urry, J. 2000. *Sociology beyond Societies: Mobilities for the Twenty-first Century*. London: Routledge.

Vogel, M. 2005. Geocache player broke all the rules of Internet treasure hunt. *Idaho's NewsChannel 7*. http://www.ktvb.com/news/topstories/stories/ktvbn-sept2805-geocaching.a28aee2c.html.

31

Online Communities on the Move: Mobile Play in Korea

Youn-ah Kang

The Internet has enabled the rapid emergence of online interactions of dispersed groups of people with shared interests. These new communities and communicative practices exhibit a wide range of characteristics and serve a variety of purposes. Core characteristics of "online communities" include: 1) members sharing goals, interests, or activities that provided the primary reason for belonging to the community; 2) members engage in repeated, active participation and there are often intense interactions, strong emotional ties, and shared activities occurring among participants; 3) reciprocity of information, support, and services among members; and 4) there is a shared context of social conventions, language, and protocols (Whittaker, Issacs, and O'Day 1997). Much existing research on computer-mediated communication (CMC) and online communities has focused on differences between CMC and face-to-face communication—CMC versus face-to-face, online versus offline, virtual versus real. In the literature on Internet communication, a debate has continued about whether online, virtual, or other types of computer-mediated communities are real or imagined (Katz et al. 2004). While the utopians say that online communities would stimulate positive change in people's lives by creating new forms of online interaction by providing a meeting space for people with common interests and overcoming limitations of space and time, the other view argues that the Internet may be diverting people from true community because online interactions are inherently inferior to face-to-face and even phone interactions (Wellman et al. 2001). According to this perspective the Internet was thought to compete for time with other activities in a twenty-four-hour day and to be a stressor that depresses and alienates people from interaction.

Blending of Online and Offline Community

While important research has been done from this dichotomous view, a growing body of research is now examining a more integrative perspective of CMC and online behavior. It is looking at the use of the Internet as well as the use of mobile communication in an individual's everyday life. Not only a rapid increase in the number of users gaining access to and using the Internet (Katz, Rice, and Aspden 2001; Ling 2004), but also users' increasing exposure and commitment to Internet-based activity indicates the increasing presence of the Internet in everyday life. These trends have resulted in new Internet-based forms of mediation, namely, blogs. With the emergence of blogs, people have become to regard online communication as a social activity in which they are intimately related. Those who participate in these environments do so through posting, reading, and commenting on each others' sites and postings. Aligned with these changes, researchers have started to examine the blending of online community and offline community. Thomsen and colleagues argued that online communities are far from the "imagined" or pseudo communities explicated by Calhoun (1991); that they are "real" in the very way in which they reflect the changing nature of human relations and human interaction (Thomsen, Straubhaar, and Bolyard 1998). However, because the limitations of virtual communities in communicating affect identifying participants and holding them accountable, the combinations of virtual and real communities were thought to be able to overcome the weaknesses of virtual communities and combine many of the strengths of both.

Multiply-Mediated Virtual Community with Mobile Communication

At the same time, with the development of wireless Internet, Web sites are becoming accessible via various other types of mobile technologies. For example, linking into online communities via small mobile devices such as mobile phones and PDAs has become popular (Preece and Maloney-Krichmar 2003). The evolution of mobile devices such as multimedia messaging services (MMS) and high-quality camera phones has further promoted the adoption of mobile communication as a supplementary form of blogging. These devices can be used in posting texts and photos (Koskinen, chapter 18 in this volume) as well as connecting into online communities, such as when bloggers or participants in virtual communities do not have access to a desktop computer.

Cyworld: Korean Online Community

An example of this type of multiply-mediated virtual community is Cyworld, a Korean social networking service. Launched in 2001, perhaps 90 percent of South Koreans in their twenties are registered users of Cyworld. It has fifteen million members, about a third of the country's population (Moon 2005). It also ranks as the seventeenth most popular site in the world (Alexa, n.d.).

The main feature of Cyworld is the service called "Minihompy" (figure 31.1), which includes 1) a photo album that allows unlimited photos, 2) a personal bulletin board that is often used for personal journaling, and 3) a guest book upon which guests—often buddies—leave messages and the owner replies to them.

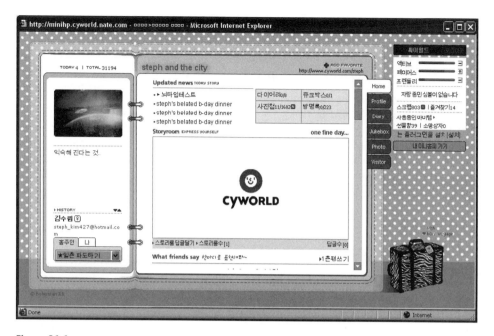

Figure 31.1
Cyworld Minihompy. Source: http://www.cyworld.com.

What sets Cyworld apart from traditional blogs such as Blogger and LiveJournal is its powerful support for customization. The users can choose to buy wallpaper and furnishings for decorating their Minihompys. Background music and other decorations such as digital furniture, art, and TVs can also be purchased. Through these new media channels, people are becoming more self-aware and have additional tools with which to demonstrate their identity. Since users of Cyworld consider the customization to reflect their own true (offline) identity (Hjorth and Kim 2005), Cyworld can be construed as a tool of self-expression and release, far from absorbing themselves through a virtual community environment (Katz and Rice 2002).

However, the most compelling feature of Cyworld arises from the relationships among people who use Cyworld. From its inception, Cyworld's developers observed closely the behavior patterns of youth. Through putting such observations into practice, Cyworld succeeded in gaining an emotional connection to its users. Users can link their Minihompy to friends' Minihompys to form buddy relationships. They then can have a chat or share memorable times with best buddies through collections of personal pictures. Music or background pictures can be purchased as a present, enabling buyers to build and enhance their friendship at a relatively small cost. To buy these, users must first exchange their real-world currency with "acorns," Cyworld money. Thus, Cyworld plays into systems of friendship and reciprocity. It closely follows one's daily life and the only difference is that it takes place in cyberspace. The use of Cyworld does not erode traditional forms of relationships, but it actually encourages people to strengthen the ties by interacting with others.

I first started Cyworld because my friends and families are all over the world, and it was easy to get in contact with them and drop by their Cyworld homepage to say hello, and share their pictures with their friends and family. I'm a very private person, so I made my pictures, family and friends pictures private—meaning only my "Cybuddies" and I can look at them. So you do have the control as to who looks at your pictures. I love Cyworld, it's great. After I joined Cyworld, i keep in contact with my friends that I haven't seen years, and my cousins who live in Korea. (from the Reader Comments of *BusinessWeek*, Moon 2005)

Mobile Cyworld

In April 2004, Cyworld started to provide "Mobile Cyworld," a wireless service through which users can manage their Minihompys. Mobile Cyworld does not limit its role to only supporting a few functions of Cyworld; in Mobile Cyworld, a user can perform most functions of Minihompy via a mobile phone. Major functions include 1) viewing and uploading pictures and postings, 2) notifying users when people ask to be added to Buddy Lists or leave messages in the guest book, 3) providing information as to whether buddies are online, 4) sending and receiving messages with buddies who are logged in, and 5) purchasing acorns. The most widely used feature in Mobile Cyworld is replying to postings in the guest book and uploading pictures taken by a camera phone directly to the photo album with texts. An average of 6.2 million photos are uploaded to Cyworld each day, many of them directly from cell phones (Moon 2005). Through the combination of online communities with mobile communication, Cyworld is intertwining users' online activities with their offline lives.

Expectations of Mobile Effects

While the impact of Cyworld and Mobile Cyworld is a communicative phenomenon particular to Korea, it is still indicative of a more general blurring between the digital and physical realms. It is important to examine how this phenomenon helps us to understand the convergence of online and offline communities. While PC access facilitates certain types of interaction, the material in this analysis shows that mobile access is employed by active users who bring new dimensions to this forum. Mobile communication, for example, is a more private medium that represents users' personal life and interpersonal relationships. Our purpose of this study is to examine how mobile communication affects social interaction in online communities through an empirical study of Cyworld. Through the comparison between Minihompys of both Mobile

Cyworld users and nonusers we make a claim that there are specific patterns that imply the blending of online and offline communities, and that mobile communication plays an important role in building social interaction in online communities.

The assertion that is examined here is that those who engage in mediated interaction via several channels (PC, mobile phone), will be more active users of the community and have a broader social network with which they are in contact. It is difficult to assert causality here with only the covariance of the two dimensions.

How the Issue Was Studied

I used multiple methods to investigate the relationship between mobile communication and interactions in online communities. Though I primarily focused on the statistical results from analysis of one hundred randomly selected Minihompys, I also did some observations and drew on Leslie Haddon's interviews of students at a Korean university. More specifically, I sampled the Minihompys of one hundred users (half of whom used both mobile and PC versions of Cyworld, the other half used only the PC-based version). The sample Minihompys were generated randomly using the Cyworld's "link to a random Minihompy" tab.

In my data collection I focused on variables that could indicate active management and interactive aspects surrounding a Minihompy. These included not only demographics (gender and age) and length of usage (since the starting date of the Minihompy), but also the number of

- photos in album (viewing is the most popular function in Minihompys and the major reason people visit others' Minihompys);
- postings on personal bulletin board;
- messages in guest book;
- monthly visitors;
- gifts from others (users can send and receive gifts such as music, decorations, background images, and even acorns);
- scrapped photos and postings by visitors ("scrap" is a similar function to blog trackbacks, which means taking a copy of a picture or a posting from others' Minihompys and using it in one's own Minihompy);
- minirooms (a miniroom is a cyber room of the user usually decorated with wallpapers, ornaments, and furniture, which are purchased by Cyworld money—acorns);
- mean response time to visitors' messages in guest book (indicates active management and involvement);
- monthly visitors (mean); and also
- the "Fame" rating given by visitors (there are quantified ratings of the user for "Erotic," "Famous," "Friendly," "Karma," and "Kindness"). (See figure 31.2.)

Figure 31.2
Minihompy fame rating. Source: http://www.cyworld.com.

To assess whether the measures of the two groups—Mobile Cyworld users and nonusers—are statistically different, we analyzed each variable using the t-test. Because the samples were chosen randomly, the analysis is appropriate to compare the means of the two groups.

What Was Found

Quantitatively, the majority of Cyworld users appear to be females; seventy-four users out of one hundred are women while twenty-six are men. Also, female users are more likely to use both the mobile and PC-based versions of Cyworld than male users. This result is consistent with an intriguing previous study, which argued that Korean females are more willing to adopt new communicative technologies than males (Hjorth and Kim 2005), and which is a striking contrast to the United States and western Europe, where the majority of early adopters are males (Katz 1999). There is no difference in the age between Mobile Cyworld users and nonusers.

The analysis of a set of one hundred Minihompys reveals that there is a clear difference between the two groups. While nonusers of Mobile Cyworld have had their Minihompys for a longer time than users of Mobile Cyworld (p = 0.04), Mobile Cyworld users are more likely to be active in using their own sites. Specifically, there were approximately twice the number of photos in the photo album (p = 0.03), and there were three times as many postings in the bulletin board (p = 0.03) when compared to the sites of nonusers. It is also significant that the interactive indicators—such as the number of visitors per month (p = 0.03), the number of gifts (p = 0.00), the number of scrapped photos (p = 0.04), and the fame rating (p = 0.03)—are considerably higher for the mobile users.

When looking at the number of messages in the guest book, there was no significant difference between groups nor was there a difference in the number of minirooms. Neither was the response time for guest books statistically different (p = 0.12). (See table 31.1.)

Table 31.1

Difference between the Two Groups (N=100)

	Variables Mobile + PC-based version (N=50)	PC-based version only (N=50)
Gender	39 females, 11 males	35 females, 15 males
Age	23.2	22.7
Duration of usage	31.5 months	35.5 months*
Visitors per month	441.8*	304.7
Photos in photo album	1279.9*	659.3
Bulletin board postings	135.0*	42.9
Messages in guest book	1381.7	1100.0
Number of gifts	30.3**	15.5
Scrapped photos	927.7*	405.6
Fame rating	1567.8*	1026.2
Minirooms	14.9	11.5
Response time	24.1 hours (N=33)	29.3 hours (N=26)

*$p<.05$
**$p<.01$

When the two groups are compared, the following differences can be discerned:

Implication 1   Mobile communication is employed by active users who bring new dimensions to the interaction in online communities.

Despite the shorter average duration of usage, Mobile Cyworld users demonstrate a higher activity level in managing their online spaces. This is seen in their archiving more photos and postings than nonmobile users. In addition, by observing the appearance of Minihompys we can see that Mobile Cyworld users are more concerned with the decoration of their Minihompys than other users. Though we cannot posit a causal relationship between the use of Mobile Cyworld and the activity level of usage, clearly mobile communication is more likely to be employed by active users who bring new dimensions to online community interactions.

Implication 2   Mobile communication plays an important role in the blending of online and offline communities.

The prevalent culture of Cyworld is that people reveal their everyday lives on their Minihompys and want many people to visit. Often, how many people visit the Minihompys is a measure of popularity of the owners. It not only reflects their popularity in online communities but also demonstrates how they are social in real world communities. The number of visitors is an indicator of face-to-face interactions in everyday life. Les lie Haddon's research at a Korean university reveals that the daily mood of Cyworld users was often influenced by what was happening online, especially how many visitors they had had (personal communication). One interviewee said,

I wanted to show myself to other people. My home page is an expression of myself to other people. And if I don't have a high number of visits, I don't feel good.

The interviewee considered the number of visitors as an indication of how popular she was. If there were few visitors she would usually post more photos or take interesting photos from others' Minihompys into her own Minihompy. If this strategy worked she could see that more people had visited her Minihompy and perhaps even some people copied her photos. According to Haddon, this made her feel good.

For many people, Cyworld is the place of self-expression. While there is the general level of portraying themselves to the world, some people use Cyworld as a tool for self-expression on a daily basis. They express their feelings indirectly on their Minihompys instead of in face-to-face communication. One of Haddon's interviewees mentioned:

When I'm not feeling happy with my boyfriend I can't tell him directly, but I can show my feelings through the music and decoration on my mini-home page.

The users of Mobile Cyworld Minihompys upload mobile photos that show everyday life. The photos taken by cell phone are neither fancy nor special; they simply show what the users' offices or schools look like, where they went on a particular day, and whom they met. By uploading these types of photos, people are able to share their everyday lives in the real world with their family and friends through an online medium. Similarly, many of the photos are titled with informal themes such as "moving by car," or "I'm in class . . ." as if the users were talking to their friends over the phone. While doing other tasks in the offline world, these users simultaneously participate in online communication through the use of their mobile phones.

The analysis here also shows that Mobile Cyworld users post more writings in the bulletin board section of their personal sites. The topics of the postings in the bulletin board are about the owners' personal thoughts and happenings in their lives and often include interesting articles or phrases. While the non–Mobile Cyworld users have few or no postings in their bulletin board, those who use the PC and mobile versions are more likely to expose their thoughts and lives to their online community members. For example, one Mobile Cyworld user considered his Minihompy the most powerful way to communicate with his friends. The title of the Minihompy was "My cell phone number has changed: xxx-xxx-xxxx," and there was an announcement saying, "Please leave your cell phone numbers in the guest book!!" This case provides evidence that this user considered his online community members to be similar to his offline community members in that people usually give updated cell phone numbers only to those with whom they have a close relationship.

These phenomena imply that the gap between the online and offline worlds is becoming narrower as the Internet is incorporated into routine practices of everyday life. As we examined, this convergence is highly supported by mobile communication in that mobile communication is more private than any other medium.

Implication 3 Mobile communication is associated with social interaction in online communities.

The apparent convergence of online and offline communities promotes social interaction in both the online and offline sphere by deepening existing relationships, facilitating reciprocity, and forming mediated relationships among users.

As shown in the observations, photos sent through mobile phones and bulletin board postings expose the user's everyday life and self-reflections to other users. When people want to find out about a friend's recent situation or get to know someone who is not a close friend, they often visit his or her Minihompy and browse the person's photo album, bulletin board, and guest book. For a relationship to gain depth and resonance, people need to move into more involved forms of interaction (Ling 2000). The more users disclose personal information such as the fact that they failed an exam or that they are having trouble with something in their Minihompys, the more possible it is for ties between the users and visitors to be strengthened. By adding another layer of mediation, mobile communication increases transparency and connection in online communities.

While the social sharing of events and emotions is crucial for sustaining the relationship, another social function that binds a group is to exclude nonmembers from certain activities and authority. Cyworld has a feature that allows users to control access to their pictures and postings. Though the feature had always existed, people rarely used it during the early history of Cyworld. At some point, however, people started to use it gradually and it became the norm of the community. It is not clear why the changeover occurred, but it implies that people started to keep a boundary between their in-group and out-group and the activity is a part of strengthening friendship. Observation of Minihompys showed this—we found few users who publicize their Phone photo folder. By prohibiting others from the Phone photo folder that has the most private photos, they are building stronger ties with friends.

The statistical results demonstrate that interactive indicators such as the number of gifts, the number of scrapped photos, and the fame rating are higher in the Minihompys of Mobile Cyworld users. This implies a higher reciprocity between users in that these indicators increase only when users interact with each other. In the real world, giving a present usually occurs during a mutual exchange between people. Similarly, in the online community when a person is given background music or a certain amount of acorns from a friend, the person usually sends a present in return. The number of times a photo has been scrapped (i.e., copied) explains how widely it has spread in Cyworld communities. If a person has a picture that is really interesting it might by seen by a number of people within a few days through multiple scrapping. The fame rating also increases through a reciprocal process; the more people visit a Minihompy and the more gifts the owner sends to other users, the faster the rating goes up. Through this reciprocity, ties between the users become strengthened. Among these reciprocal activities, the best "gift" that users perceive is visiting a friend's Minihompy, which adds to the counter of visits, and leaving a message. A male Cyworld user mentioned that about twenty people, mostly his friends, visited his Minihompy every day, and he reciprocated visiting theirs, looking at their postings and photos to see what they did that day. The norm in Cyworld is "if I visit your Minihompy, you will visit mine, and vice versa."

In addition, the fact that Mobile Cyworld users are more likely to have a guest book in their Minihompys than are non-Mobile Cyworld users indicates that Cyworld users are more open to a mediated relationship with other users. A guest book is accessible to absolute strangers and users can use the stranger's message to see who they are and visit that person's Minihompy. The messages in a guest book are often used to gauge the owner's reputation and as mediation for friendship. The following is a review written by a new user of Cyworld:

When I first got on Cyworld, I got all my friend's Cyworld names so that I could see their Minihompys. The next day, my friend asked if I went to his Minihompy. I told him yes. He asked me if I left a message. I didn't, as I didn't know how to do that. He was very upset that I didn't leave a message. He was waiting for my message. Of course I can call him on the phone any time, or leave him an email, or even write him an IM. You see, these messages are public on Cyworld. When someone leaves you a message, everyone else that goes to your Minihompy sees how cool you are. They read the messages to see what type of a person you are.

However, the interaction in Cyworld has drawbacks. In terms of sociability, any activity that involves the work of maintaining social ties requires time and effort. Sometimes, people may be obligated to attend social events, even when they do not want to do. In case of Cyworld, the time and effort to maintain one's own Minihompy may be seen as a burden. A female interviewee who would have to decorate her pictures using the Photoshop facility on Cyworld stated,

Because everyone does it (laughs) but for me it's totally annoying...it takes a long time. It can take 3–5-minutes to decorate a single picture. Therefore if you take lots of pictures you have to do a lot of work!

For those who desire to keep their everyday lives on their Minihompys, they have to make an effort to capture the most memorable moments—taking and uploading pictures. The work is not trivial if the pictures are to be posted on one's Minihompy. An interviewee shows that one sometimes may avoid bringing a camera to an event:

When we have picnics I sometimes don't take a camera on purpose. Then the person who has camera has the job of taking pictures, decorating them and posting them for the others to access!

Peer pressure is another aspect of strengthening social ties through Cyworld experiences. Because Cyworld has already crossed a critical mass, those who do not have their own Minihompy or do not manage it actively are hindered from participation in social life.

If I do not put the pictures up my friends will say...what are you doing...you can update your home page...so...OK...Almost all of my friends had one...if you didn't have one you can't make common things with my friends and other people...nowadays I'm addicted to Cyworld.

This final comment highlights the often compelling nature of social networking when combined with mobile communication. This underscores the need to understand in a full way how people interact with and through their technologies.

In sum, mobile communication can be another mode of interaction in online communities among members who are already in established relationships. Without necessarily positing causality, we found significant differences between the mobile users and nonusers of Cyworld. Use of mobile communication in an online social networking sphere is positively correlated to social interactions. These differences may stem from the role mobile communication plays in promoting the convergence of online and offline worlds. The adoption of mobile communication in online communities may help them evolve into another more intensely involving place for mediated interaction. By supporting the blending of online and offline communities, mobile communication opens a possibility of facilitating social interaction between people.

References

Alexa Traffic Rankings. n.d. http://www.alexa.com/data/details/traffic_details?range=2ysize =mediumy=rurl=http://cyworld.nate.com/#top.

Calhoun, C. 1991. Indirect relationships and imagined communities: Large-scale social integration and the transformation of everyday life. In *Social Theory for a Changing Society*, edited by P. Bourdieu and J. S. Colemen. Boulder, Colo.: Westview Press.

Cyworld. n.d. http://wiki.pikkle.com/index.php/CyWorld.

Hjorth, L., and H. Kim. 2005. Being there and being here: Gendered customising of mobile 3G practices through a case study in Seoul. *Convergence* 11(2): 49.

Katz, J. E. 1999. *Connections: Social and Cultural Studies of the Telephone in American Life*. New Brunswick, N.J.: Transaction Publishers.

Katz, J. E., and R. E. Rice. 2002. *Social Consequences of Internet Use*. Cambridge, Mass.: The MIT Press.

Katz, J. E., R. E. Rice, and P. Aspden. 2001. The Internet, 1995–2000: Access, civic involvement, and social interaction. *American Behavioral Scientist* 45(3): 404–418.

Katz, J. E., R. E. Rice, S. Acord, K. Dasgupta, and K. David. 2004. Personal mediated communication and the concept of community in theory and practice. In *Communication and Community: Communication Yearbook 28*, edited by P. Kalbfleisch. Mahwah, N.J.: Lawrence Erlbaum Associates.

Ling, R. 2004. *The Mobile Connection*. San Francisco: Morgan Kaufmann.

Ling, R. 2000. Direct and mediated interaction in the maintenance of social relationships. In *Home Informatics and Telematics: Information, Technology and Society*, edited by A. Sloane and F. van Rijn. Boston: Kluwer.

Moon, I. 2005. E-Society: My World Is Cyworld. *BusinessWeek*, Sept. http://www.businessweek.com/magazine/content/05_39/b3952405.htm.

Preece, J., and D. Maloney-Krichmar. 2003. Online communities. In *Handbook of Human-Computer Interaction*, edited by J. Jacko and A. Sears. Mahwah, N.J.: Lawrence Erlbaum Associates Inc.

Rheingold, H. 2002. Mobile virtual communities. http://economia.unipv.it/marketing_high_tech/high_tech_letture/global%20VO/06_02.pdf.

Thomsen, S. R., J. D. Straubhaar, and D. M. Bolyard. 1998. Ethnomethodology and the study of online communities: Exploring the cyber streets. *Information Research* 4(1). http://informationr.net/ir/4-1/paper50.html.

Wellman, B., A. Q. Haase, J. Witte, and K. Hampton. 2001. Does the Internet increase, decrease, or supplement social capital? *American Behavioral Scientist* 45: 437–456.

Whittaker, S., E. Issacs, and V. O'Day. 1997. Widening the Net. Workshop report on the theory and practice of physical and network communities. *SIGCHI Bulletin* 29(3): 27–30.

Conclusions and Future Prospects

James E. Katz

This chapter probes selected issues raised by the studies in this volume. The chapter seeks not to summarize the book's contents, a daunting task given the already compressed nature of the handbook, but to highlight points that have arisen across several chapters to give them additional analysis. It also seeks to make some broad conclusions. Of course any general statement about mobile communication will have numerous exceptions. After all, about one person in two on the planet has a mobile phone. Plus, many of those who do have a mobile phone also have more than one mobile electronic device. Nonetheless, general trends may be seen about mobile communication in ordinary ways of life. These may be clustered into overlapping topics: adoption, uses, and consequences.

Adoption and Access in a Mobile World

To make a wordplay on "An Internet of Things," the title of a White Paper report tabled at the 2005 World Information Summit, it seems the world is also coevolving into a mobile network of people and resources. Evidence for this view is presented in Srivatava's chapter 2 (she coincidentally was also the author of the White Paper). The technological trajectory is toward ever more mobile devices doing more things and being owned by more people from every demographic and age category.

High-end or luxury-oriented users often seek to make a fashion statement with their mobile devices, and are regularly willing to pay hundreds, or even thousands, of dollars to do so. In most cases they already have less sophisticated versions of the device in question and seek to move up in power and prestige. It is different for those without money to spend; countless others would like to be able to afford to pursue these same technologies. But not having the resources, they will instead settle for less expensive models. Indeed, they sometimes will even settle for a nonworking device that merely looks like the prestige model (as for instance, in Namibia, a hand-carved wooden mobile phone).

Yet on the other side of the equation, there is also a strong drive toward making simpler, low-tech phones, due to economics and demographics. As Chipchase and Portus show (chapters 7 and 9, respectively), even at the lowest economic strata there is strong interest in having a mobile phone. The chapter on India pointed out that its telecom minister believes that very-low-cost handsets (in the range of USD 20) could be a boon not only to the world's poor but also to those companies that seek to market to them. Apparently acknowledging the merit of his argument, international as well as indigenous firms are building plants in India to manufacture just such models.

In terms of the demographics of distribution, adoption is spreading from rich to poor areas, from urban to rural, and from young adults to even younger and older cohorts. As Lai suggested, the mobile phone is evolving into an all-purpose entertainment portal for kids. It also is becoming a monitoring device for parents concerned about their children and wanting to keep tabs on them.

A mobile digital divide remains of concern despite the fact that, as Srivastava shows, the digital divide for mobiles is far less severe than it is for the Internet. It is relevant to note that even though the Internet digital divide is narrowing, it is still enormous. This remains the case despite strenuous efforts by donor nations and organizations to provide the poor with Internet access. The situation with the Internet contrasts with that of the mobile phone. The latter is enormously popular even without donors having to provide a kick start. Moreover, it is impressive to see how many of the world's poorest people who despite their poverty are able to get their hands on a mobile. Certainly rural areas in the poor parts of the world remain far too isolated from mobile services, and in many developing countries tariffs are unfairly high. But the mobile phone may deserve the subtitle or service mark of "the real world's Internet."

Nonetheless, despite the obvious popularity of the mobile phone, there continue to be many nonadopters due to reasons of poverty or absence of infrastructure and because of limitations of their physical condition or lack of education. Yet there is another form of "nonadoption," or to put it more precisely, disadoption or involuntarily dropping out. These are people who, usually for financial reasons, do not currently have a mobile phone even though they had one at an earlier time. Numbers in this category are surprisingly large, as the survey in Mexico suggests (chapter 6), especially in light of the oft-stated, and clearly hyperbolic, opinion of users that the mobile phone is vital to their continued existence ("I couldn't live without it!"). Yet this dropout phenomenon remains as persistent as it is surprising. (The Mexican survey, incidentally, is one of the first confirmations in developing countries of the "dropout" effect that Ron Rice and I discovered in our 2000 U.S. national surveys.)

Rejection

Little has been said about individuals and groups who make an outright decision not to adopt mobile communication technology even though they could do so if they wished. One reason little has been said is because there are so few individuals who fall into this category. But still there are some who elect not to have mobiles because of group or individual beliefs. There are a few groups that explicitly forbid its members to have mobile phones under any circumstances; these are usually religious orders. Yet even those groups that have traditionally not allowed wireline phones seem to be less stringent about mobiles. This is the situation for groups such as the Amish of Indiana in the United States and a few Kibbutzim in Israel.

There are also individuals who make the choice not to have a mobile phone for other than economic or physical reasons. However, their numbers are very small, and ever diminishing. Students I have worked with, Nina Aversano and Shenwei Zhao, have done some research on purposeful nonadopters in the United States. Part of their problem even conducting research in the first place was to find such people (nonelderly) to interview. But one of their more intriguing findings was that many of the physically and economically capable people who said that they had chosen not to have their own mobile phone only appeared to be nonadopters. Technically they were nonadopters. However, that did not strictly mean they were nonusers. Rather, their use might be more accurately if unflatteringly described as parasitic. That is, even though they did not have a mobile phone of their own, their spouse or other frequently accompanying companions did. Nor were they inexperienced in mobile phone use: in all cases the students were able to identify, the nonuser had previously experienced using a mobile phone and was familiar with its operation.

However, some nonusers were quite stubborn in their unwillingness to become a mobile phone owner, even at the grave cost of being without one. For instance, Nina Aversano (personal communication) encountered in her study a businessman from Mexico City who refused to get a mobile phone even though he had been kidnapped and held for ransom—twice!

More generally, refusal to adopt often stems from individual (and self-described) archaic views of manners, standards, and good public behavior. The decision is indicative of a life-philosophy that asserts that one must live by one's own rules, regardless of what others think. Another prominent reason for nonadoption was that by having a mobile phone they would be reachable by others and have to deal with situations they might not like. So both for value and (oddly enough) convenience reasons, some small number have declined to get a mobile phone. At the same time, many later adopters reported that they got one only because of immense pressure from family or friends. This pressure forced them, albeit reluctantly, to accept a mobile phone into their lives.

Uses

Given the trends discussed by chapter authors, it seems clear that talk and text will remain preeminent mobile applications. They will be followed in popularity by music, gaming, mobile photography and video-recording, navigation, and Internet access.

Enhanced services and add-on features, such as the ringtones that Licoppe discussed (chapter 11), will continue to be important in terms of overall market size if not necessarily breadth of demographic penetration. Mobile technology will serve as an interface to an array of informational and commercial resources. These will include multimedia mail, portable entertainment players, and identification and financial transaction. Mobile technology will also reconfigure the nature of public space and what it means to be at various cultural and family events.

Significantly, from a social perspective, it will link us more tightly to our interpersonal networks even as it allows us to wander further from them physically. In essence, mobile technology will act as something like a fanciful homunculus, slowly commanding more of our lives, knowing what we should know better than we ourselves do, and entertaining us (and our friends) endlessly. Continuing in a hyperbolic vein, it can be said that mobile technology will have a mind of its own that can turn on us in a most unpleasant way. (More will be said about this later.)

Politics

Mobile communication has helped overcome many barriers to political mobilization. Yet while mobile technology can allow speedy assemblage of demonstrators, and coordinate their reaction to security forces, mobiles do not of themselves seem capable of producing spontaneous eruptions or stimulating otherwise torpid groups to take vigorous action, as Rheingold shows (chapter 17).

On the other hand, mobiles can work to send messages and sentiments through networks to lay the groundwork for future action. This preparatory function seems to be reflected in activities currently taking place in the Middle East, according to Ibahrine (chapter 19).

Concomitantly, SMS messages are used for disinformation and demoralization. Sükösd and Dányi (2003) have reported that barrages of SMS messages were dispatched during Hungary's national elections asking supporters of certain parties to meet at various inconvenient times and noncentral locations. However, these were false messages meant to distract, exhaust, and demoralize party elements. Their source was unknown, but the practice of false messages borne by agent provocateurs and double and triple agents has been long-standing. Such concerns continue and have become more important as more people use mobile technology. A recent case was in March 2006 when mysterious text messages circulated through Minsk, the capital of Belarus, on the eve of an allegedly fraudulent national election. Its dire warning was: "On the evening of the 19th on Oktyabrskaya Square, provocateurs are preparing bloodshed. Look after your life and health." The source of the SMS message, and whose interests it served, remained unconfirmed, although opposition figures claimed it was government-originated (Associated Press 2006).

At the level of interpersonal and small group politics, the cell phone is also significant, certainly in terms of migrant workers in China. Workers are able to build links to their compatriots to establish coordinated efforts to advance their rights collectively, as was shown by Law and Peng (chapter 5).

Economic Development

The mobile appears to have a powerful effect on promoting economic opportunity. Yet it seems to have an even more powerful effect on promoting social need. In terms of the first, Donner (chapter 3) shows the tremendous economic benefits in developing countries made possible by mobile communication technology. As is the tendency of academics, we look for the dark side or unintended consequences (preferably negative and ironic) of attempts to improve the world. Mobile phones should not be, nor are they, exceptions to the skeptical examination of social scientists. Here the big question is that once mobile communication technology gets into the hands of the world's poor, what will they do with it. Certainly tremendous economic benefits can be forthcoming as Molony suggests (chapter 25). But Molony also suggests that the most important ramifications may be in terms of social networks. For his part, Donner suggests that mobile phones in the poorest villages might exacerbate the divide been the poor and the poorest, a question that requires further investigation.

Tracking and Location Monitoring

Location is where it is at for many mobile services, of course, and an important aspect of their use. This for instance was shown in Lai's study of Japan (chapter 20) and White and White's study of tourism in New Zealand (chapter 15). Orientation to space is important to both the caller and called party. But beyond the mechanics of establishing a framework for conducting a phone call, location is important in other ways including the offering of location-based services and navigation aides. Other services coming to the fore include location-based ad delivery and billing for tolls. Retail mobile marketing and recommender services should grow along with multichannel shopping functionality.

Location has the corollary of tracking and implied-content monitoring. Worldwide tracking of anybody who has a mobile in her or his possession now seems possible and may be on the verge of the probable. Certainly this is the implicit promise of a commercial company operating out of the UK that offers to pinpoint any GSM user's location. The World Tracker supposedly uses cell tower GPS data to track the phone's location. Whether it can work as claimed (I was unable to use it), the service presumably relies on existing technological capabilities, which easily could be exploited behind the scenes. Another company (U.S.-based) was demonstrably able to provide (for a fee) Americans' private mobile phone calling records. (After a national scandal, it was put out of business.) Tracking and spying on mobile conversations has a long pedigree, including during the late twentieth century when recordings surfaced of Prince Charles and Princess Di chatting via mobiles to their respective lovers. News of illicit monitoring continues to break out across the globe ranging from Korea to Greece. Licit monitoring, on the other hand, has allowed some serious terrorist plots and criminal conspiracies to be broken.

Everyday Social Reproduction

Photography is a frequent activity for many users. This includes trading photos and other images. But much less used are the video capabilities of mobile devices (Koskinen's chapter 18). This is not merely a result of limited storage and frame size. Even when people are able to record copiously their activities and special events, they seem to become quickly satisfied, and a few hundred photos are often more sufficient than hours of full-motion video and sound.

There are exceptions. These include low-frequency, high-expense rites, events, and transitions (e.g., births and weddings). However, events at the other end of the spectrum, including children's sports events and birthday parties, over time become less frequently recorded and thereafter less frequently viewed. Homemade mobile videos, ringtone concerts, and other adventurous experiments will doubtless continue but will generally not command large audiences for individual productions.

Massively parallel multiplayer mobile games are proliferating in sophistication and degree of involvement. Jane McGonigal has formulated these as Ludic Networking and Massively Collaborative Play. One mobile game genre, geocaching, is a form of the old-fashioned treasure hunt. A leading site claims there is a quarter-million active caches distributed among most of the world's countries, and has more than twenty-six thousand active players (geocaching.com). There should be continued growth as some people seek games that are reality-based and use the computer and mobile as a springboard to physical action in social space.

Increasingly, there will be artistic endeavors using mobile devices, especially mobisodes (mobile phone–watchable video materials or "mobile episodes"). While these will proliferate via Web sites and moblogs (mobile Web logs or blogs), this promises to remain a largely specialized endeavor involving networks of friends. The proposition that they will become mainstream remains dubious for a lot of people.

Everyday Multimedia Consumption

The level of mass interest appears much higher when it comes to consuming professionally produced multimedia, especially music, mobile games, and video entertainment. Currently the main interest is in portable audio entertainment, as evidenced by iPods, but this interest is likely to spread rapidly to a variety of other audio and visual material. Although much of the new content will be sheer entertainment, audio information and "podcasts" should attract considerable attention, especially among older and more-educated segments. Already many positive educational and cultural uses are being found. As but one example, museums are increasingly offering via mobile phones free audio lectures and tours of their collections.

Initially there was skepticism over the viability of commercial video programming being delivered via cell phones. This was in part because it was thought that if people were walking or driving, it would be hard for them to also watch a small screen. Skeptics are likely to be wrong in this instance.

In 2005, I had the opportunity to witness Seoul's TV programming available through enhanced cell phones. The images were clear and smooth with excellent sound. Korea's program is the most advanced in the world, and is free (due to government subsidies). Although other countries also have video broadcasting, the images are less clear and usually cost money to access. The service is popular and growing rapidly; by 2010, TV phone services may well be generating USD 30 billion annually on a worldwide basis. These expectations are, if anything, modest. Korean researchers tell me that the small screen, rather than a disadvantage, provides a way to watch programming privately. Due in part to high-fidelity sound, young people will happily watch movies of ninety minutes or longer on the small screen. So what was initially a shortcoming of mobility—the tiny screen—is actually a plus since the user can easily move the "screen" and watch at convenient times.

The Korean case can also point to where services are not likely to be growing anytime soon, and put in relief some of the nontechnical barriers to certain multimedia services. During that same visit to Korea I had dinner with the head of SK Telecom's division in charge of the TV-based services. He told me that real-time, two-way video conferencing was a capability that could in fact be deployed on his system but that there was not much of a market for it. Yes, it did work, he assured me. In fact he himself had tried it on an experimental basis. He related that after the first day of the trial, he went out in the evening to a bar with business colleagues (a common practice in Korea). While in the bar he received a call from his wife. After chatting a bit, and hearing the background noise, she asked him to slowly rotate 360 degrees his new video camera–equipped cell phone so that she could confirm his location and companions. As it turned out, shortly after that first day, the executive decided to terminate the experiment. Not to put too fine a point on experience, it does show that two-way video telephony is no more popular when it is mobile than it was as a stationary service repeatedly offered in the United States by AT&T, branded as the Videophone service.

Consequences

Can People Handle Multitasking?

Fears once were voiced about whether people could handle multiple cognitive and operational tasks involving their technology seem to have been definitively answered: Yes, they can. It also seems, according to Ling (chapter 13), that they are able to do so while still discharging ordinary social rituals without substantial loss of "face," as Goffman defined it. From Baron (chapter 14), we learn that not only can people handle multitasking but they also can use it as part of their strategic and tactical communication considerations. It appears there is an important generational dimension to usage, and that over time we can expect multitasking to be much more frequent and developed.

Contested Space and Norms

The process of defining and enforcing appropriate behavior in public places has a long history, intermittently punctuated by conflict. A factor precipitating conflict has often been a change in technology. Historically, management of public space was done through repressive means including laws and armed suppression. Occasionally, it has been done by altering the built environment. More rarely, it was done through incentives. (An example of the latter was the practice of tossing coins to a crowd when an emperor or pope would pass in the teeming, narrow streets of Renaissance cities. This was not a gesture of munificence but instead was done so the pope could pass through the crowds and so that the people would not press "too closely upon the Holy Father" [King 2000, p. 137]). When it comes to systems to deal with mobile communication in public, systems vary and many incentives and disincentives may be deployed.

There is the possibility I raised elsewhere that exposure to the mobile phone calls of others can be unpleasant and can degrade the public environment (Katz 2006), a likelihood that has been confirmed in various public opinion surveys including one done by Rich Ling in the United States via a national random sample that showed about 70 percent of the population is disturbed by others' use of mobile phones in public. While this may be a problem for low-density industrialized countries, it may not be universal. Yet for those who are bothered by "teledensity," which here means the widespread use of mobile communication technology in public places, the experience can be unpleasant and even depressing and anger-inducing.

Numerous formal and informal responses have been undertaken to impose order on the tumult of contested terrain over use and "use-free" locations. A popular approach has often been signage. I am reminded of a sign posted in a Broadway theater forbidding clientele from using their mobile phones even while standing in line to buy tickets. Public address announcements are also among the techniques used to impose order.

Going beyond formal written requests, stricter sanctions have been considered. For mobile phone users who disrupt a cultural performance in New York City, a small fine may be levied. In other venues, more severe penalties are possible. In fact, several U.S. football players have received substantial fines for using cell phones during games, reaching in one case to USD 30,000.

Another avenue for tackling the problem of public mobile communication behavior is to apply technical interventions. In terms of cell phones, these have been imposed in some settings including some national houses of parliament, as mentioned in chapter 29 on India. They have also been put into use in some churches in Mexico. The idea of jamming cell phone signals is expected to improve a deteriorating situation in U.S. movie theaters, where some claim that declining attendance is due to patrons frequently using their mobile phones during the screenings. To fight this problem, movie theater owners are asking federal authorities for permission to jam cell phone reception. "I don't know what's going on with consumers that they have to talk on phones in the middle of theaters," declared the president of the United States National Association of Theater Owners in a speech. He said that plans were afoot to remove federal regulations forbidding the jamming of cell phones by private parties, in this case theaters (*China Post* 2006).

Despite the various attempts to regulate public use of mobile communication technology, it is unlikely that many boundaries will remain unpierced. In part this is because of the irresistible sweetness of using the technology, and the fact that over time people become inured to such practices. The experience of being on the street or in public places is quantitatively changing as more people have and use mobile communication while on the move. It may even be the case that mobile communication technology is redrawing public space in ways that are, speaking phenomenologically and sociologically, at least as dramatic as that of the automobile.

Apparatgeist and Perpetual Contact: Do Mobile Technologies Erode Individual Senses of Identity and Promote a Collective One?

An important question is whether mobile communication technology can serve as an effective substitute for, or even enhancement of, traditional activities through which social cohesion has been achieved. It may be the case, especially given the heavily social uses of mobile communication technology, that it can help people redevelop a sense of group integration as well as reduce anomie.

Gergen has extended the prospect that mobile interpersonal communication technology could restore some of the lost sense of cohesion and belongingness people presumably felt a century ago. In his chapter (22) he also outlines how mobile phones could give voice to people to relink themselves to the political agencies responsible for policy. He sees that they could operate in clusters to restore active political participation. His view is complemented indirectly by those of Licoppe, Ling, and Ibahrine in the social sphere. As well, the theme comes through in the chapters of Molony, Kang, Campbell, Portus, Mesch and Talmud, and Law and Peng. Rheingold lends support from the perspective of political mobilization (chapter 17). Combined, these analyses add weight to the view that mobile communication may have a substantial impact on reconfiguring social and political lives of adults every bit as much as it has impacted the lives of teens in industrialized nations.

As seen by the analysis of Kang; Mesch and Talmud; and Miyata, Boase, and Wellman (chapters 31, 23, and 16, respectively), the mobile has the effect of engaging users in psychologically important communities. These groups can easily become vital sources of meaning. Sherry Turkle has observed in her studies of teens that mobile technology may be used less to convey feelings than to determine one's own feelings, which is done through communication (chapter 10). She finds that people in general, but teens in particular do not necessarily contact a friend to tell them how they are feeling. Instead, they initiate contact in order to gain cues as to how they should be feeling. In effect, they are not saying, "Here I am and this is how I feel" (which is a typically understood rationale for communication). Rather, Turkle asserts, the teen says, "Here I am—how should I feel?" (Turkle 2006; my paraphrasing.)

Certainly the intensive use of mobile technology for social relationships has been a theme that many chapter authors have highlighted. Mark Aakhus and I have attempted to spotlight this role in our development of Apparatgeist theory (Katz and Aakhus 2002). We came to appreciate a subtle but nonetheless critical idea that the tools of mobile communication elicit and further develop human feelings about what interpersonal communication should really achieve, namely transcending the personal physical limits of being human. Through technology, the hollow promise is extended by joining in a transcendent state of connectedness with others. Mobile communication offers the tantalizing prospect of crossing the boundary of individual barriers to merge the self with a higher sense of place and group. (This sense of the individual blending with the unified sense of authentic home is captured by the German term *Heimat*, which means home-place but also much more in that it connects several layers of the emotional aspects of belonging.) These devices extend the unfulfillable promise of letting a person slip loose from the severe cognitive and physical limits of human existence. Or, in Ernest Becker's evocative terminology, they could set free the angel bound in heavy armor (Becker 1969; these ideas are discussed further in Wynn and Katz 1997).

Perpetual contact is a view of what communication ought to be. Mark Aakhus and I have described it as a logic that informs choices about what practices and technologies for communicating are desirable. In terms of the autonomy-connectedness dialectic, perpetual contact gives priority to connectedness to the detriment of autonomy. This sociologic also has its limits because in positing the ideal form of communication as perpetual contact another dimension is minimized—that people as physical beings are bound by their material nature. This they cannot escape despite the seemingly compelling powers of personal communication technology. So, a dialectic springs up between the promise of what a machine can seem to deliver in terms of the depth of communication and how the material world in which it acts actually responds. As communication technology tools allow users to aspire toward perpetual contact and control, the environment responds with counter-practices and counter-technologies. This does not suggest that there is no progress, or that users do not get much of what they bargain for, but rather that movement is complex and may introduce new barriers and dampers that minimize or even reverse the sought-after effects.

In light of the studies presented in this volume, we can conclude that even mobile communication, while increasing the range of physical and social possibilities, more tightly links the individual to a group and set of responsibilities and outlooks, thus also acting to limit and socially integrate one's own individuality. In essence, then, mobile communication appears to have a net integrative effect on society.

Mobile Network Society: M-Networks Out-Compete

The work of Manuel Castells (2000; Castells et al. 2006; and his afterword for this volume) informs some of the conclusions that can be drawn from the analyses. He posits that networks out-compete other forms of social organization, and if one accepts economic and evolutionary principles as important in human behavior, one can expect people increasingly to involve themselves in networks. Humans are natural communicators, and language expressed through voice is one of the paramount aspects of humanity. Speaking, and increasingly reading and writing, have become rather indispensable parts of what it means to be human in modern society.

And so it is becoming with electronic communication in industrialized and many developing countries. If anyone but the most august or elderly university professor decided to forego indulging in any e-mail communication, there would be a serious question about the sanity of that individual. Indeed, that professor would be perceived, to the extent labeling theory holds, to be "crazy." In wealthy industrialized societies, the conscious decision for anyone under sixty to not "go mobile" is worthy of critical comment bordering on opprobrium. The group views such a decision as the imposition of an inconvenience bordering on social dereliction.

So it is unsurprising that mobile communication, especially voice- and text-based modes, has become so widespread and popular. But Castells notes that people in networks also act as switches, moving information around in certain directions, electing to block its flow in certain other directions or even stopping its flow entirely (Castells 2004, p. 224). This observation seems confirmed by many of the studies in this volume. Noteworthy examples include Singapore's spreading of rumors and the interpersonal mediation that Molony, White and White, and Law and Peng discuss. Capabilities are increasingly allowing people to take their social networks with them. Their social networks are in their pockets, as it were, and are readily accessible. This gives a greater reach of information resources and becomes a management interface with the physical and geographic world. As mobiles become mainstream, Castells's network society takes on another dimension (which I cannot resist describing as the "M-dimension" as in *M*anuel). It is not simply N + 1 or N × 1, but N quickly becoming raised by M [N^M], achieving a vastly higher power of dynamic information, connectivity, and reciprocity.

Conclusion

Mainstreamed mobile communication is a new accompaniment to physical and social freedom. In terms of the physical dimension, mobile communication has allowed people to go farther and yet stay effectively closer; it gives them an enhanced sense (and reality) of safety and control. It allows distant resources to be managed while on the move. It has thus extended physical freedom. In terms of the social dimension, mobile communication has allowed people to modify their immediate social environment while on the move, adjust their networked relationships in detail, and dynamically reorganize their schedules and activities. As such, its net effect seems to be to enhance several dimensions of freedom and to increase their choices in life.

But social scientists are, or at least should be, skeptical. By inclination and training, social scientists want to scratch through positive gloss to seek evidence of unfair treatment, negative trends, and depressing risks. In this spirit, it should be noted—as many others have—that the positive and self-empowering features of mobile communication can also be turned against the user. This occurs at levels ranging from sophisticated monitoring by security or criminal organizations to the casual snooping of family members. Use of mobile communication technology leaves traces and data that can directly or indirectly restrict or even deny freedom. At the very least, they can be used to invade personal privacy and cause emotional distress. These threats, real and emerging, require continuing vigilance and thoughtful protections that balance competing interests at the level of both policy and technology.

Yet the word *freedom* also connotes less predictability and more chaos. Mobile communication can be abused with destructive consequences, as when they are used to effect terror attacks. Mobile communication has also added new complexity to the management of personal relationships, commercial ventures, and political processes. It also confronts and confounds traditional guidelines for social interaction. Such challenges require reconfiguration of institutional arrangements.

Mobile communication remains a double-edged sword. It can be used both to achieve one's own freedom as well as to bind one more tightly to systems of sociopolitical commands and economic demands. It can add complexity to politics but also can add complexity to shared systems of public life. Thus it is incumbent upon scholars of communication to examine in detail the ways in which new frontiers of mobile technology both disencumber individual action as well as entwine coordinated lives. In this way, scholars can dispassionately analyze human communication processes as technologically mediated, scholars' first and foremost goal. Perhaps naïvely, it may also be hoped that through their illuminating efforts, they can help empower citizens and guide leaders to use mobile communication technology to extend human freedom and welfare.

References

Associated Press. 2006. Text messages warn of violence in Belarus. http://news.yahoo.com/fc/ World/belarus.

Becker, E. 1969. *Angel in Armor: A Post-Freudian Perspective on the Nature of Man.* New York: The Free Press.

Castells, M. 2000. *The Rise of the Network Society*, 2nd ed. Oxford: Blackwell.

Castells, M. 2004. Why networks matter. *Demos* 20: 219–225.

Castells, M., J. L. Qiu, M. Fernández-Ardèvol, and A. Sey. 2006. *Mobile Communication and Society: A Global Perspective.* Cambridge, Mass.: The MIT Press.

China Post. 2006. Theaters bullish on prospects. *China Post Online.* http://www.chinapost.com.tw/ art/detail.asp?ID=78735&GRP=h.

Katz, J. E. 2006. *Magic in the Air.* New Brunswick, N.J.: Transaction Publishers.

Katz, J. E., and M. Aakhus, eds. 2002. *Perpetual Contact: Mobile Communication, Private Talk, Public Performance.* Cambridge: Cambridge University Press.

King, R. 2000. *Brunelleschi's Dome.* New York: Walker.

Sükösd, M., and E. Dányi. 2003. M-Politics in the making: SMS and e-mail in the 2002 Hungarian election campaign. In *Mobile Communication: Essays on Cognition and Community*, edited by K. Nyíri. Vienna: Passagen Verlag.

Turkle, S. 2006. Cyberplaces in our lives: Technology and society. Eastern Sociological Society Annual meeting. Boston.

Wynn, E., and J. E. Katz. 1997. Hyperbole over cyberspace: Self-presentation in Internet home pages and discourse. *The Information Society* 13(4): 297–329.

Afterword

Manuel Castells

At the end of 2007, the number of mobile phone subscriptions reached the mark of 3.2 billion, which represents a rate of penetration of fifty percent in the population of the planet as a whole. Thus, mobile telephony is the fastest diffusing communication technology in human history. Of course, there are vast disparities in penetration rates (that is, standardized by population) between rich countries and developing countries (with the European Union exceeding 86 percent of penetration; Australia and New Zealand at 82 percent; Japan and South Korea at 75 percent; the United States and Canada lagging at 66 percent; and Latin America, the rest of Asia, and Africa trailing, respectively, at 31 percent, 17 percent, and 9 percent). Furthermore, in the developed world, there is a high proportion of two or more subscribers per household, while there are still a number of households without mobile phone access. On the other hand, we know that one subscriber number can be used by many, particularly in the developing world, where community-owned mobile phones is a common practice. This is to say that mobile communication has become the prevalent mode of communication in our world, and that its digital divide, at least in terms of access, is much less pronounced than the one we find with the Internet. Because of fast diffusion of broadband, mobile telephony is becoming the support of multimodal communication (including audio, video, and text) and it is widely distributing Internet through wireless networks, thus freeing the Internet from some of the limitations of the fixed-lines infrastructural layout.

As with the Internet, mobile communication became a new medium of self-directed mass communication to the surprise of its designers and business owners. What Motorola conceived as a high-end tool of communication for business professionals, and what Nokia first designed at the request of the Finnish army, was appropriated by youth and their families for their own personal needs. Only then did some firms, Nokia being the earliest and foremost player in this field, discover the extraordinary potential of mobile telephones as a consumer good, while telecommunication operators scrambled to follow the lead of their customers avid to weave communication capability into the diversified fabric of their lives. The result, that we can already fully observe in our social practice, is that the communication process, which is the heart of human experience, is built around wireless communication circuits that amplify the human mind. This, at the same time, empowers people and makes them dependent on the owners and regulators of the basic tools of their existence.

Because of the speed of diffusion of this new communication pattern, social research has lagged in the analysis of its causes, modalities, and consequences. Yet, in recent years, a number of scholars, and some perceptive observers, have started to set the record of the transformation and have produced a good deal of knowledge that cuts through the unreliable accounts that populate the media. This volume, building on the collective effort of some of the pioneers in mobile communication research, brings together a series of analyses that, in their diversity, both cultural and methodological, provide a basis for the understanding of an evolution that I personally consider to be a turning point in the patterns of social organization and practice. The studies speak for themselves, and James Katz has distilled, with his usual mastery, the essence of their contribution, while providing the original framing of the handbook. Thus, I would only add a few remarks to highlight what, in my personal view as a researcher in this field, are some of the critical issues that derive from our current knowledge, both from the works collected here, and from other sources.

The first one is that mobile communication is not about mobility but about autonomy.

In fact, research shows that most calls made through wireless networks are made from the usual places where people are: their homes, their workplaces, their schools. And this is so even if they have fixed lines to call from. This is why there is a general trend toward the rapid substitution of wireless for fixed telephone lines, only slowed down by abusive rates that do not have a technological justification. But now people can also call from any other place, where there is no alternative. And this includes transportation sites (cars, trains, stations, airports—not yet airplanes) and the whole variety of waiting rooms, and ultimately all time/spaces that supersede the prearranged time/spaces. In other words, we now have a wireless skin overlaid on the practices of our lives, so that we are in ourselves and in our networks at the same time. We never quit the networks, and the networks never quit us; this is the real coming of age of the network society. But these networks are self-centered networks. Every individual has her definition of her networks and becomes the connecting node of these networks. Our lives are mapped in our address books. But we write the address books. The consequences, as observed by research, are extraordinary. For instance, wireless communication has come to the rescue of the postpatriarchal family. This is the family formed by individuals who assert their autonomy, including children, and that at the same time needs constant coordination, monitoring, support, and backup systems. The possibility to reach any one at any time anywhere provides this safe autonomy pattern that characterizes the daily life of millions of families around the world. Autonomous communication extends as well to the sociopolitical realm. People can now build their own information systems. They can disintermediate the media, not only via the Internet, but with their mobile phones, including instant video reporting of "citizen journalists." Researchers have documented many instances in which the spontaneous social movements and political mobilizations based on this autonomous communication capacity have profoundly impacted power relationships including the change of governments in several countries in the past five years.

However, the argument is not that technology makes possible a less-centrally planned family or triggers grassroots mobilizations. Rather, it is the other way around: because our societies demand increasing levels of individual and collective autonomy, wireless communication fits perfectly this demand, and thus its use spreads like fire in the prairie of our desires. New family relationships result from the challenge, in the past three decades, to patriarchal authority, to the confinement of women in their exclusive role as household managers, and to the reduction of children to passive subjects of their regimented socialization. These structures of authority are irreversibly shaken in many societies and in many families, but people still like to be together and have children and cherish them. And the children want to be loved and protected, yet they also want to be human subjects as well as active consumers. Thus, it is the complex process of negotiation, managed consensus, and schedule compatibility that makes wireless communication a precious tool for perpetual contact and the relentless adjustment of the daily lives of multiple subjects. Or, in the case of sociopolitical mobilization, what is behind the increasing use of mobile phones and the Internet is the worldwide crisis of legitimacy of political parties and governments, the growing lack of credibility of subservient media, and the conviction of many citizens, as documented by surveys everywhere, that they have to take matters into their own hands, using moments of outrage to establish insurgent politics as a new component in the political system.

The second issue that, in my view, weighs considerably in the process of transformation is the blurring of time and space, inadequately characterized as multitasking. I have identified in my research the emergence of a new form of space, the *space of flows*, and a new form of time, *timeless time*, as characteristic of our society, the network society. Without repeating these analyses here, I emphasize their direct connection to what we learn from the observation of wireless communication. The *space of flows* means simply that simultaneity of social interaction can be achieved without territorial contiguity. Most of the dominant activities in our society (from financial transactions to broadcast media) are performed in the space of flows. But now, through wireless communication, everything migrates to the space of flows. Not that places disappear, but they are folded into the logic, infrastructure, and dynamics of the space of flows. Thus, cities, as shown by William Mitchell in his remarkable studies, do not disappear, but become hybrid spaces of physical settings and communication networks so that people have to constantly manage the transition between these two dimensions of their experience. Similarly, the structural trend toward time compression (the acceleration of life) and toward the blurring of sequences (first home, then work, then transportation, then home, and so on), which define what I conceptualized as *timeless time*, is materially supported by a technology that allows us to "kill time" in free moments by calling someone or checking the BlackBerry, so filling in with activity every second of our existence as well as blurring working time, family time, leisure time, and any content time around the networks that converge in us in relentless waves from the ocean of communication. Not that we become slaves of the technology. Rather we select the technology to enslave our freedom because we are free to do so (freedom can be used for self-destruction, as history shows).

Thirdly, wireless communication has become an essential tool for economic and social development in our world, both for individuals and for countries. There is ample evidence that information and communication are essential components of the development process in the global knowledge economy, the core of our economy. Yet, the layout of the communication infrastructure in the planet has followed the geography of colonization and international domination. As a result, vast areas of the world, including China and India as well as Latin America and Africa, have a very deficient fixed-line communication infrastructure that, together with poor transportation, make connectivity (or the lack of it) a key factor in keeping shared development out of reach for hundreds of millions of people. Yet, there is a potential reserve of entrepreneurship to be tapped in many of these areas, and enough capital (public or private) to be invested in spurring new competitive strategies in the global economy. Institutional reform is a precondition for such development strategy, but appropriate communication infrastructure is a must to establish the connection. The global economy is an economy of networks. But networks connect and disconnect at the same time, creating a new geography of inclusion and exclusion. Satellite-based wireless communication and networks of connection between wireless networks and the fixed communication infrastructure offer new possibilities to alleviate the connectivity gap. Thus, studies at the village level, such as those conducted by Araba Sey in Ghana, or those coordinated by Hernan Galperin in Latin America, show the potential impact of access to wireless networks for local communities. Here again, technology cannot substitute for development and for community control over this development. But given the social and institutional conditions to engage in a developmental process, wireless connectivity is an essential medium to leapfrog toward full participation in the global economy—on the condition that governments and telecommunication providers play a fair game.

Which brings me to the last issue that I consider strategically important. The more wireless communication becomes the fabric of our lives the more its regulation on behalf of the public interest raises to the highest priority of public policy. In recent years, around the world, fast technological change and the coming of digital networking have made obsolete not only our communications infrastructure, but the regulations and policies that were built in the analog era. Governments and corporations have responded in a piecemeal approach. Corporations, naturally, vie for competitive advantage without broader vision, often damaging the expansion of opportunities for everybody, as is the case with cell phone operators in the United States, lagging vis-à-vis their European competitors that unified and expanded their market using a more cooperative approach. On the other hand government regulators more often than not have been more responsive to pressure groups and to their own vested interests than to the challenges of shaping a new communication era on behalf of their citizens. The result is a chaotic system of standards, technologies, pricing structures, and international agreements that is leaving the tidal wave of mobile telephony free to move in whatever direction is indicated by imperfect market signals. Once again, our technological over-development is hampered by our institutional underdevelopment.

In the last resort, however, if the lessons of history are of any use, people will shape the new communication system, largely based on wireless communication. They will do it as users, and they will do it as citizens. And researchers should be attentive to rigorously follow what they do, and to report it to society at large, so that our personal choices and our public strategies will become better informed than they currently are.

About the Editor and Authors

James E. Katz is Chair of the Communication Department at Rutgers University and Director of the Center for Mobile Communication Studies. He is the author of *Magic in the Air* (Transaction 2006), coauthor of *Social Consequences of Internet Use* (The MIT Press 2002), and editor of *Machines that Become Us: The Social Context of Personal Communication Technology* (Transaction 2003).

Sophia Krzys Acord is completing her PhD at the University of Exeter. Her thesis is on the cultural control and reproduction of art as performed by curators.

Bart Barendregt has his doctorate in anthropology and currently is at the Institute of Social and Cultural Studies, Leiden University, the Netherlands.

Naomi S. Baron is Professor of Linguistics and Director of the TESOL Program in the Department of Language and Foreign Studies at American University in Washington, DC. A Guggenheim Fellow and former president of the Semiotic Society of America, she is the author of six books including *Alphabet to Email: How Written English Evolved and Where It's Heading*.

Jeffrey Boase is Postdoctoral Researcher in the Department of Social Psychology at the University of Tokyo. Using a social network approach, his current research explains why the use of mobile phones and PC email varies so greatly between Japan and America. His dissertation focused on the social implications of PC email in America by drawing on data collected from a national survey that he codesigned with Barry Wellman and the Pew Internet & American Life Project. During his short career, he already has coauthored more than twelve scholarly publications and has received awards from the Social Sciences and Humanities Research Council of Canada and the National Center for Digital Government at Harvard University.

Carla Marisa Bonina is Associate Researcher in the Telecommunications Research Program, Telecom-CIDE, at the Centro de Investigación y Docencia Económicas (CIDE) in Mexico City. She holds a BS in economics from the University of Buenos Aires and master's degrees from CIDE and the University of Texas's Lyndon B. Johnson School of Public Affairs.

Scott Campbell is Assistant Professor and Pohs Fellow of Telecommunications at the University of Michigan, Department of Communication Studies. His research explores the social implications of new media, with an emphasis on mobile communication practices. With Richard Ling, he is coeditor of Transaction Publishers' Volume I of a new series of annuals on mobile communication research.

Manuel Castells is Professor of Communication and the holder of the Wallis Annenberg Chair in Communication Technology and Society at the Annenberg School for Communication, University of Southern California, Los Angeles. He is also Research Professor at the Open University of Catalonia in Barcelona, and Professor Emeritus of Sociology and of City and Regional Planning at the University of California, Berkeley. He is coauthor of *Mobile Communication and Society: A Global Perspective* (The MIT Press 2006).

Yi-Fan Chen earned her doctorate from Rutgers University and is Assistant Director of its Center for Mobile Communication Studies.

Jan Chipchase is Principal Researcher in the User Experience Group of Nokia Research. He graduated with an 'MSc' in user interface design from London Guildhall University, and worked at the Institute for Learning and Research Technology in the UK. Portions of his chapter first appeared in *Advance*, Nokia's internal research magazine.

Mira Desai is Reader in Communication Technology at the Postgraduate Department of Extension Education, Shreemati Nathibai Damodar Thackersey Women's University, Mumbai. She has also authored two books in Gujarati.

Jonathan Donner is Researcher at Microsoft Research India in Bangalore, where he studies the social and economic impacts of mobile communication technologies in developing countries. Between 2003 and 2005, he was a Postdoctoral Research Fellow at the Earth Institute at Columbia University, where he helped develop a nationwide information system to support HIV/AIDS care and treatment in Rwanda. In 1999, he earned a doctorate in communication research from Stanford University.

Kenneth J. Gergen is the Mustin Professor of Psychology at Swarthmore College, and President of the Taos Institute. Among his major works are *Realities and Relationships, The Saturated Self,* and *An Invitation to Social Construction.* Of focal interest in his work is the relationship between transformations in communication technology and patterns of cultural life.

Gerard Goggin is Professor of Digital Communication and Journalism, and Deputy Director of the Centre for Social Research in Journalism and Communication, at the

University of New South Wales. He is also an ARC Australian Research Fellow, completing a project on mobile phone culture. His books include *Internationalizing Internet Studies* (2008), *Mobile Phone Cultures* (2007), *Cell Phone Culture* (2006), *Virtual Nation: The Internet in Australia* (2004), and *Digital Disability* (2003). He is editor of the journal *Media International Australia*. This chapter was written while he held a position in the Department of Media and Communications at the University of Sydney.

Mohammed Ibahrine is Assistant Professor at the School of Humanities and Social Sciences, Al-Akhwayn University in Ifrane (AUI), Morocco. He is the author of *New Media and Neo-Islamism* and some fifteen peer-reviewed journal articles. He serves on the editorial committee of *Global Media Journal* and *Journal of Arab and Muslim Media Research*.

Youn-ah Kang is a PhD student at the School of Interactive Computing at the Georgia Institute of Technology. She earned her master's degree from the School of Information at the University of Michigan. With emphasis on human-computer interaction (HCI), her research interests include user interface (UI) design and evaluation of systems and products, information visualization, and social computing.

Ilpo Koskinen Professor of Industrial Design at the University of Art and Design, Helsinki, has presented his research findings at numerous international conferences. Trained in sociology and conversation analysis and coauthor of the widely admired book *Mobile Image*, Koskinen also is the author of *Mobile Multimedia in Action*.

On-Kwok Lai is Professor at the Graduate School of Policy Studies, Kwansei Gakuin University, Japan, with an honorary professorship (social policy) and fellowship (urban planning and environmental management) at the University of Hong Kong. This research was funded by Kwansei Gakuin University and an honorary professorship (social policy) at the University of Hong Kong.

Pui-lam (Patrick) Law received his PhD in sociology from the University of New South Wales, Australia, and is currently Assistant Professor in the Department of Applied Social Sciences of the Hong Kong Polytechnic University. He coauthored *Marriage, Gender, and Sex in a Contemporary Chinese Village* (New York: M. E. Sharpe 2004).

Katie M. Lever is a PhD student at Rutgers University in the School of Communication, Information, and Library Studies.

Christian Licoppe is Professor in the Department of Economics, Management, and Human Sciences, École Nationale Supérieure des Télécommunications, Paris. He earlier headed the social science laboratory at France Telecom R&D, whom he thanks for their support of the research contained in his chapter. He also thanks Romain Guillot for his contribution to the chapter's empirical work.

Rich Ling   is Sociologist and Senior Researcher at Telenor's research institute located near Oslo, Norway, and has been the Pohs Visiting Professor of Communication Studies at the University of Michigan in Ann Arbor. He is the author of *The Mobile Connection: The Cell Phone's Impact on Society* and, with Per E. Pederson, editor of *Mobile Communications: Renegotiation of the Social Sphere*. He is completing his next book, entitled *Mediated Ritual Interaction*.

Judith Mariscal   is Director of the Telecommunications Research Program from the Centro de Investigación y Docencia Económica's (CIDE), an independent research and educational institution based in Mexico City.

Patricia Mechael   is Research Associate at the Rutgers Center for Mobile Communication Studies. She completed her PhD at the London School of Hygiene and Tropical Medicine, focusing on the use of mobile phones for health in Egypt. She currently is working with the World Health Organization to develop an mHealth strategy for developing countries.

Gustavo Mesch   is Senior Lecturer at the Department of Sociology and Anthropology at the University of Haifa, Israel. He has been analyzing the domestication of the Internet.

Kakuko Miyata   is a professor at Meiji Gakuin University in Tokyo. She received her PhD in social psychology from the University of Tokyo. *Social Capital at the Age of Internet* (NTT Press 2005), one of her several books, won the 2005 Okawa Publications Prize and an award from the Japanese Society of Social Psychology.

Thomas Molony   is Research Fellow at the Centre of African Studies, University of Edinburgh and Postdoctoral Researcher at the School of Geography, Archaeology, and Environmental Studies, University of the Witwatersrand, Johannesburg, South Africa. He thanks Josipa Petrunic (Science Studies Unit, University of Edinburgh) for her helpful comments.

Ragnhild Overå   holds a PhD in geography. She carried out the research on telecommunications in Ghana while she was Senior Researcher at the Christian Michelsen Institute in Bergen. Overå is currently Associate Professor in the Department of Geography at the University of Bergen.

Yinni Peng   obtained her bachelor's and master's degrees in sociology at Peking University. She is currently a PhD student in the Department of Sociology at the Chinese University of Hong Kong.

Raul Pertierra   holds a doctorate in anthropology and currently is a visiting professor at the Asian Center, University of the Philippines and the Ateneo de Manila University. His latest book is *Transformative Technologies: Altered Selves—Mobile Phone and Internet Use in the Philippines* (2006).

Lourdes M. (Odette) Portus is Associate Professor in the College of Mass Communication of the University of the Philippines, from which she earned her PhD in communication. She is the author of *Streetwalkers of Cubao* (Giraffe Books 2005) and numerous articles on communication.

Nimmi Rangaswamy is Associate Researcher at Microsoft Research India in Bangalore, where she studies consumption patterns of urban middle-class Indians. All views expressed in chapter 24 are those of the authors alone.

Madanmohan Rao a consultant and writer from Bangalore, is a research consultant at the Asian Media Information and Communication Centre (AMIC), Singapore. Among his edited books are *The Asia Pacific Internet Handbook* and *The Knowledge Management Chronicles*. He graduated from the University of Massachusetts at Amherst, with an MS in computer science and a PhD in communication.

Howard Rheingold an independent author, has been extraordinarily influential in forming and commenting upon the evolution of the modern digital world. His books include *Tools for Thought*, *The Virtual Community*, and *Smart Mobs*. He has taught at the University of California, Berkeley, and at Stanford University.

Shahiraa Sahul Hammed previously was associated with the School of Communication and Information and Singapore Internet Research Centre, Nanyang Technological University, Singapore. She was conferred a master of psychology degree from the National University of Singapore.

Lara Srivastava is Senior Policy Analyst and Project Director of the New Initiatives Programme at the Strategy and Policy Unit (SPU) of the International Telecommunications Union (ITU). She writes and edits ITU publications such as "The Internet of Things" (2005) and "digital.life" (2006).

Molly Wright Steenson is a doctoral student in the Department of Architecture at Princeton University. She completed her master's degree in environmental design at Yale School of Architecture in 2007 and was a research intern at Microsoft Research India.

Ilan Talmud is Senior Lecturer in the Department of Sociology and Anthropology, University of Haifa, Israel. His main interests include network models of social structure and online social capital.

Sherry Turkle is Abby Rockefeller Mauzé Professor of the Social Studies of Science and Technology in the Program in Science, Technology, and Society at MIT. She is author of *Life on the Screen*. Portions of her chapter appeared as "Tethered," in *Sensorium: Embodied Experience, Technology, and Contemporary Art*, edited by Carolyn A. Jones (Cambridge: List Visual Art Center and The MIT Press 2006); "Tamagotchi Diary" in the April 20, 2006, *London Review of Books*; and "Can You Hear Me Now" in the May 7, 2007, *Forbes*. Research reported in this essay was funded by an NSE ITR grant, "Relational

Artifacts" (Turkle 2001) award number SES-0115668, by a grant from the Mitchell Kapar Foundation, and by a grant from the Intel Corporation.

Carolyn Wei   is User Experience Researcher at Google. She received her doctorate in technical communication at the University of Washington in 2007 and was a research intern with Microsoft Research India.

Barry Wellman   directs NetLab, and is S. D. Clark Professor in the Department of Sociology at the University of Toronto. He is an expert on social network analysis and is widely published in the field of Internet studies.

Naomi Rosh White   is associate professor in the sociology program of the School of Political and Social Inquiry at Monash University, Melbourne, Australia. Her current work examines the social psychology of travel and tourism.

Peter B. White   is associate professor in the Media Studies Program at La Trobe University, Melbourne, Australia. His current interests are in the ways new media and communications services are incorporated into people's lives. Parts of chapter 15 by White and White incorporate material previously published in *Convergence*.

Index

habits, 406–407
intimate life and, 411
mobisodes and, 413–414
online versus, 125
public space and, 411–412
race, ethnicity and, 405
social modalities of, 407–408
sociological trends through, 408–409
Garfinkel H., 410
Gay, G., 183
Gender and mobile communications
in India, 331–333
in Japan, 326
in Mexico, 67–68, 75–76
mobile gaming and, 404–405
MP3 players and, 374
in the Philippines, 108–109, 111–113
in Tanzania, 346–347
General Packet Radio Service (GPRS), 82
Gergen, Kenneth, 241–242, 243, 245, 251, 252,
278, 302
Germany, 392
Geser, H., 412
Ghana, 343, 450
benefits of mobile communications in, 52–
53
coordination, monitoring, and timing of trade
in, 50–51
new trading practices in, 47–48
political collective action in, 226, 228
reconfigurations of power in, 51–52
rural-urban market chain and traders'
information needs in, 45–47
telecommunications infrastructure in, 43–45
Ghost culture, 385–386
Gibson, James, 177
Global positioning systems (GPS), 274, 280
Global System for Mobile Communications
(GSM), 15, 355
Goffman, E., 165, 166, 167, 169, 170, 410
Grameen phone program, 33, 36
Great Britain. See United Kingdom, the
Green, N., 160

Growth of mobile communications
compared to the Internet, 15
in the developing world, 22–24, 389–390,
434
numbers in, 16–17, 447
GSM Association, 24
Guangdong, China. See Migrant workers
Gundolf, A., 246

Haddon, Leslie, 97, 325–326, 329
Hall, E. T., 411
Hamon, Patrick, 235
Hardcore gamers, 407
Harper, R., 160, 327
Hasselberger, Sepp, 228
Health services and mobile communication
applications, 91–94
combined with fixed-lines, 94–95
coordinating care, 95–96, 98–101
informal networks in, 98–101
in Singapore, 292
Heidegger, M., 380
Hembrooke, H., 183
Henry, John, 233, 234
Herd immunity, 97–98
Hewlett-Packard, 187
Hirsch, Tad, 233, 234
Huizinga, J., 412
Hungary, 226, 227
Hutchby, I., 273
Hyper-coordination, 264–267

Icons, 81–82
Identity and mobile communication, 124–125
ringtones in, 146–150, 286–287
social interactions and, 165–167, 172–174,
441–443
Inada, Y., 409
India, 23–24, 33, 325, 326–328, 333–335, 441
civic and political impact of mobile
communications in, 390–393
courtship in, 330–331, 392
domestic space in, 331–333, 334